WITHDRAWN
WRIGHT STATE UNIVERSITY LIBRARIES

NOVEL THERAPEUTIC AGENTS *for the* TREATMENT *of* AUTOIMMUNE DISEASES

NOVEL THERAPEUTIC AGENTS
for the
TREATMENT
of
AUTOIMMUNE DISEASES

edited by

VIBEKE STRAND
Stanford University School of Medicine
Stanford, California

DAVID L. SCOTT
King's College School of Medicine and Dentistry
and King's College Hospital
London, England

LEE S. SIMON
Harvard Medical School
Deaconess Hospital
and Dana–Farber Cancer Institute
Boston, Massachusetts

Marcel Dekker, Inc.   New York•Basel•Hong Kong

Library of Congress Cataloging-in-Publication Data

Novel therapeutic agents for the treatment of autoimmune diseases /
edited by Vibeke Strand, David L. Scott, Lee S. Simon.
  p. cm.
Includes index.
ISBN 0-8247-9748-5 (hardcover : alk. paper)
  1. Autoimmune diseases—Chemotherapy. 2. Autoimmune diseases-
-Immunotherapy. I. Strand, Vibeke. II. Scott, David L.
[DNLM: 1. Arthritis, Rheumatoid—therapy. 2. Autoimmune Diseases-
-therapy. 3. Antibodies, Monoclonal—therapeutic use.
4. Cytokines—antagonists & inhibitors. 5. Receptors, Antigen, T
-Cell—immunology. WE 346 N937 1997]
RC600.N68 1997
616.97 ' 8061—dc20
DNLM/DLC   96-18671
  for Library of Congress   CIP

The publisher offers discounts on this book when ordered in bulk quantities. For more information, write to Special Sales/Professional Marketing at the address below.

This book is printed on acid-free paper.

Copyright © 1997 by MARCEL DEKKER, INC. All Rights Reserved.

Neither this book nor any part may be reproduced or transmitted in any form or by any means, electronic or mechanical, including photocopying, microfilming, and recording, or by any information storage and retrieval system, without permission in writing from the publisher.

MARCEL DEKKER, INC.
270 Madison Avenue, New York, New York 10016

Current printing (last digit):
10 9 8 7 6 5 4 3 2 1

PRINTED IN THE UNITED STATES OF AMERICA

To our families, without whom there would be
no place to go during both the good times and the bad.
We know we may often not be there in body,
but we are always with you in mind.

Preface

The idea for this book arose late one night, during a meeting, when the three of us lamented the difficulties in finding a useful source for clinical information on the novel therapies quickly coming into vogue in clinical research, and those just as quickly becoming "passé." Although the biyearly "Biologics Meeting" publishes proceedings, these are often released more than a year after the meeting and are limited to those topics included in the program. Several basic science reviews have discussed a variety of promising biological treatments, but are, of course, limited to preclinical data in mice and rats. For this reason, we believe there is great benefit to a compendium of potential new therapies in autoimmune diseases, emphasizing the experience gained in humans, however preliminary. We have focused on "innovative" therapies, including promising new pharmaceutical as well as biological agents.

Our goal has been to make this book an invaluable single source for information on newer interventions in the treatment of autoimmune diseases. Some of these agents have yielded disappointing clinical results, or demonstrated toxicity, and have passed on, without hope for approval. For others, data are still too preliminary to predict their future. In the 14 months since this volume was conceived, promising agents have come and gone, and we have come to appreciate the challenging task we undertook. Data are often interim or just about to be published, and therefore cannot be included in this book. Individual life cycle changes, corporate politics, and other events beyond our control have thwarted some of our best attempts to secure critical manuscripts. As a result, we have been forced to overlook several promising agents. Nonetheless, the chapters in this book reflect a great deal of patience and hard work on the part of the contributing authors. If this volume serves a critical need, then a second edition may well expand significantly the knowledge base presented in this edition.

Each of the innovative therapies discussed in this book has incrementally helped us further understand the complicated pathophysiologic processes underlying autoimmunity, which have daunted mankind for hundreds of years. For example, we now recognize the significant role TNFα plays in the inflammatory cascade, as was previously recognized for IL-1. And the clinical experiences with each product have helped us better refine the process of "rational" drug development, especially difficult when definitive treatments are lacking.

Although no treatment has yet offered a cure, we have made significant progress. Since this book was conceived, two forms of Interferonβ have received approval for the treatment of multiple sclerosis (MS). Clinical trials evaluating three promising products are now under way in systemic lupus erythematosus (SLE), a difficult disease to treat or even evaluate experimental treatment therapies for. Some of the agents discussed in this book, as well as newer products, are now being studied in psoriasis and/or Crohn's disease. An international collaboration is under way to develop a registry for autologous and allogeneic stem-cell transplantation in patients with severe manifestations of autoimmune disease such as SLE, MS, and systemic sclerosis. The first gene therapy trial in arthritis has just been initiated.

There is every reason to believe that these innovative approaches, including those discussed in this book, will yield better, more rational and well-tolerated treatments for a variety of diseases heretofore inadequately addressed by our therapeutic armamentarium.

Vibeke Strand
David L. Scott
Lee S. Simon

Contents

Preface v
Contributors xi

I. INTRODUCTION

1. The Evaluation of Biological Agents　　　　　　　　　　　1
   Vibeke Strand and David L. Scott

II. CELLULAR TARGETED THERAPIES

2. Anti-CD5/Ricin A Chain Immunoconjugate Therapy in
   Rheumatoid Arthritis　　　　　　　　　　　　　　　　　11
   John J. Cush

3. Early Clinical Studies of IL-2 Fusion Toxin in Patients with
   Severe Rheumatoid Arthritis, Recent-Onset Insulin-Dependent
   Diabetes Mellitus, and Psoriasis　　　　　　　　　　　　25
   Thasia G. Woodworth and Karen Parker

4. Chimeric Anti-CD4 Antibody as a Potential Therapeutic Agent
   for Rheumatoid Arthritis　　　　　　　　　　　　　　　41
   Larry W. Moreland and William J. Koopman

5. CD4 Monoclonal Antibody Therapy in Rheumatoid Arthritis 55
   F. C. Breedveld

6. The Use of CE9.1, a Primatized Monoclonal Anti-CD4, in the
   Treatment of Rheumatoid Arthritis 65
   David E. Yocum, Alan M. Solinger, and John A. Lipani

7. CAMPATH-1H Therapy in Autoimmune Diseases 75
   Richard A. Watts and John D. Isaacs

8. CAMPATH-1H in Rheumatoid Arthritis: United States
   Experience 83
   David E. Yocum and Jeffrey M. Johnston

III. CYTOKINE TARGETED THERAPIES

9. Interferon-Gamma in the Treatment of Rheumatoid Arthritis 95
   Eric M. Veys, Herman Mielants, and Gust Verbruggen

10. Tumor Necrosis Factor Blockade in Rheumatoid Arthritis 107
    Michael J. Elliott, Marc Feldmann, and Ravinder N. Maini

11. Engineered Human Anti-Tumor Necrosis Factor-Alpha (TNFα)
    Antibody, CDP571, in Rheumatoid Arthritis 121
    Ernest H. S. Choy and Gabriel S. Panayi

12. Clinical Experience with Recombinant Human Interleukin-1
    Receptor Type I (Rhu IL-1RI) in Patients with Rheumatoid
    Arthritis 131
    *Richard M. Pope, Barbara Drevlow, Jennifer Capezio, Rosa Lovis,
    and Alan Landay*

13. Treatment of Rheumatoid Arthritis with Soluble Tumor
    Necrosis Factor Receptor 141
    Gary R. Margolies, William J. Koopman, and Larry W. Moreland

IV. ADHESION MOLECULE TARGETED THERAPIES

14. Treatment of Rheumatoid Arthritis with a Monoclonal Antibody
    to Intercellular Adhesion Molecule-1 155
    Arthur F. Kavanaugh and Peter E. Lipsky

V. POTENTIAL ANTIGEN-SPECIFIC THERAPIES

15. T-Cell-Receptor Peptide Therapy for Multiple Sclerosis   173
Arthur A. Vandenbark, Dennis N. Bourdette, Ruth H. Whitham, and Halina Offner

16. T-Cell-Receptor Peptide Vaccination Studies in Rheumatoid Arthritis   189
Louis W. Heck, Larry W. Moreland, and William J. Koopman

17. Oral Tolerance for the Treatment of Autoimmune Disease   201
David A. Hafler and Howard L. Weiner

18. Oral Tolerance   221
David E. Trentham

19. Intravenous Immunoglobulin (IVIg) in the Treatment of Autoimmune Diseases   235
Vibeke Strand and Martin L. Lee

VI. PURINE AND PYRIMIDINE SYNTHESIS INHIBITORS

20. Inhibitors of De Novo Nucleotide Synthesis in the Treatment of Rheumatoid Arthritis   257
Robert I. Fox and Randall E. Morris

21. Leflunomide: A New Immunosuppressive Drug   287
David L. Scott and Vibeke Strand

VII. FUTURE DIRECTIONS

22. An Assessment of Novel Agents in the Treatment of the Rheumatic Diseases   295
Vibeke Strand and Lee S. Simon

Index   *301*

Contributors

Dennis N. Bourdette, M.D. Acting Chief, Neurology Service, Portland Veterans Affairs Medical Center, and Association Professor, Department of Neurology, Oregon Health Sciences University, Portland, Oregon

F. C. Breedveld, M.D., Ph.D. Professor, Department of Rheumatology, Leiden University Hospital, Leiden, The Netherlands

Jennifer Capezio, M.D. Department of Medicine, Northwestern University Medical School, Chicago, Illinois

Ernest H. S. Choy, M.D., M.R.C.P. Lecturer in Rheumatology, Rheumatology Unit, Division of Medicine, Guy's Hospital, London, England

John J. Cush, M.D. Associate Professor of Internal Medicine, Rheumatic Diseases Division, University of Texas Southwestern Medical Center at Dallas, Dallas, Texas

Barbara Drevlow, M.D. Department of Medicine, Northwestern University Medical School, Chicago, Illinois

Contributors

Michael J. Elliott, Ph.D., F.R.C.P Department of Clinical Immunology, Kennedy Institute of Rheumatology, London, England

Marc Feldmann, Ph.D., F.R.C.P. Department of Cytokine Immunology, Kennedy Institute of Rheumatology, London, England

Robert I. Fox, M.D., Ph.D. Member, Departments of Rheumatology and Immunology, Scripps Clinic and Research Foundation, La Jolla, California

David A. Hafler, M. D. Director, Laboratory of Molecular Immunology, Center for Neurologic Diseases, Brigham and Women's Hospital, Boston, Massachusetts

Louis W. Heck, M.D. Associate Professor, Department of Medicine, University of Alabama at Birmingham, Birmingham, Alabama

John D. Isaacs, M.B., B.S., M.R.C.P. MRC Research Fellow, Department of Pathology, Addenbrooke's Hospital, Cambridge, England

Jeffrey M. Johnston, M.D. Senior Clinical Program Head, Rheumatology and Immunology Clinic Research, Glaxo Wellcome, Inc., Research Triangle Park, North Carolina

Arthur F. Kavanaugh, M.D. Associate Professor of Internal Medicine, Rheumatic Diseases Division, University of Texas Southwestern Medical Center at Dallas, and Chief of Rheumatology, Dallas Veterans Affairs Medical Center, Dallas, Texas

William J. Koopman, M.D. Professor and Chairman, Department of Medicine, University of Alabama at Birmingham, Birmingham, Alabama

Alan Landay, Ph.D. Department of Immunology/Microbiology, Rush-Presbyterian–St. Luke's Medical Center, Chicago, Illinois

Martin L. Lee, Ph.D., C.Stat. Lecturer, School of Public Health, University of California, Los Angeles, Los Angeles, California

John A. Lipani, M.D. Group Director, Inflammation and Tissue Repair, SmithKline Beecham Pharmaceuticals, Collegeville, Pennsylvania

Peter E. Lipsky, M.D. Professor of Medicine and Microbiology, Department of Internal Medicine, University of Texas Southwestern Medical Center at Dallas, Dallas, Texas

Contributors

Rosa Lovis Research Technician, Department of Medicine, Northwestern University Medical School, Chicago, Illinois

Ravinder N. Maini, M.B., B.Chin., F.R.C.P. Professor, Department of Clinical Immunology, Kennedy Institute of Rheumatology, London, England

Gary R. Margolies, M.D. Lexington, Kentucky

Herman Mielants, M.D., Ph.D. Department of Rheumatology, Ghent University Hospital, Ghent, Belgium

Larry W. Moreland, M.D. Associate Professor of Medicine, Division of Clinical Immunology and Rheumatology, Department of Medicine, University of Alabama at Birmingham, Birmingham, Alabama

Randall E. Morris, M.D. Director, Transplantation Immunology, and Research Professor, Department of Cardiothoracic Surgery, Stanford University School of Medicine, Stanford, California

Halina Offner, Ph.D. Professor, Department of Neurology, Oregon Health Sciences University, Portland, Oregon

Gabriel S. Panayi, M.D., Sc.D., F.R.C.P. ARC Professor of Rheumatology, Rheumatology Unit, Division of Medicine, Guy's Hospital, London, England

Karen Parker Manager, Department of Medical Communications, Seragen, Inc., Hopkinton, Massachusetts

Richard M. Pope, M.D. Professor, Department of Medicine, Northwestern University Medical School, Chicago, Illinois

David L. Scott, M.D. Reader in Rheumatology and Honorary Consultant Rheumatologist, King's College School of Medicine and Dentistry, and King's College Hospital, London, England

Lee S. Simon, M.D. Associate Professor of Medicine, Harvard Medical School; Assistant to the Chairman, Department of Medicine, for Undergraduate Medical Education; Assistant to the President, Deaconess Hospital, for Medical Education; and Director of Rehabilitative Services, Deaconess Hospital, and Dana-Farber Cancer Institute, Boston, Massachusetts

Alan M. Solinger, M.D. Director, Clinical Therapeutics, IDEC Pharmaceuticals Corporation, and Clinical Professor of Medicine, Division of Rheumatology, Department of Internal Medicine, University of California, San Diego, San Diego, California

Vibeke Strand, M.D. Clinical Associate Professor, Division of Immunology, Stanford University School of Medicine, Stanford, and Consultant in Clinical and Regulatory Affairs, San Francisco, California

David E. Trentham, M.D. Chief, Division of Rheumatology, Beth Israel Hospital, and Associate Professor of Medicine, Harvard Medical School, Boston, Massachusetts

Arthur A. Vandenbark, Ph.D. Career Scientist, Department of Neuroimmunology, Portland Veterans Affairs Medical Center, and Professor, Department of Neurology, Oregon Health Sciences University, Portland, Oregon

Gust Verbruggen, M.D., Ph.D. Department of Rheumatology, Ghent University Hospital, Ghent, Belgium

Eric M. Veys, M.D., Ph.D. Department of Rheumatology, Ghent University Hospital, Ghent, Belgium

Richard A. Watts, D.M., M.R.C.P. Senior Registrar, Rheumatology Research Unit, Addenbrooke's Hospital, Cambridge, England

Howard L. Weiner, M.D. Co-Director, Center for Neurologic Diseases, Brigham and Women's Hospital, Boston, Massachusetts

Ruth H. Whitham, M.D. Staff Physician, Neurology Service, Portland Veterans Affairs Medical Center, and Associate Professor, Department of Neurology, Oregon Health Sciences University, Portland, Oregon

Thasia G. Woodworth, M.D. Senior Associate Director, Department of Experimental Medicine, Pfizer Central Research, Groton, Connecticut

David E. Yocum, M.D. Associate Professor and Director, Department of Medicine, University of Arizona College of Medicine, Tucson, Arizona

I
Introduction

1
The Evaluation of Biological Agents

VIBEKE STRAND
Stanford University School of Medicine, Stanford, California
DAVID L. SCOTT
*King's College School of Medicine and Dentistry and
King's College Hospital, London, England*

I. INTRODUCTION

Many biological products are under investigation for the treatment of rheumatoid arthritis (RA). The number of agents that will be evaluated is likely to increase considerably in the next few years. The agents vary from monoclonal antibodies to cell receptors such as CD4, cytokines such as TNF-α, and recombinant products such as IL-1 receptor antagonist. Other classes of biological products will eventually be evaluated for their therapeutic potential. This evaluation requires defined standards for clinical trials and determining efficacy of these agents.

Many uncontrolled trials of biological agents have been conducted using various methods to assess their efficacy. While such approaches have a place in the investigation of a new area of therapeutic interest, too many open-label, poorly designed trials waste time and resources. Utilizing different methods to evaluate biological therapies makes comparison across studies difficult and weakens the argument in favor of using any of these products. Clinical trials of new immunotherapies in RA should meet three key objectives: (1) patients need symptomatic relief and potential induction of disease control; (2) investigators require standard methods to allow comparison of different agents across patient groups; (3) sponsors require early and efficient assessment of the clinical activity and safety of their products. These aims need an agreed framework, and firm recommendations have been made for work in this area (1).

II. STUDY DESIGN

There have been too many reports of open-label uncontrolled studies with small numbers of patients. While investigators are correctly enthusiastic about initial findings in open studies, these should not receive excessive emphasis. Clinical trials should be prospective, randomized, and blinded. The classic placebo control has limitations. It can be difficult to maintain placebo therapy for prolonged periods in patients with active disease (2). This issue is made more complex by associated polypharmacy and variability of supportive therapies. It is possible to initiate active therapy in placebo-treated cases when there is a documented failure to respond. The case for having active treated controls is relatively strong. However, it has some drawbacks: there are no biological agents of proven activity at present, and only conventional slow-acting agents such as methotrexate can be used as comparators; in addition, power of the studies deceases when an active control is used, so the number of patients required to achieve significance is increased.

The development of biological therapies should follow a path from early placebo-controlled studies, which are relatively small, to larger, multicenter trials employing active controls. Detailed dose schedule work is typically required to discern the most appropriate treatment regimen to achieve maximal efficacy. There may be an additional need to consider treatment combining two or more agents. While studies of combinations of these agents are attractive to visualize, they are difficult to evaluate. Indeed, the problem of evaluating combinations of drugs of the same class is well recognized with studies of slow-acting drugs (3,4).

Only patients with active disease should be evaluated. These patients will have: six or more swollen joints; nine or more tender joints; morning stiffness lasting at least 45 min; and an erythrocyte sedimentation rate of at least 28 mm/hr. The greater the activity of RA, the more likely it is that a biologic will be shown to be effective. A major concern is disease duration. There is no dilemma in trying a biological therapy in an RA patient with disease who has failed many slow-action drugs and still has active disease. But such patients do not represent an ideal population to evaluate the activity of a new therapeutic agent. There may also be differential responses in early disease with late disease being more refractory than early disease (5,6). On the other hand, administering experimental agents to patients with early RA of less than 2 years' duration who have not yet had sufficient exposure to slow-acting drugs may be difficult to justify. There is a reasonable case for giving biological agents in established disease of more than 2 years' duration when there has been a failure to respond to one or more slow-acting drugs and the disease is persistently active. As treatment is offered to patients with early disease, there must be careful appraisal of the potential risks, including a host immune response to the biologic, infection, autoimmune manifestations, and the more remote possibility of malignancy. Such limitations must be weighed against the potential benefits of controlling disease activity and possibly inducing remission.

III. DISEASE ASSESSMENT

Outcome measures should be based on the core data set proposed at Maastricht (7) and accepted by WHO, ILAR, ACR, and other groups (8). This core data set is shown in Table 1. It includes tender joint counts, swollen joint counts, patient's assessment of pain, global assessments of disease activity, and laboratory evaluation of an acute-phase reactant. Radiology is recommended for long-tern analysis.

A. Articular Indices

The various articular indices have been compared in a longitudinal study by Prevoo et al. (9). They examined the validity and reliability of seven indices: the Ritchie articular index (10), a modified Ritchie articular index, the Thompson-Kirwan index (11), the 28-joint index, the 36-joint index, total tender joints, and total swollen joints. The validity and reliability of traditional joint indices do not differ substantially. Weighted joint indices are less valid and reliable. No joint index was superior for measuring disease activity. Taking simplicity into account, the 28-joint index, not graded and not weighted, was preferred. Other studies also agree with the use of a reduced joint count (12).

There is a problem with variability in clinical assessments. This was examined in two studies of comparisons across European centers (13,14). Determining the number of swollen joints and determining the number of tender joints were the most stable clinical methods. Nevertheless there was considerable intercenter variation. Several other studies highlight the extent of disagreement between

Table 1 Core Data Set for Assessing Rheumatoid Arthritis

Core assessments
Number of swollen joints
Number of tender joints
Patient's assessments of pain
Patient's global assessments of disease activity
Physician's global assessments of disease activity
One laboratory evaluation of an acute-phase reactant
Self-administered functional assessment
Plain joint radiology (long-term evaluations)

assessors. Hart et al. (15) evaluated the results from three physicians who assessed the Ritchie articular index in 18 patients, examined in random order. Interrater agreement hardly exceeded the chance level when degree of tenderness alone was assessed. At individual centers with collaboration between observers there is less variation. Lewis et al. (16) studied this using 42 RA patients assessed by two metrologists over 14 weeks. Results showed close agreement within and between metrologists.

B. Patients' Views

Concordance between clinicians' and patients' assessments of the patients' physical and mental functioning was examined in 166 RA patients by Kwoh et al. (17). Clinicians and their patients with RA often disagreed in their assessment of the degree of physical and mental impairment that the patient experiences. Various demographic, medical, and psychological factors influence patients' perceptions. This was examined by Parker et al. (18) in 135 patients with RA. Patients were examined using the systemic index, articular index, McGill Pain Questionnaire, Symptom Checklist-90-R, and a pain visual analog scale. Multiple regression analyses found no significant relationships between pain and the medical variables. Age, income, and selected psychological variables were significantly correlated with pain. The greatest management challenge occurred in patients who were middle-aged, living on limited incomes, and experiencing major stresses in everyday life. These patients were worry-prone and felt isolated and lacking in social support.

Variations in the articular index could have a cultural basis. The features of RA were compared in 108 Greek and 107 British patients with RA by Drosos et al. (19). They found that British patients had more severe articular involvement than did Greeks, as judged by the duration of morning stiffness, grip strength, and the number of swollen and tender joints. Although genetic and environmental factors may be responsible for such differences in disease expression between these two European populations with RA, some of the variation may be related to the clinical methods of measuring joint involvement.

Do self-report measures of symptoms and functional health status provide unique outcome information, or are they no more than a proxy for self-reported arthritis symptoms? Mason et al. (20) examined this in 138 patients with RA. Information from the Rapid Assessment of Disease Activity in Rheumatology (RADAR) measure was compared with same-day functional health scores collected with the recently revised Arthritis Impact Measurement Scales (AIMS2). Self-assessed arthritis symptoms, physical function and work impact, psychological status, and social health each made independent contributions to outcome. Satisfaction with health status was not independent of symptoms, functional capacity, or psychological status. Self-reported symptom and health status information provides complementary rather than duplicative information.

C. Simple Clinical Endpoints and Combined Indices

In clinical trials on disease-modifying drugs it is common to apply a large number of endpoint measures. This has several disadvantages, and the number of measures can be reduced by using only those that are generally considered important, are sensitive to change, and are able to differentiate between drugs. Van der Heijde et al. (21) proposed that a joint count, assessment of pain, a questionnaire on functional status, and measurement of erythrocyte sedimentation rate are sufficient. An alternative approach is to look at changes in the core data set for RA disease activity. For example, changes in ≥20% in five of the seven criteria may indicate a satisfactory response (22). This approach is likely to be widely accepted, though the precise definition of responders needs further clarification. A similar approach can be based on the disease activity score (23).

D. Imaging

There are several approaches to joint imaging including plain radiology, magnetic resonance imaging (MRI), bone scans and isotope labeling methods, and DEXA scans for periarticular osteoporosis. At present conventional X-rays are best for studying drug efficacy, while MRI is best for defining changes in synovitis.

Current X-ray scoring methods are predominantly those of Sharp et al. (24) and Larsen et al. (25). Both methods are reproducible (26,27). The scores are composite indices combining joint space loss, erosions, and other changes. They have highlighted the value of standardization. But the methods combine diverse changes in a single score and assign numerical values to qualitative changes. The 1993 WHO/ILAR Task Force Meeting recommended developing a new approach based on juxta-articular erosions and joint destruction. Until new methods are developed and evaluated, the Sharp and Larsen scores should be used.

A debate exists about the long-term effects of slow-acting antirheumatic drugs (SAARDS) and whether they influence the progression of the disease. Early placebo-controlled studies show reduced radiological progression with gold (28,29) and cyclophosphamide (30). Recent work shows sulfasalazine also reduces radiological progression (31). But several large studies of injectable gold, penicillamine, and similar drugs failed to detect effects on X-ray progression. In general, studies have been too small, too brief, have evaluated only patients remaining on specific therapies, and have used too many different methods to determine joint damage.

Digital image analysis of standard radiographs theoretically increases sensitivity of measurement, but has limited value for assessing erosions (32).

Microfocal radiology increases sensitivity (33). The specialized equipment and technical support needed will not become widely available, making it unsuitable as a standard method.

MRI may give more information than plain radiology, define clinicopathological interrelationships, and increase the sensitivity of assessing damage. This would help in assessing new therapeutic approaches. There is growing familiarity with MRI of peripheral joints. MRI can show effusions, synovial inflammation, synovial proliferation, bone erosions, bone marrow edema, and cartilage changes. Most reports have described changes in one or a few joints at single time points. Experience in early RA (34,35), in a limited number of cases, suggests MRI is more sensitive than X-rays in demonstrating bone erosions. In established disease MRI may also be more accurate for detecting soft tissue changes and minimal skeletal lesions (36). There is some information on MRI changes during therapy with potential disease-modifying drugs. Jevtic et al. (37) studied 45 patients over 12 months. They imaged one carefully selected and clinically active hand joint in each case. MRI showed 27 joints had highly active synovitis and/or destructive pannus, 12 had synovial proliferation, and six had no demonstrable inflammatory activity. Although only carefully selected joints with active synovitis clinically were studied, there was no homogeneous appearance on MRI. Instead a spectrum was seen from presumed markedly active synovitis to total normality. This indicates heterogeneity of MRI appearances in joints thought clinically to have relatively uniform disease severity.

IV. SURROGATE LABORATORY MARKERS

There is less certainty about surrogate laboratory measures of the efficacy of biological agents. In vitro indicators of immune function such as mitogen assays and flow cytometric analyses of peripheral lymphocyte subsets have shown little relationship to clinical effects. This may be due in part to sampling difficulties and our limited knowledge of the key pathophysiological events in RA. There is a need to resolve what to measure and where to sample, as well as the relevance of the proposed mechanisms of action of the agents under study. One might envision the use of certain assays specific to the disease process and others to the agent, indicating its successful delivery and duration of its biological effect.

Beckham et al. (38) evaluated the relationship of a variety of immune and inflammation variables with traditional disease severity measures in RA. They examined plasma tumor necrosis factor-α, soluble interleukin-2 receptor, sCD4 and sCD8 (and the sCD4/sCD8 ratio), neopterin, and fibrin D-dimer in relationship to the erythrocyte sedimentation rate, physician assessment of disease activity, joint pain count, grip strength, and Arthritis Impact Measurement Scale scores. RA patients had higher mean levels of all these immune and inflammation variables except sCD4 compared to healthy subjects. These variables were also related to most of the traditional clinical measures, but they showed no special advantage in assessing disease activity. Wolf et al. (39) found similar relationships in a study of cytokines (IL-1α and IL-2) and soluble IL-2 receptors, but again did not conclude that there were any special advantages. Measuring IL-6 may be more appropriate as this cytokine stimulates hepatocytes to produce acute-phase reactants and B lymphocytes to produce immunoglobulin. Dasgupta and his colleagues (40) have provided some evidence to support the use of IL-6 in disease assessment. But its role as a surrogate marker remains under debate.

V. DATA ANALYSIS

Assessment of efficacy should include an intent-to-treat analysis whenever possible. There are advantages in evaluating those patients who complete a course of therapy, but they should not be studied exclusively. In controlled trials where dropout to allow open-label therapy occurs after documented treatment failure, a time to treatment failure analysis should be included. When possible, patients should be followed for 12 months to further define the safety and activity of longer-duration therapy and permit X-ray assessment of disease progression. Although interest is often generated by studies of the mechanisms of action for a biological therapy, it is essential to ascertain activity by pragmatic changes in clinical measures.

Wells et al. (41) examined the point at which differences in clinical assessment scores on physical ability, pain, and overall condition are sufficiently large to correspond to a subjective perception of a meaningful difference from the perspective of the patient. The patients rated themselves as "somewhat better" when they had a (mean) 7% better score on the HAQ, 6% less pain, and 9% better global assessment. Anderson and Chernoff (42) examined the efficiency of measures that may be used to make statistical comparisons of sensitivities of trial outcome measures. Efficiency was expressed as the mean change divided by the standard deviation of change. Variability and correlations of efficiencies for typical RA trial outcomes are described. From among a variety of joint assessments and other clinical measures, pain and global measures, and health status and laboratory measures, they found that joint tenderness and pain measures were the most sensitive in trials of various antirheumatic drugs.

VI. CONCLUSIONS

The development of new therapies in RA requires both that new biological products be available and that they be tested in well-organized clinical trials. Some general principles for study design are summarized in Table 2. Whether

Table 2 Outline of Clinical Trials Methodology with Biologics

Key components of clinical trial strategy

1. Prospective randomized controlled clinical trials of central importance
2. Detailed dosing and dose scheduling work needed early in clinical development
3. Careful selection of a patient population with active, responsive disease
4. Use of the WHO/ILAR core set of 7 clinical outcome measures
5. Performance of an intent-to-treat analysis as well as a completer analysis
6. Regular consideration of long-term open-label treatment and follow-up of patients after protocol completion

all these proposals have been met will be discussed in the ensuing chapters of this volume.

REFERENCES

1. Strand V, Scott DL, Panayi GS. Evaluating biologic agents in rheumatoid arthritis: a framework for clinical trials. J Rheumatol 1994; 21:1390–1392.
2. Pullar T, Capell HA. A rheumatological dilemma—is it possible to modify the course of rheumatoid arthritis? Can we answer the question? Ann Rheum Dis 1985; 44:134–140.
3. Dawes PT, Tunn PD, Scott DL. Combination therapy in rheumatoid arthritis. Br J Clin Pract 1987; 41(Suppl 52):28–35.
4. Porter DR, Capell HA, Hunter J. Combination therapy in rheumatoid arthritis—no benefit of addition of hydroxychloroquine to patients with a suboptimal response to intramuscular gold therapy. J Rheumatol 1993; 20:645–649.
5. Strand V, Lee ML. Differential patterns of response in patients with rheumatoid arthritis following administration of an anti-CD5 immunoconjugate. Clin Exp Rheumatol 1993; 11:S161–163.
6. Legerton C, Callahan L, Marcum SB, et al. Shorter duration of disease is associated with a higher likelihood of response in a clinical trial of rheumatoid arthritis. Arthritis Rheum 1991; 34:S36.
7. Tugwell P, Boers M. OMERACT conference on outcome measures in RA clinical trials: conclusion. J Rheumatol 1993; 20:590–594.
8. Felson DT, Anderson JJ, Boers M, et al. The American College of Rheumatology preliminary core set of disease activity measures for rheumatoid arthritis clinical trials. The Committee on Outcome Measures in Rheumatoid Arthritis Clinical Trials. Arthritis Rheum 1993; 36:729–740.
9. Prevoo ML, van Riel PL, van-'t Hof MA, van Rijswijk MH, van Leeuwen MA, Kuper HH, van de Putte LB. Validity and reliability of joint indices: a longitudinal study in patients with recent onset rheumatoid arthritis. Br J Rheumatol 1993; 32:589–594.

10. Ritchie DM, Boyle JA, McInnes JM, et al. Clinical studies with an articular index for the assessment of joint tenderness in patients with rheumatoid arthritis. Q J Med 1968; 37:393–406.
11. Thompson PW, Silman AJ, Kirwan JR, Currey HLF. Articular indices of joint inflammation in rheumatoid arthritis. Arthritis Rheum 1987; 30:618–623.
12. Fuchs HA, Pincus T. Reduced joint counts in controlled clinical trials in rheumatoid arthritis. Arthritis Rheum 1994; 37:470–475.
13. Scott DL, Panayi GS, van Riel PLCM, van de Putte LBA, Smolen J. Disease activity in rheumatoid arthritis: preliminary report of the Consensus Study Group of the European Workshop for Rheumatology Research. Clin Exp Rheumatol 1992; 10:521–525.
14. Scott DL, Panayi GS, van Riel PLCM. Variations between centres when assessing disease activity. Clin Rheumatol 1993; 12:37.
15. Hart LE, Tugwell P, Buchanan WW, Norman GR, Grace EM, Southwell D. Grading of tenderness as a source of interrater error in the Ritchie articular index. J Rheumatol 1985; 12:716–717.
16. Lewis PA, O'Sullivan MM, Rumfeld WR, Coles EC, Jessop JD. Significant changes in Ritchie scores. Br J Rheumatol 1988; 27:32–36.
17. Kwoh CK, O'Connor GT, Regan-Smith MG, Olmstead EM, Brown LA, Burnett JB, Hochman RF, King K, Morgan GJ. Concordance between clinician and patient assessment of physical and mental health status. J Rheumatol 1992; 19:1031–1037.
18. Parker J, Frank R, Beck N, Finan M, Walker S, Hewett JE, Broster C, Smarr K, Smith E, Kay D. Pain in rheumatoid arthritis: relationship to demographic, medical, and psychological factors. J Rheumatol 1988; 15:433–437.
19. Drosos AA, Lanchbury JS, Panayi GS, Moutsopoulos HM. Rheumatoid arthritis in Greek and British patients: a comparative clinical, radiologic, and serologic study. Arthritis Rheum 1992; 35:745–748.
20. Mason JH, Meenan RF, Anderson JJ. Do self-reported arthritis symptom (RADAR) and health status (AIMS2) data provide duplicative or complementary information? Arthritis Care Res 1992; 5:163–172.
21. van der Heijde DM, van-'t Hof MA, van Riel PL, van Leeuwen MA, van Rijswijk MH, van de Putte LB. Validity of single variables and composite indices for measuring disease activity in rheumatoid arthritis. Ann Rheum Dis 1992; 51:177–181.
22. Felson DT, Anderson JJ, Boers M, et al. American College of Rheumatology preliminary definition of improvement in rheumatoid arthritis. Arthritis Rheum 1995; 38:727–735.
23. Van Gestel AM, Prevoo MLL, van der Putte LBA, van Riel PLCM. Development of response criteria for rheumatoid arthritis based on the disease activity score. Arthritis Rheum 1994; 37:S334.
24. Sharp JT, Lidsky MD, Collins LC, Morland J. Methods of scoring the progression of radiological changes in rheumatoid arthritis. Arthritis Rheum 1971; 14:706–720.
25. Larsen A, Dale K, Eek M. Radiographic evaluation of rheumatoid arthritis and related conditions by standard reference films. Acta Radiol (Diagn) 1977; 18: 481–491.
26. Sharp JT, Young DY, Bluhm GB, et al. How many joints in the hands and wrists should be included in a score of radiologic abnormalities used to assess rheumatoid arthritis? Arthritis Rheum 1985; 28:1326–1335.

27. Grindulis KA, Scott DL, Struthers GR. The assessment of radiological changes in the hands and wrists in rheumatoid arthritis. Rheumatol Int 1983; 3:39–42.
28. Co-operating Clinics Committee of the American Rheumatism Association. A controlled trial of gold salt therapy in rheumatoid arthritis. Arthritis Rheum 1973; 16: 353–358.
29. Sigler JW, Bluhm GB, Duncan H, et al. Gold salts in the treatment of rheumatoid arthritis: a double-blind study. Ann Intern Med 1974; 80:21–26.
30. Co-operating Clinics Committee of the American Rheumatism Association. A controlled trial of cyclophosphamide in rheumatoid arthritis. N Engl J Med 1970; 282: 883–889.
31. Van Der Heijde DM, van Riel PL, Nuver-Zwart IH, et al. Effects of hydroxychloroquine and sulphasalazine on progression of joint damage in rheumatoid arthritis. Lancet 1989; 1:1036–1039.
32. Richmond BJ, Powers C, Piraino DW, et al. Diagnostic efficacy of digitized images vs plain films: a study of the joints of the fingers. Am J Roentgenol 1992; 158:437–441.
33. Dacre JE, Buckland-Wright JC. Radiological measures of outcome. Clin Rheumatol (Int Pract Res) 1992; 6:39–68.
34. Foley Nolan D, Stack JP, Ryan M, Redmond U, Barry C, Ennis J, Coughlan RJ. Magnetic resonance imaging in the assessment of rheumatoid arthritis—a comparison with plain film radiographs. Br J Rheumatol 1991; 30:101–106.
35. Rominger MB, Bernreuter WK, Kenney PJ, Morgan SL, Blackburn WD, Alarcon GS. MR imaging of the hands in early rheumatoid arthritis: preliminary results. Radiographics 1993; 13:37–46.
36. Corvetta A, Giovagnoni A, Baldelli S, Ercolani P, Pomponio G, Luchetti MM, Rinaldi N, De Nigris E. MR imaging of rheumatoid hand lesions: comparison with conventional radiology in 31 patients. Clin Exp Rheumatol 1992; 10:217–222.
37. Jevtic V, Watt I, Rozman B, Kos-Golja M, Rupenovic S, Logar D, Presetnik M, Jarh O, Demsar F, Musikic P, et al. Precontrast and postcontrast (Gd-DTPA) magnetic resonance imaging of hand joints in patients with rheumatoid arthritis. Clin Radiol 1993; 48:176–181.
38. Beckham JC, Caldwell DS, Peterson BL, Pippen AM, Currie MS, Keefe FJ, Weinberg JB. Disease severity in rheumatoid arthritis: relationships of plasma tumor necrosis factor-alpha, soluble interleukin 2-receptor, soluble CD4/CD8 ratio, neopterin, and fibrin D-dimer to traditional severity and functional measures. J Clin Immunol 1992; 12:353–361.
39. Wolf RE, Brelsford WG, Hall VC, Adams SB. Cytokines and soluble interleukin 2 receptors in rheumatoid arthritis. J Rheumatol 1992; 19:524–528.
40. Dasgupta B, Corkill M, Kirkham B, Gibson T, Panayi G. Serial estimation of interleukin 6 as a measure of systemic disease in rheumatoid arthritis. J Rheumatol 1992; 19:22–25.
41. Wells GA, Tugwell P, Kraag GR, Baker PR, Groh J, Redelmeier DA. Minimum important difference between patients with rheumatoid arthritis: the patient's perspective. J Rheumatol 1993; 20:557–560.
42. Anderson JJ, Chernoff MC. Sensitivity to change of rheumatoid arthritis clinical trial outcome measures. J Rheumatol 1993; 20:535–537.

II
Cellular Targeted Therapies

The central role of T cells in the initiation and perpetuation of disease has been clearly shown in murine models of autoimmunity. Monoclonal antibodies (MAbs) were developed to target cell surface antigens on T cells, attempting either to remove the "offending" subset or interfere with its function. Although initial pilot studies utilizing murine antibodies were positive, concern over immunogenicity and effectiveness with retreatment led to the development of chimeric and subsequently primatized and humanized MAbs. Toxin conjugates (immunoconjugates) were also employed in an attempt to more effectively remove the targeted cell population.

An early biologic agent used in the treatment of rheumatoid arthritis (RA), and in pilot studies in type I diabetes (IDDM), systemic lupus (SLE), and systemic sclerosis (SS), was an immunoconjugate of murine IgG1 MAb against CD5, coupled to ricin, a ribosomal inhibitory protein (Chapter 2). Although open label trials in 76 patients reported clinical benefit, a subsequent placebo controlled trial in 104 did not. Of significance was the placebo response, which exceeded that in the active treatment groups at days 60 and 90. Adverse effects included constitutional symptoms and myopathy. Transient T cell depletion did not correlate with clinical response.

Other agents were developed to target activated T cells. Treatment with both murine and chimeric MAbs against the T cell activation antigen, CD7, failed to show benefit in a pilot study in RA (1). A humanized MAb to CD25, the high affinity IL-2 receptor, anti-TAC, is under evaluation in transplantation and autoimmune disorders. A fusion protein of IL-2 and the enzymatically active portion of diphtheria toxin, DAB_{389} IL-2, was initially evaluated in a short-term trial in RA as well as IDDM (Chapter 3). Adverse events included constitutional symptoms and elevated hepatic transaminases. Currently DAB_{389} IL-2 is being

II: Cellular Targeted Therapies

studied in psoriasis, where clinical benefit in a placebo randomized controlled trial (RCT) was reported.

Monoclonal antibodies to CD4 comprise the largest clinical experience with T cell targeted agents. Initial reports utilizing several different murine MAbs in open label trials in RA, juvenile RA, SLE, and multiple sclerosis (MS) were positive (2). Subsequently, a chimeric anti-CD4 MAb, cM-T412, was evaluated in two RCTs in RA in patients with early disease as well as those receiving methotrexate (Chapters 4 and 5). Neither showed significant clinical responses in comparison to placebo, despite profound CD4 T cell depletion of as long as 30 months. Adverse events included constitutional symptoms. Two "non depleting" and anti-CD4 MAbs are now under evaluation in RA: a primate/human IgG1 MAb, CE9.0 (Chapter 6), and a humanized IgG4 MAb. Preliminary reports are positive, but further study will be required to determine if these agents can offer benefit without associated T cell depletion.

A humanized MAb to the CD52 antigen, CAMPATH 1H, has been studied in RA, systemic vasculitis, and MS (Chapters 7 and 8). It has been demonstrated to effectively deplete lymphocytes, causing longstanding CD4 cytopenia for as long as 36 months after administration. Acute treatment-associated infections have been common, and accounted for two deaths in a total of 140 RA patients treated in pilot or single blind trials. Adverse effects included fever, rigors, nausea, and hypotension, attributed to lympholysis. Despite these significant biological effects, clinical responses were only of several months duration.

In addition to the newer, "non depleting" anti-CD4 MAbs, which may inhibit T cell activation, several other agents have been designed to interfere with the interaction between activated T cells and B cells. These include a MAb to the gp39 ligand on activated T cells, and CTLA-4 Ig, which blocks the binding of B7 to CD28. Further work will be necessary to determine whether targeting a T cell population can offer significant clinical benefit in the treatment of autoimmune diseases.

REFERENCES

1. Kirkham BW, Thien G, Pelton B, et al. Chimeric CD7 monoclonal antibody therapy in rheumatoid arthritis. J Rheumatol 1992; 48:428–433.
2. Burmester GR, Hiepe F, Horneff G, Emmrich F. The experience with murine anti-CD4 monoclonal antibodies. In Strand V, Johnson K, Simon L, eds. Biologic Agents in Autoimmune Diseases IV. Arthritis Foundation, 1996:125–136.

2
Anti-CD5/Ricin A Chain Immunoconjugate Therapy in Rheumatoid Arthritis

JOHN J. CUSH
University of Texas Southwestern Medical Center at Dallas, Dallas, Texas

I. INTRODUCTION

Rheumatoid arthritis (RA) is characterized by the intense infiltration of activated T lymphocytes into the synovium. These activated T cells are thought to play an instrumental role in the initiation and perpetuation of chronic rheumatoid synovitis (1–3). Although the traditional use of slow-acting antirheumatic drugs (SAARDS) intends to down-regulate the chronic inflammatory response within the synovium, such agents are limited by their incomplete, unpredictable, and poorly sustained clinical benefits (4,5). Moreover, these agents are largely nonspecific in their alteration of effector immune responses (4). For these reasons alternative therapies have emerged to selectively interfere with leukocyte number and function. Therapies such as thoracic duct drainage, total lymphoid irradiation, leukapheresis, and cyclosporine have been studied because of their ability to abrogate circulating T-cell/leukocyte number or activity (6). Unfortunately, the clinical utility of such therapies has been hampered by the limited clinical responses and/or unacceptable toxicities. Newer approaches to therapy in RA have largely focused on the quantitative or qualitative impairment of T-cell function with the hope of achieving greater disease control.

The utilization of monoclonal antibody (MAb) technology has permitted the selective depletion or functional impairment of cellular subsets by selectively targeting specific surface antigens. Immunoconjugates, or immunotoxins, have been engineered to target specific cellular subsets. An immunoconjugate is comprised of a cytotoxin coupled to a tissue-specific, cell-binding ligand. Commonly used ligands include monoclonal antibodies (MAb), hormones, or growth factors. Most cytotoxins are either plant (e.g., ricin) or bacterial (e.g., diphtheria) modified toxins that lead to intracellular toxicity and/or cell death (7–11). A murine monoclonal anti-CD5/ricin A chain immunoconjugate (Fig. 1) has been developed as an immunosuppressive agent for use in a variety of autoimmune disorders (12). This CD5 immunoconjugate (CD5IC) has been evaluated in patients with RA and other autoimmune states including graft-versus-host disease, systemic lupus erythematosus, type I diabetes mellitus, inflammatory bowel disease, and psoriasis (12–16). This report will focus on the biological and clinical effects of CD5IC in RA.

II. ANTI-CD5/RICIN A CHAIN IMMUNOCONJUGATE

CD5 is the human homolog to the murine Lyt-1 surface determinant that is expressed on a subset of B cells. In mice the Lyt-1 subset has been implicated as a primary source of certain IgM autoantibodies. In humans, CD5 is present on greater than 90% of mature T lymphocytes and up to 10% of all circulating B cells (6). However, the production of autoantibodies in humans is not limited to CD5+ B cells alone. Instead CD5 identifies a subset of memory, activated B cells (17,18). Furthermore, CD5 is capable of functioning as a costimulatory molecule on proliferating T cells (19,20). A murine MAb against CD5 was therefore chosen as an immunosuppressive agent because of its ability to target mature T cells and

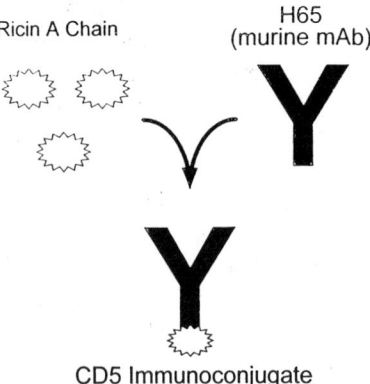

Figure 1 Schematic diagram of CD5 immunoconjugate construct.

a subset of B cells, its binding affinity, and its ability to be covalently bound to ricin A chain as an immunoconjugate.

CD5IC couples an anti-CD5 IgG1 murine MAb (H65) with ricin A chain by a disulfide bond (Fig. 1). Ricin is a 65-kDa heterodimeric glycoprotein from the plant *Ricinus communis,* composed of an A chain and B chain. While the A chain is inactive extracellularly and possesses the ribosome-inactivating protein, the B chain possesses a lectin-binding domain that attaches to cells manifesting appropriate glycoprotein receptors. The cytotoxic ricin A chain is dependent on the B chain or other conjugated carrier molecules (e.g., MAb) to access the cytosol. After binding its ligand, the toxin is endocytosed and is routed via clathrin-coated vesicles to acidic endosomes and subsequently to the Golgi network and cytoplasm where it acts as a potent inhibitor of protein synthesis (7–9). The appeal of CD5IC as an immunotoxin is based on its tissue specificity, high affinity for its cellular ligand, ability to be endocytosed, and in vitro cellular effects.

In vitro studies have shown that once bound to CD5+ cells, the immunoconjugate is internalized and leads to cell death (10,11). The biological effects of the CD5IC are evident from studies demonstrating significant reductions in mitogen and allogeneic induced T-cell proliferation in vitro (10,21) and the rapid reduction of circulating CD5+ T cells with in vivo administration (13,14). Anti-CD5 MAbs without ricin A chain have also been shown to augment IL-2-induced proliferation of circulating peripheral blood T cells (19,20) and conversely to suppress IL-2-induced proliferation of synovial fluid T cells (20).

A. CD5IC—Mechanism of Action

The rationale for a pan-T-cell immunotoxin has been suggested by the limited success of other T-cell-targeted therapies and sporadic reports of clinical improvement in RA patients infected with HIV (1,6,22). It has also been shown that anti-CD5 MAb therapy decreases synovitis and disease severity in mice with established collagen-induced arthritis (23). Moreover, the presence of CD5 on all mature T cells and a subset of activated B cells and the in vitro demonstration of cytotoxicity and antiproliferative effects of CD5IC underscore the potential of target CD5 with a murine MAb-ricin A chain immunoconjugate. Nonetheless, several findings shed some doubt on this approach or may contribute to its limited success. This includes the lack of CD5 B cells in RA synovium (24), disparate functional roles for CD5 in human and mouse (19), and the inability of CD5IC to consistently alter serum RF titers (14,25).

Previous in vitro and in vivo experience therefore suggests that any clinical benefits appreciated using CD5IC therapy are likely to result from selective targeting and depletion of T cells with an immunoconjugate. In vivo cytotoxicity has been presumed because of the 3.5-hr drug half-life, a nearly month-long depletion of CD3+CD5+ T cells from the circulation, and the observed in vitro cytotoxic effects of ricin A chain. Such claims, however, are presumptive at best, and in vivo studies to support such a mechanism are lacking. Also unknown is whether previously used doses of clinically administered CD5IC exert any effect on the synovium or if this agent merely binds to circulating rather than intraarticular T cells. It is possible that the ricin moiety of CD5IC may not be internalized after binding and only occupy the receptor without resultant cytotoxicity. It is also possible that CD5IC may opsonize circulating immunoreactive T cells and B cells that are then taken up by the reticuloendothelial system for clearance. The net effect may be a down-regulation in the number and function of T cells, sufficient to produce clinical improvement.

Verwilghen and co-workers have also shown that MAb against CD5 alone (without conjugation to ricin) may exert differing effects in vivo (20). They have demonstrated that the addition of anti-CD5 MAb inhibits IL-2-induced synovial fluid T-cell proliferation, but tends to augment IL-2-induced proliferative responses in peripheral blood T cells from RA patients. Thus engagement of this costimulatory molecule may have a distinct "switching off" effect on activated T cells within the joint. Moreover, these studies suggested that CD5 may have differential regulatory roles for resting and activated T cells by affecting naïve and memory and naïve T cells differently (20). These authors suggested that the efficacy of CD5-directed therapies may not depend on T-cell depletion, but rather on the functional impairment of immunoreactive cells within the synovium.

B. Immunoconjugate Toxicity

The toxicities associated with the use of immunoconjugates have been well described (Table 1) (8,9,14) and relate to either the toxin alone or the host's immune response to the tissue-specific ligand (i.e., murine MAb), toxin, or both (immunoconjugate). A common immunological sequela to immunotoxin therapy includes the development of human anti-mouse (HAMA) or human anti-immunoconjugate (HAIA) antibody responses. Although HAMA and HAIA commonly occur with CD5IC treatments and retreatment, their presence does not seem to predict toxicity or dampen clinical efficacy (14,26).

The *cytokine release syndrome* is often associated with CD5IC administration and is characterized by flu-like symptoms such as fever, chills, malaise, myalgias, headache, or nausea (27). The cytokine release syndrome has also been ascribed to other anti-T-cell MAb preparations (i.e., anti-CD3, anti-CD4, anti-CD7, anti-CD52 MAb) and is thought to result from the MAb-mediated T-cell activation and release of cytokines such as TNF-α and possibly IFN-γ (27). Whether a similar mechanism underlies these manifestations in CD5IC therapy is unknown.

Anti-CD5/Ricin A Chain Immunoconjugate

Table 1 Side Effects Associated with Anti-CD5/RTA Immunoconjugate Therapy

Common (>15%)	Uncommon (5–15%)		Rare (<5%)
Weight gain	Abdominal pain	Cough	Anaphylactoid reaction
Edema	Anorexia	Conjunctivitis	Vasculitis
Hypoalbuminemia[a]	Diarrhea	Visual blurring	(palpable purpura)
Rash/urticaria	Dyspepsia	Insomnia	Pulmonary edema
Pruritus	Flatulence	Somnolence	Synovitis
Fever	Arthralgia	Dizziness	Chest "tightness"
Headache	Chills	Tachycardia	Pericardial effusion
Nausea/vomiting	Flushing	Palpitations	Dehydration
Fatigue/malaise	Diaphoresis	Paresthesia	CPK elevation
Myalgias	Telangiectasias		Myopathy
Elevated hepatic enzymes			Rhabdomyolysis

[a]Defined as >20% decrease from baseline values.

Clinical toxicity may also manifest as anorexia, exanthems, pruritus, weight gain, edema, hypoalbuminemia, or other features of the *capillary leak syndrome* (27). The capillary leak syndrome is characterized by low serum albumin, weight gain, edema, and uncommonly hypotension, wheezing, or pulmonary edema. It has also been observed with other biological therapies, such as IL-2, GM-CSF, and murine anti-CD3 MAb therapy. The etiology of the capillary leak syndrome is unclear, but a recent report suggests that ricin A chain is capable of rapidly inducing changes in endothelial cell morphology and permeability that might account for the capillary leak syndrome (28). Finally, ricin A is rich in mannose- and fucose-containing oligosaccharides that are recognized by receptors on Kupffer cells. If this toxin were to become unconjugated, it might potentially result in hepatotoxicity (7–9,29). Myalgias are commonly observed and there have been rare reports of myopathy or rhabdomyolysis in association with CD5IC use. Mechanisms underlying these manifestations have not been elucidated (8).

III. CLINICAL EXPERIENCE WITH CD5IC

A. Uncontrolled Clinical Trials in RA

The results of initial clinical trials utilizing CD5IC first appeared in 1989 (13). Sixteen patients were treated in a phase I, open-label, dose-escalating trial for 5–9 days. Depletion of CD5+ T cells was accompanied by clinical improvement in nearly half of the patients and 25% of patients demonstrated sustained clinical responses for 3 months. This encouraging efficacy was associated with limited toxicity and suggested that further trails should be performed. Importantly, this early trial helped to establish the doses used in subsequent trials.

Two phase II open-label trials in RA were initiated in 1989 and the combined results were reported by Strand et al. in 1993 (14). Seventy-nine patients received open label CD5IC at a dose of either 0.20 or 0.33 mg/kg/day for 5 consecutive days. Patients were enrolled after demonstrating they had active disease and had failed at least one SAARD. Twenty-five patients had early disease with a mean disease duration of 1.8 years and the remaining 54 patients had longstanding disease with a mean disease duration of 10.7 years and were largely refractory, having received a mean of 4.3 SAARDS prior to enrollment. In the latter group, 32 patients continued to receive maintenance methotrexate or azathioprine during the CD5IC trial. Clinical responses were ascertained using the response criteria established by Paulus et al. (30). The Paulus criteria for clinical response required at least 20% improvement in at least four of six predefined variables.

Although 55% of patients were improved during the first week following treatment, maximal clinical responses were observed 1 month postinfusion with CD5IC, such that 68% of patients were "responders" by the Paulus criteria (Fig. 2). However, subsequent monthly evaluations demonstrated a rapid decline in response rates, and only 32%, 33%, 25%, and 15% of patients remained responders for 2 months, 3 months, 6 months, and 12 months, respectively. Although nearly one-third of patients demonstrated a 50% reduction in acute-phase reactants (erythrocyte sedimentation rate and C-reactive protein) or serum rheumatoid factor titers, these results were not consistent.

Regardless of clinical status, all patients demonstrated a significant lymphopenia and a median maximal depletion of CD3+ T cells to 21% of baseline values between day 2 and 5 of treatment. Circulating CD3+ T cell numbers returned to greater than 50% of pretreatment values in 57% of patients 1 month posttreatment. CD5+ B cells were also depleted but returned to normal by day 15 or day 29. Circulating neutrophils and monocytes were not altered as a result of therapy. Thus, a 5-day infusion of CD5IC produced a disproportionate depletion of CD3+CD5+ T cells from the circulation. Although the absolute numbers of other T-cell subsets (e.g., CD4, CD8, CD4/CD29, CD4/CD45RA, CD4RO, HLA-DR, TCR$\alpha\beta$, and TCR$\gamma\delta$) demonstrated a parallel decline in lymphocyte counts (21,26), the percentage of lymphocytes expressing these surface markers was not significantly altered as a result of CD5IC therapy. Neither the clinical results nor phenotypic data were significantly affected by concomitant therapy with methotrexate or azathioprine (14).

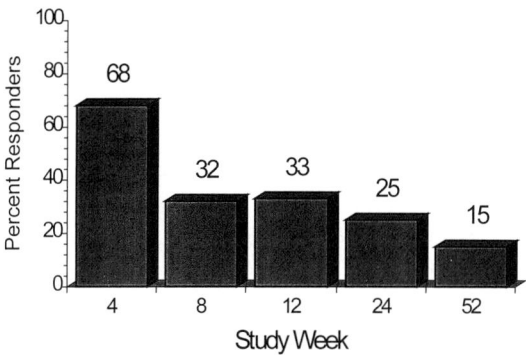

Figure 2 Clinical results of CD5IC administration in RA. (Adapted from Ref. 14.)

B. Toxicity

Adverse effects were observed in the majority of patients reported by Strand et al. (14). However, for most, the side effects were mild to moderate, self-limited, and seldom required interventions or dose adjustments. Moreover, the vast majority of patients who went on to receive a second course of CD5IC did so despite prior adverse events. Only six of 79 patients (7.6%) exited the study because of an adverse event (14). Table 1 indexes the side effects observed in this and other clinical trials of CD5IC (13,14,31). Constitutional complaints and edema were most common. Constitutional complaints included fever, fatigue, malaise, myalgias, or a "flu-like" syndrome and were observed during or immediately following the 5-day infusion. Self-limiting edema, weight gain, and/or hypoalbuminemia were frequently seen 7–10 days following treatment. Also common were reports of rash, urticaria, pruritus, and elevation of hepatic enzymes.

Weight gain and edema were common manifestations, suggesting that the capillary leak syndrome may result from CD5IC administration (8). Infrequent, rare, or serious complications of CD5IC have been described (18) and may include pulmonary edema, anaphylactoid reactions, palpable purpura, vasculitis, immune-complex-mediated phenomenon (i.e., acute postinfusion synovitis, urticaria, and chest tightness), myopathy, and rhabdomyolysis (Table 1).

A dose-related increase in constitutional symptoms, edema, rash, or myalgias was more common in those receiving CD5IC 0.33 mg/kg/day compared with the 0.20 mg/kg/day treatment group (14). Nonetheless, for most patients increased toxicity was not correlated with retreatment, the presence of HAMA, human anti-immunoconjugate antibodies (HAIA), or the use of concomitant SAARDS. In fact, patients receiving a second course of CD5IC demonstrated less toxicity compared with the first course of therapy (26). It is possible that the presence of HAIA may have limited the tissue exposure to infused immunotoxin, thereby limiting the number of adverse events due to CD5IC.

C. Retreatment

The study reported by Strand et al. (14) suggests that the success of CD5IC may be limited by a relatively short duration of clinical efficacy. Another objective of this study was to assess the safety and efficacy of repeated CD5IC administration in those patients consenting to retreatment. Of the 79 patients reported, retreatment with CD5IC was undertaken in 15 patients (26). These patients received comparable or lower doses of CD5IC (0.20–0.33 mg/kg/day) for an additional 5 days and were assessed in a similar manner. Clinical efficacy similar to that observed during the first course was seen. Although nearly two-thirds of patients were "responders" according to the Paulus criteria at 1 month, only one-third were still responding 3 months following retreatment. In this small group of patients, it appeared that those who responded to the first course of CD5IC were most likely to respond to retreatment. Clinical improvement during retreatment with CD5IC was not related to the degree of T-cell depletion, HAMA or HAIA titers, or use of concomitant azathioprine or methotrexate (unpublished observation).

One patient (#349) received three courses of open-label CD5IC therapy. For this patient, CD5IC and low-dose azathioprine were the mainstay of therapy for nearly 17 months. The first two courses produced beneficial effects and were without significant toxicity (Table 2). Interestingly, during the second course of CD5IC she demonstrated clinical improvement with little toxicity despite the presence of high-titer HAMA, HAIA (data not shown), and anti-idiotypic antibodies. The third course of CD5IC was aborted after 4 days because of postinfusion urticaria, pruritus, chest tightness, and acute synovitis in joints previously affected by RA on days 2, 3, and 4. An immune complex etiology for such findings was surmised from the presence of high-titer HAMAs and HAIAs, temporal onset with infusion, and resolution of symptoms within 1 hr of treatment with 200 mg of IV hydrocortisone. These events suggest that a considerable amount of murine immunoconjugate must be given repeatedly to elicit an immune response by the host that might result in toxicity.

Although none of the CD5IC-treated patients experienced opportunistic infections, this same patient developed an acute staphylococcal (*Staphylococcus aureus*) septic arthritis of the wrist following the abbreviated third course of immunoconjugate therapy. This event was judged to be a likely result of bacteremia induced by repeated venipuncture, as both RA and repeated instrumentation are known risk factors for septic arthritis.

Table 2 Patient #349: Clinical, Pharmacokinetic, and Immunological Effects of Three Courses of CD5IC

	1st course			2nd course			3rd course[b]		
Study week	0	1	4	12	13	16	36	37	40
Tender joint score	46	19	25	35	34	13	49	16	23
CD5+CD3+ cells[a]	860	0	493	365	254	257	322	257	326
Peak CD5+ levels	1520	ND	ND	100	ND	ND	32	ND	ND
HAMA (reciprocal titer)	100	6.9 $\times 10^3$	3.8 $\times 10^3$	100	5.2 $\times 10^5$	2.1 $\times 10^5$	100	2.2 $\times 10^6$	1.9 $\times 10^6$
Blocking antibody (reciprocal titer)	NS	NS	NS	NS	<25	NS	ND	ND	ND
Adverse drug reactions		Edema, weight gain, tachycardia			Elevated liver function tests[c]			Rash, flushing, pruritus, edema, acute synovitis, chest tightness	

[a]Number of cells/mm^3, normal range 805–1765 cells/mm^3.
[b]Therapy withdrawn after 4 days of therapy secondary to ADRs.
[c]Possibly secondary to concomitant therapy with azathioprine.
ND = not performed; NS = not significant.

D. Immunological Response to Therapy

A humoral response to both the murine antibody (HAMA) and whole immunoconjugate (HAIA) was observed in nearly all of the patients receiving open-label CD5IC (14,26). Marked anamnestic responses, with very high HAMA titers, were observed in those patients who received a second or third course of CD5IC (Table 2). HAIA titers were similarly present upon retreatment (unpublished observation). Anti-idiotypic antibodies (blocking antibodies) were assessed in nine patients treated in the study by Strand (32). None of the patients developed blocking antibodies after the first course of CD5IC and only two patients developed markedly elevated anti-idiotypic antibodies upon retreatment. The remaining patients developed variably low titers of blocking antibodies with retreatment. Anti-idiotypic antibodies did not alter the efficacy or toxicity of the readministered immunoconjugate (32).

E. Pharmacokinetics

Peak serum drug levels were observed immediately postinfusion and were reduced by nearly 90% 8 hr later. Thus the early peak in CD5IC levels was associated with the subsequent decline in circulating CD3+CD5+ T cells. The mean terminal half-life of CD5IC has been shown to be 1.5 hr in graft-versus-host disease and 3.5 hr in patients with RA (14). Retreatment with CD5IC was associated with markedly lower peak drug levels, especially in those patients with marked HAMA titers. The presence of elevated HAMA titers and low levels of CD5IC precluded the reliability of pharmacokinetic determinations during retreatment. For both the initial and retreatment cycles, no correlation between drug levels and therapeutic efficacy was observed.

F. Controlled Trial of CD5IC

The results of a double-blind, placebo-controlled trial of CD5IC in RA were reported in 1994 at the 58th Annual Meeting of the American College of Rheumatology in Minneapolis (32). A total of 109 patients with early, active RA were enrolled in a multicenter trial. Patients who failed more than one SAARD or who had a disease duration more than 8 years were excluded. Because of results obtained in earlier trials and the increased toxicity observed with the 0.33 mg/kg/day regimen, the dosage was reduced and the schedule was altered to minimize the frequency, quantity, and immunogenicity of CD5IC, yet allow for maximal depletion of CD5+ T cells. Thus, a treatment "cycle" consisted of CD5IC administered intravenously on days 1, 2, 29, 30. Patients received either placebo, 0.2, 2.0, 8.0 mg/m^2/day in the blinded portion or cycle 1. Whereas the 0.2 mg/m^2/day dose was given as a "very low" or inconsequential dose, the 8.0 mg/m^2/day or "high" dose was calculated to be equivalent to the 0.20 mg/kg/day dose used in earlier trials. Patients with an insufficient response or disease flare were eligible to receive open-label CD5IC at 8 mg/m^2/day in cycle 2 ($n = 51$) or cycle 3 ($n = 6$). Patients enrolled in this placebo-controlled trial differed from those reported by Strand et al. (14) because of a lower mean disease duration (32 months), and only half of the patients had previously received and/or failed one SAARD.

During cycle 1 CD3+ T-cell depletion to <50% of baseline was observed in 77% of the high-dose (8.0 mg/m^2/day) group and 28%, 20%, and 11% of the middose, very-low-dose, and placebo-treated patients, respectively. Regardless of the level of the T-cell depletion, clinical response rates at day 43 were greatest in the placebo-treated group (49%). Placebo response rates were significantly greater than those in the very-low-dose (37%), middose (42%), and high-dose (33%) treatment groups (32). During subsequent follow-up, none of the CD5IC-treated patients demonstrated efficacy superior to the placebo-treated patients. Whereas adverse effects were generally similar to those previously described (Table 1), a decreased frequency and severity of adverse events was noted. Moreover, edema and features of the capillary leak syndrome were rarely found in CD5IC-treated patients. Finally, the number of side effects did not increase for those receiving retreatment.

This well-controlled trial of "cyclic" CD5IC failed to demonstrate significant clinical efficacy and further questions rationale of biological therapies directed against T cells (33). Moreover, the magnitude of placebo responses in this biological trial suggests that controlled trials requiring parenteral therapies and frequent outpatient or inpatient evaluations may be associated with very high placebo response rates that may hinder new efforts to develop biological therapies for RA. This most recent trial of CD5IC failed either because of insufficient dosing, because trial design errors (e.g., sample size, dose, and frequency of administration) may have obscured the potential benefits of CD5IC, or simply because this modality is ineffective in patients with RA. Finally, this trial and others failed to show any correlation between the level of T-cell depletion and the patient's clinical response to CD5IC. Such data support the claims of some that the ricin A chain may not be necessary and may only add to the toxicity profile of the drug. Thus, the deliverance of sufficient antibody levels with subsequent intra-articular immunosuppression may be more desirable than systemic T-cell depletion via cytotoxicity.

IV. FUTURE DIRECTIONS FOR CD5IC

The ideal biopharmaceutical agent should be one that is easy to administer, produces sustained clinical benefits with acceptable toxicity, possesses a measurable tissue or target specificity, and is nonimmunogenic. The chronic nature of autoimmune diseases, such as RA, demands that such an agent have the capacity for repeated administration with reproducible and safe results (27). Thus, our prior experience with CD5IC therapy in RA has been disappointing because of its limited clinical efficacy, unacceptable immunogenicity, and an uncertain mechanism of action. For these reasons and others, further testing of CD5IC has ceased in RA.

Further development of this agent and other similar therapies may benefit from the lessons learned in these studies and the continued advances in biotechnology. One desirable modification might include the development of a humanized anti-CD5 MAb, one that would be less immunogenic yet retain the binding avidity of the murine MAb H65. It also seems likely that further conjugation with plant or bacterial toxins might be unnecessary if a humanized anti-CD5 were available. Finally, the limited duration of effect and potential immunogenicity may restrict further use of these therapies to selected circumstances or possibly as an accompaniment to traditional SAARD or immunosuppressive therapies. Thus, future trials may utilize immunoconjugates or anti-CD5 therapies as part of a multidrug disease-induction regimen or as a single agent to be used systemically or intraarticularly for rheumatoid flares.

REFERENCES

1. Panayi GS, Lanchbury JS, Kingsley GH. The importance of the T cell in initiating and maintaining the chronic synovitis of rheumatoid arthritis. Arthritis Rheum 1992; 35:729–735.
2. Cush JJ, Lipsky PE. Phenotypic analysis of synovial tissue and peripheral blood lymphocytes isolated from patients with rheumatoid arthritis. Arthritis Rheum 1988; 31:1230–1238.
3. Sewell KL, Trentham DE. Pathogenesis of rheumatoid arthritis. Lancet 1993; 341: 283–290.
4. Kavanaugh AF, Lipsky PE. Gold, penicillamine, antimalarials, and sulfasalazine. In: Gallin JI, Goldstein IM, Snyderman R, eds. Inflammation: Basic Principles and Practice, 2nd ed. New York: Raven Press, 1992:1083–1101.
5. Wolfe F. 50 years of antirheumatic therapy: the prognosis of rheumatoid arthritis. J Rheumatol 1990; 17:24–32.
6. Cush JJ, Lipsky PE. Cellular basis for rheumatoid inflammation. Clin Orthop Rel Res 1991; 265:9–22.
7. Hertler AA, Frankel AE. Immunotoxins: a clinical review of their use in the treatment of malignancies. J Clin Oncol 1989; 7:1932–1942.
8. Vitetta ES, Thorpe PE, Uhr JW. Immunotoxins: magic bullets or misguided missiles? Immunol Today 1993; 14:252–259.
9. Vallera DA. Immunotoxins: will their clinical promise be fulfilled? Blood 1994; 83:309–316.
10. Fishwild DM, Staskawicz MO, Wu HM, Carroll SF. Cytotoxicity against human peripheral blood mononuclear cells and T cell lines mediated by anti-T cell immunotoxins in the absence of added potentiator. Clin Exp Immunol 1991; 86:506–513.
11. Collinson AR, Lambert JM, Liu Y, O'Dea C, Shah SA, Rasmussen RA, Goldmacher VS. Anti-CD6 blocked ricin: an anti-pan T-cell immunotoxin. Int J Immunopharmacol 1994; 16:37–49.
12. Kernan NA, Byers V, Scannon PJ, Mischak RP, Brochstein J, Flomenber N, Dupont B, O'Reilly RJ. Treatment of steroid-resistant acute graft-vs-host disease by in vivo administration of an anti-T cell ricin A chain immunotoxin. JAMA 1988; 259: 3154–3157.
13. Byers VS, Caperton E, Ackerman S, Shepard J, Scannon PJ. Modification of the immune system in patients with rheumatoid arthritis treated with anti-CD5 ricin A chain immunotoxin. FASEB J 1989; 3:A1122.
14. Strand V, Lipsky PE, Cannon GW, Calabrese LH, Wisenhutter C, Cohen SB, Olsen NJ, Lee ML, Lorenz TJ, Nelson B, The CD5 Plus Rheumatoid Arthritis Investigators Group. Arthritis Rheum 1993; 36:620–630.
15. Wacholz MC, Lipsky PE. Treatment of lupus nephritis with CD5 plus, an immunoconjugate of an anti-CD5 monoclonal antibody and ricin A chain. Arthritis Rheum 1992; 35:837–839.

16. Skyler JS, Lorenz TJ, Schwartz S, Eisenbarth GS, Einhorn D, Palmer JP, Marks JB, Greenbaum C, Saria EA, Byer V. Effects of an anti-CD5 immunoconjugate (CD5-plus) in recent onset type I diabetes mellitus: a preliminary investigation. The CD5 Diabetes Project Team. J Diabetes Complications 1993; 7:224–232.
17. Hardy RR. Variable gene usage, physiology and development of Ly-1+ (CD5+) B cells. Current Opin Immunol 1992; 4:181–185.
18. Lydyard PM, Lamour A, MacKenzie LE, Jamin C, Mageed RA, Youinou P. CD5+ B cells and the immune system. Immunol Letters 1993; 38:159–166.
19. Spertini F, Stohl W, Ramesh N, Moody C, Geha RS. Induction of human T cell proliferation by a monoclonal antibody to CD5. J Immunol 1991; 146:47–52.
20. Verwilghen J, Kingsley GH, Ceuppens JL, Panayi GS. Inhibition of synovial fluid T cell proliferation by anti-CD5 monoclonal antibodies. Arthritis Rheum 1992; 35:1445–1452.
21. Fishwild DM, Strand V. Administration of an anti-CD5 immunoconjugate to patients with rheumatoid arthritis: effect of peripheral blood mononuclear cells and in vitro immune function. J Rheumatol 1994; 21:596–604.
22. Orenstein MH, Kerr LD, Spiera H. Reexamination of the relationship between active rheumatoid arthritis and the acquired immunodeficiency syndrome. Arthritis Rheum 1996; 38:1701–1706.
23. Plater-Zyberk C, Taylor PC, Blaylock MG, Maini RN. Anti-CD5 therapy decreases the severity of established disease in collagen type II-induced arthritis in DBA/1 mice. Clin Exp Immunol 1994; 98:442–447.
24. Smith MD, O'Donnell J, Highton J, Palmer DG, Rozenbilds M, Roberts-Thompson PJ. Immunohistochemical analysis of synovial membranes from inflammatory and non-inflammatory arthritides: scarcity of CD5 positive B cells and IL2 receptor bearing T cells. Pathology 1992; 24:19–26.
25. Olsen NJ, Teal GP, Strand V. In vivo T cell depletion in rheumatoid arthritis is associated with increased in vitro IgM–rheumatoid factor synthesis. Clin Immunol Immunopathol 1993; 67:124–129.
26. Cush JJ, Nichols LA, Strand V, Lipsky PE. Treatment of rheumatoid arthritis patients with an anti-CD5 immunoconjugate. Arthritis Rheum 1991; 34(Suppl):S159.
27. Cush JJ, Kavanaugh AF. Biologic interventions in rheumatoid arthritis. Rheum Dis Clin North Am 1995; 21:797–816.
28. Soler-Rodriguez AM, Ghettie MA, Oppenheimer-Marks N, Uhr JW, Vitetta ES. Ricin A-chain and ricin A-chain immunotoxins rapidly damage human endothelial cells: implications for vascular leak syndrome. Exp Cell Res 1993; 206:227–234.
29. Baldwin RW, Byers VS. Monoclonal antibody immunoconjugates for cancer treatment. Curr Opin Immunol 1989; 1:891–894.
30. Paulus HE, Egger MJ, Ward JR, Williams HJ, and the Cooperative Systematic Studies of Rheumatic Diseases Group. Analysis of improvement in individual rheumatoid arthritis patients treated with disease-modifying antirheumatic drugs, based on the findings in patients treated with placebo. Arthritis Rheum 1990; 33:477–484.
31. LeMaistre CF, Rosen S, Frankel A, Kornfeld S, Saria E, Meneghetti C, Drajesk J, Fishwild D, Scannon P, Byers V. Phase I trial of H65-RTA Immunoconjugate in patients with cutaneous T-cell lymphoma. Blood 1991; 78:1173–1182.

32. Olsen NJ, Cush JJ, Lipsky PE, St. Clair EW, Matteson E, Cannon G, McCune WJ, Strand V, Lorenz T. Multicenter trial of an anti-CD5 immunoconjugate in rheumatoid arthritis (RA). Arthritis Rheum 1994; 37(Suppl):S810.
33. van der Lubbe PA, Dijkmans BAC, Markusse HM, et al. Lack of clinical benefit of CD4 monoclonal antibody therapy in early rheumatoid arthritis. Arthritis Rheum 1994; 37(Suppl):S94.

3
Early Clinical Studies of IL-2 Fusion Toxin in Patients with Severe Rheumatoid Arthritis, Recent-Onset Insulin-Dependent Diabetes Mellitus, and Psoriasis

THASIA G. WOODWORTH
Pfizer Central Research, Groton, Connecticut

KAREN PARKER
Seragen, Inc., Hopkinton, Massachusetts

I. BACKGROUND

As a result of its limited distribution on activated lymphocytes and monocytes, the interleukin-2 receptor (IL-2R) is a target for potential treatment of those diseases in which activated lymphocytes or monocytes may play a pathogenic role. In these instances, elimination of IL-2 receptor-expressing activated cells should result in selective immunosuppression, while resting, memory, and other nonactivated cells should be spared. In addition, the IL-2R is a therapeutic target for selected leukemias and lymphomas where malignant cells may constitutively express IL-2R.

IL-2 fusion toxin ($DAB_{486}IL\text{-}2$ and $DAB_{389}IL\text{-}2$), the first of a new class of targeted biologicals called fusion toxins, is specifically cytotoxic for IL-2R-expressing cells. Naturally occurring bacterial and plant toxins often possess specific structure-function activity and, by using genetic engineering techniques, these structure-function relationships can be manipulated to create biological agents that specifically target and kill disease-causing cells. Typically, there is a toxic region, a membrane translocating portion, and a receptor-binding domain that mediates cells targeting. In the case of $DAB_{486}IL\text{-}2$ or $DAB_{389}IL\text{-}2$ (Fig. 1), the receptor-binding domain of diphtheria toxin has been replaced by a specific targeting ligand, human IL-2, creating a recombination protein that kills activated IL-2R expressing lymphocytes at 10^{-10} M to 10^{-11} M concentrations (1).

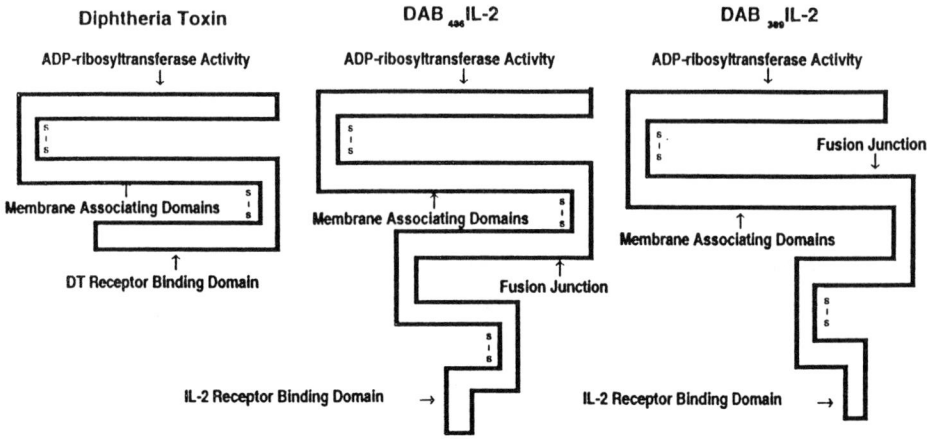

Figure 1 Schematic drawings of diphtheria toxin, DAB$_{486}$IL-2 and DAB$_{389}$IL-2 showing the regions of enzymatic activity (ADP-ribosyltransferase), membrane associating domains, receptor-binding domains and, in the case of DAB$_{486}$IL-2 and DAB$_{389}$IL-2, the fusion junction. Disulfide bonds are also indicated $\begin{pmatrix} s \\ | \\ s \end{pmatrix}$.

DAB$_{486}$IL-2 (67 Kd) was the first fusion toxin to be evaluated clinically; DAB$_{389}$IL-2 (58 Kd) is approximately 10-fold more potent compared to DAB$_{486}$IL-2 with an enhanced therapeutic index in animal disease models. Potency is measured according to the IC$_{50}$ corresponding to 50% inhibition of protein synthesis in a standardized bioassay in which 10 U/ml is approximately 10^{-10} M. Specific activity is expressed in kU/mg.

In contrast to monoclonal antibody (MAb)-targeted immunotoxins (such as the CD5 ricin A conjugate), which are passively and slowly internalized, IL-2 fusion toxin, once bound to the cell membrane IL-2 receptor, is actively and rapidly internalized into an acidic endocytic vesicle. The toxic fragment then reaches the cell cytosol, initiating a cytotoxic event within 5–15 min. Protein synthesis is irreversibly inhibited as a result of ADP ribosylation of elongation factor-2, and cell death occurs over 36–72 hr.

In vitro, nonspecific toxic effects are detectable at concentrations $>10^{-7}$ M; specific effects are noted at concentrations $\leq 10^{-9}$ M. At the highest doses (10-fold higher than doses evaluated in humans) evaluated in 14-day toxicology studies in mice and monkeys, there were hepatocellular and renal tubular effects, all of which were reversible during a 2-week recovery period.

Initial safety and pharmacokinetic analyses as well as evidence for IL-2R-specific cytotoxicity were obtained in patients with IL-2R-expressing malignancies

IL-2 Fusion Toxin in Autoimmune Diseases

(2–7). These studies served as a basis for initiating evaluation of $DAB_{486}IL\text{-}2$ in patients with severe methotrexate refractory rheumatoid arthritis (RA) and phase I/II evaluation in patients with recent-onset insulin-dependent diabetes mellitus (IDDM).

II. EVALUATION OF IL-2 FUSION TOXIN IN AUTOIMMUNE DISEASES

A. Overview

The clinical strategy was to gain early safety and pharmacokinetic, and pharmacodynamic (evidence of IL-2R-specific targeting) evaluation of $DAB_{486}IL\text{-}2$ in patients with refractory IL-2R-expressing malignancies. IL-2 fusion toxin dose levels administered to patients with autoimmune diseases were typically less than 25% of those administered in lymphoma studies. An open-label phase I/II dose-escalation study of $DAB_{486}IL\text{-}2$ in patients with methotrexate refractory severe RA was conducted, followed by a double-blind, placebo-controlled study in patients with active RA refractory to ≥2 DMARDS. In addition, a phase I/II dose-escalation study was conducted to evaluate $DAB_{486}IL\text{-}2$ in patients with recent-onset IDDM. The results of these studies, together with comparative $DAB_{389}IL\text{-}2/DAB_{486}IL\text{-}2$ preclinical toxicology and data from animal models of lymphoma and arthritis, served as a basis for evaluating $DAB_{389}IL\text{-}2$ in phase I/II clinical protocols in patients with RA, recent-onset IDDM, and psoriasis. Table 1 summarizes these studies.

B. RHEUMATOID ARTHRITIS

1. RA Study Methods and Populations

Entry criteria for all four clinical trials in RA were similar, requiring patients with active RA (≥6 swollen joints, ≥9 tender joints, with either ≥45 min of morning stiffness or WESR ≥28 mm/hr), uncontrolled by DMARD therapy. Prednisone ≤10 mg/day was permitted, along with stable NSAIDS. The demographics of patients enrolled in each of the studies are shown in Table 2. Except for the trial of $DAB_{389}IL\text{-}2$ and methotrexate, all trials included a 3–4-week DMARD washout period. Each course consisted of five daily intravenous doses administered as a brief infusion. Subjects were treated as outpatients. Subsequent courses were repeated every 4 weeks (except in the original pilot study). Evaluations of arthritis activity were performed by the same observer at entry, baseline, and 4 weeks after each course. Additional evaluations at weeks 1, 2, and 3 were performed in the original pilot trial of $DAB_{486}IL\text{-}2$, and at weeks 2, 6, and 8 in the trial with concomitant methotrexate. Safety was monitored with biweekly laboratory studies.

Table 1 Clinical Studies of IL-2 Fusion Toxin in Autoimmune Disease

Molecule	Protocol indication	Study design	Primary objective	No. of patients	Status
DAB$_{486}$IL-2	Refractory, severe RA	Open, uncontrolled, dose-escalation	Safety, PK, biological effects	19	Complete (8)
	Severe RA despite ≥2 DMARDS	Double-blind (third-party blinded observer) placebo-controlled, one dose level	Pilot antiarthritic effects	45	Complete (9)
	Recent-onset IDDM	Open, uncontrolled, dose-escalation	Safety, PK, biological effects	43	Complete (10)
DAB$_{389}$IL-2	Refractory severe RA	Double-blind (third-party blinded observer) placebo-controlled; 3 dose levels	Safety, PK, biological effects	55	Complete (11)
	Active RA in patients taking methotrexate	Open, uncontrolled, 2 dose levels	Safety, PK, biological effects	20	Complete (12)
	Maintenance in patients with active RA (previously treated with IL-2 fusion toxin)	Open, uncontrolled (one dose level) repetitive administration up to 4 times/year for 2 years	Safety, biological effects	45	Ongoing[a]
	Recent-onset IDDM	Double-blind, placebo-controlled dose-ranging (2 dose levels)	Safety, biological effects	48	Ongoing[a]
	Recalcitrant psoriasis	Open, uncontrolled dose-escalation	Safety, PK, biological effects	20	Ongoing (13)

[a]Information from this study is not included in this report.

Table 2 RA Patient Demographics and Study Design

	DAB₄₈₆IL-2		DAB₃₈₉IL-2	
	Open-label, uncontrolled, dose-escalation	Double-blind, placebo-controlled, one dose level	Double-blind, placebo-controlled, 3 dose levels	Open, uncontrolled, 2 dose levels, with concurrent Mtx
Number (F, M)	19 (16 F, 3 M)	45 (31 F, 14 M)	55 (42 F, 13 M)	20 (15 F, 5 M)
Mean age (years)	44 (23–73)	51 (17–74)	50 (25–75)	47 (32–62)
Mean duration of RA (years, range)	10 (2–34)	10 (2–26)	12 (1–42)	20 (2–21)
Mean number of prior DMARDS	4.6	4.7	4.9	2.6
RF seropositive (%)	68%	67%	78%	80%
Mean number of swollen joints at entry	12	25	18	23
Mean number of tender joints at entry	19	30	29	26
Dose levels evaluated	0.04, 0.07, 0.1 mg/kg/day (50, 150, 300 kU/kg/day)	Placebo, 0.07 mg/kg/day (150 kU/kg/day)	Placebo, 75, 150, 300 kU/kg/day	75, 150 kU/kg/day
Number of patients per group	6–7	22, 23	12–14	4, 16
Number of courses	1–5	3 active 1 placebo, 3 active	2 active 1 placebo, 2 active	1
Interval between courses	At time of disease flare	4 weeks	4 weeks	4 weeks
Criteria for response (index)	≥25% improvement in swollen and tender joints and 2 of 6 other parameters (modified Paulus)	≥25% improvement in swollen and tender joints and 2 of 6 other parameters (modified Paulus)	≥25% improvement in swollen and tender joints and 2 of 8 other parameters (modified consensus)	≥25% improvement in swollen and tender joints and 2 of 8 other parameters (modified consensus)

Mtx = methotrexate.

2. Safety Observations in RA

a. $DAB_{486}IL$-2 Phase I/II Results. The maximum tolerated dose (MTD) was established by transient, dose-related hepatic transaminase elevations, greater than 5 times normal, in two of seven patients receiving the 0.1 mg/kg/day dose. Some patients experienced fever, nausea, and/or fatigue.

b. Phase II Evaluation of $DAB_{486}IL$-2 in Active Severe RA. Overall, $DAB_{486}IL$-2 was reasonably well tolerated. Adverse effects, which were significantly more common in the treatment group, included transiently elevated hepatic transaminases (≤ 3 times normal), chills and/or fever, nausea and/or vomiting. Hepatic transaminase elevations were dose-related and less frequently elevated during subsequent courses. Twelve patients (eight placebo, four active) dropped out of the study—eight for lack of efficacy, two because of adverse events, one was lost to follow-up, and one died of myocardial infarction. Fifty-six percent of patients had antibodies to $DAB_{486}IL$-2 (titer 1:25–1:625) prestudy, with 100% of patients developing antibody or increased titers (1:3125–1:15,625) by the end of the study. Low titers of antibody to IL-2 were present (titer 1:25–1:625) in 16% of patients prestudy and approximately 80% of patients poststudy.

c. Phase I/II Dose-Ranging Evaluation of $DAB_{389}IL$-2 in Active, Methotrexate-Refractory RA. $DAB_{389}IL$-2 was generally well tolerated at all dose levels: 75, 150, and 300 kU/kg/day. The most frequent adverse events are described in Table 3. All adverse events were transient; dose-related effects included only nausea/emesis and transaminases, which were reduced in frequency and magnitude on readministration of $DAB_{389}IL$-2. Eleven of 55 patients dropped out prior to completing two active courses, five because of lack of benefit, and six because of adverse affects (including one placebo patient).

d. Phase I/II Study of $DAB_{389}IL$-2 in Patients with Active RA Despite Concurrent use of Methotrexate. This study was conducted primarily to assess the safety of the combination of $DAB_{389}IL$-2 and methotrexate. Two dose levels were evaluated: low-dose and middose levels used in the placebo-controlled, dose-ranging trial. Adverse experiences were similar to those reported in trials of

Table 3 Adverse Experiences for the First Course

Adverse event	Dose group			
	Placebo (%)	Low (%)	Mid (%)	High (%)
Elevated transaminases ($<5 \times$ nl)	0	29	33	60
Nausea/emesis	23	14	46	60
Rash ($\pm$ pruritus)	8	36	33	33
Fever/chills	15	29	25	47

$DAB_{389}IL-2$ alone. A transient increase in transaminase enzymes was reported for 45% of patients (1.5–2.5 times normal—one patient; 2.6–5 times normal—four patients; 6 times normal—one patient). The weekly dose of methotrexate was temporarily suspended or decreased in eight patients due to these elevations. Rash was reported in only one patient.

e. Immune Function Effects. No broad immunosuppressive effects were noted. Infections occurred similarly in placebo and active patients, either requiring no treatment or responsive to standard antibiotic therapy.

In each of these studies, repetitive lymphocyte subset analyses were performed. No lymphocyte depletion was observed, and there was no persistent decrease in the number of CD4 lymphocytes, even with repeated courses of fusion toxin administration. The number or percent of IL-2R-expressing CD4 lymphocytes did not consistently change with changes in disease activity, suggesting that peripheral blood lymphocytes may not directly reflect changes in activity in an individual joint(s). Further investigation is underway.

3. Antiarthritic Activity

a. $DAB_{486}IL-2$ Phase I/II Results. This was an open-label, dose-escalation trial. After a single course, no significant improvement in arthritis parameters was noted in patients receiving the 0.04 mg/kg/day dose, whereas three of six patients receiving the 0.07 mg/kg/day dose and two of seven patients receiving the 0.1 mg/kg/day dose experienced ≥25% (most >50%) improvement in both joint counts and at least two other parameters of arthritis activity. Onset of action occurred as early as 7 days and appeared to persist for 4–30 weeks beyond the treatment period. Additional 5-day courses at the 0.07 or 0.1 mg/kg/day dose level were administered to five of six patients in the low-dose group and eight of those in the two higher dose groups when they experienced a recurrence of arthritis symptoms. Eight of these 13 patients experienced additional improvement, despite the development of antibodies to $DAB_{486}IL-2$. Figure 2 illustrates the response in one patient who received three courses and experienced prolonged improvement in both swollen and tender joint counts as well as the other parameters of arthritis activity. Erythrocyte sedimentation rate decreased from 90 mm/hr to 27 mm/hr during the observation period.

b. Phase II Evaluation of $DAB_{486}IL-2$ in Active Severe RA. This was a double-blind, placebo-controlled (for the first course), trial with a blinded third-party observer. By the predefined response criteria, four of 22 (18%) $DAB_{486}IL-2$-treated patients responded; none of the placebo patients were responders. Thirty-three patients subsequently received a total of three 5-day courses of $DAB_{486}IL-2$.

As shown in Table 4, the number of responses increased with subsequent courses; after three courses were completed, approximately one-third of the patients were responders. Typical responses included ≥50% improvement in both joint counts (Table 5) and four of six other parameters (Table 6).

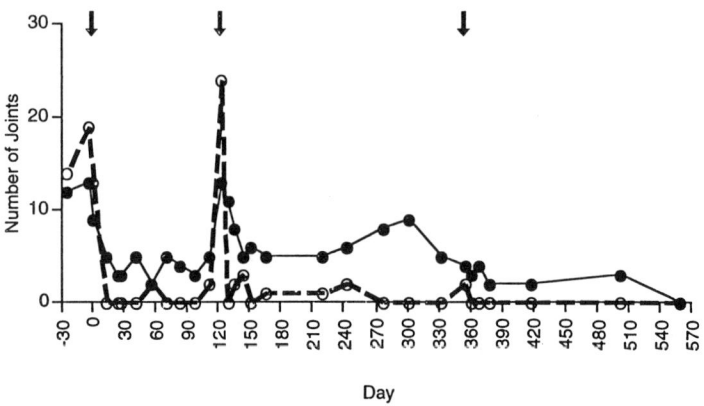

Figure 2 Swollen (———●———) and tender (– -◯- –) joint counts after $DAB_{389}IL-2$ administration in a 53-year-old female with RA for 25 years. The patient's previous antiarthritic therapies included methotrexate, gold, azathioprine, penacilliamine and hydroxychloroquine. The time of $DAB_{486}IL-2$ administration is indicated by ↓.

While there was a trend for improvement in each disease activity parameter in the treated group, only the response index provided statistical significance ($p = .05$) for the first course. Thus, the use of such a predefined response index provided early insight regarding the potential antiarthritic effects of $DAB_{486}IL-2$.

  c. Phase I/II Dose-Ranging Evaluation of $DAB_{389}IL-2$ in Active, Methotrexate-Refractory RA. There was no significant difference in the number of responders between the four treatment groups at the end of the first course (placebo and $DAB_{389}IL-2$), first active course, or second active course. Two patients in the placebo group qualified as responders after the 4-week blinded period and there was one responder in each of 75 kU/kg and 300 kU/kg dose groups after the first course.

  A responder analysis was conducted for this exploratory study based on the patients in the intent-to-treat population who completed two active courses and met the criteria for response. Given the sample sizes in each group, there was 80% power to detect a 55% difference between any two randomly selected treatments under a two-sided test of equality of response rates. A total of seven patients had ≥25% improvement in both joint counts and met response criteria: first course placebo ($n = 2$), 75 kU/kg ($n = 2$), 150 kU/kg ($n = 2$), and 300 kU/kg ($n = 1$). Table 7 lists improvement after two active courses.

Table 4 Response Distribution by Course

	Total no. of patients	No. of responders
Course 1	42	7 (17%)
Course 2	36	11 (31%)
Course 3	33	11 (33%)

Table 5 Joint Counts (Mean ± SE) in Patients Receiving Three Courses

	Screen/ day 21	Baseline	Post CRS 1	Post CRS 2	Post CRS 3
Painful/tender					
All ($n = 11$)	30.8 ± 3.7	35.2 ± 3.4	27.0 ± 4.1	13.0 ± 2.3	12.2 ± 2.7
All ($n = 33$)	32.5 ± 2.7	33.1 ± 2.9	29.9 ± 3.1	24.9 ± 3.2	24.2 ± 3.3
Swollen					
responders ($n = 11$)	24.2 ± 3.7	28.0 ± 4.6	19.4 ± 3.7	11.1 ± 2.0	10.8 ± 2.4
All ($n = 33$)	26.0 ± 2.6	26.6 ± 3.1	22.5 ± 2.9	18.7 ± 2.6	18.7 ± 2.9

CRS = Course.

Table 6 Improvement over Baseline at End of Third Course in Patients Receiving Three Courses

Parameter	All patients ($n = 33$)	Responders ($n = 11$)
Swollen joints	31%	63%
Painful joints	29%	68%
Morning stiffness	30%	69%
Grip strength	30%	60%

Table 7 Improvement After Two Active Courses

	Responders ($n = 7$)
Number of swollen joints	52%
Number of painful joints	52%
Morning stiffness	66%
50-min walk time	16%
Grip strength	39%
Observer assessment of pain	59%
Observer assessment of disease activity	45%
Patient assessment of pain	50%
Patient assessment of disease activity	51%
WESR	10%

Only one patient experienced a ≥25% improvement in WESR. In general, the seven responders experienced a broad degree of improvement in a majority of the 10 response categories. The majority of responders had >40% improvement for at least five of 10 outcome parameters, suggesting treatment-associated benefit in this subset of patients.

d. Phase I/II Study of DAB$_{389}$IL-2 in Active RA Despite Concurrent Use of Methotrexate. Four of 20 patients experienced a decrease in arthritic activity after DAB$_{389}$IL-2 administration. Twelve patients chose to enroll in an open-label maintenance protocol to receive additional courses of DAB$_{389}$IL-2. Of those 12 patients, six are still active in the study, five discontinued after one additional course, and two patients received four additional courses.

e. Antibodies. Nearly all patients with RA develop antibodies to IL-2 fusion toxin and to IL-2 (low titers) after treatment. The titer of antibodies to IL-2 fusion toxin (preexistent at low levels in 30–40% of patients at entry) increases to a maximum by the end of a second course (median titer 1:625–1:3125). Anti-IL-2 titers are also maximal by the end of a second course, with a lower median titer (1:25–1:125). The impact of antibodies to IL-2 is unknown.

The presence of antibody prior to IL-2 fusion toxin administration did not preclude the potential for antiarthritic effects, nor did the development of antibody preclude additional antiarthritic effects in subsequent courses.

C. Phase I/II Evaluation of DAB$_{486}$IL-2 in Recent-Onset IDDM

Based on the safety and apparent benefit in patients with severe RA, and also on effects demonstrated in an accelerated model of autoimmune diabetes in the nonobese diabetic mouse (14), a phase I/II evaluation of $DAB_{486}IL$-2 in patients with recent-onset IDDM was conducted. As a result of experience in clinical studies of cyclosporine A and other immunoregulatory agents such as azathioprine in recent-onset IDDM, we defined a clinical model in which the immunological effects of new agents can be assessed in patients with recent-onset IDDM.

This study involved in "induction phase" (with $DAB_{486}IL$-2) and a "maintenance phase" (with cyclosporine A, 5 mg/kg/day), recognizing that patients with symptoms of type I diabetes have <10–20% of their islet cell mass remaining, and an underlying continuing stimulus to lymphocyte destruction of islet cells. Eligible patients were otherwise healthy with: HLA tissue typed DR3 and/or 4; anti-islet cell antibodies; a history of symptoms and signs of hyperglycemia of <4 months duration and insulin therapy of <8 weeks duration. Patients were assigned consecutively to cohorts of six, each of whom received 0.025 (~50 kU/kg/day), 0.05 (~100 kU/kg/day), or 0.07 (~150 kU/kg/day) mg/kg/day of $DAB_{486}IL$-2 as a 1-hr infusion daily for 4 or 7 days. All patients maintained insulin therapy according to good glycemic control. Safety and immune function parameters were evaluated twice weekly for 2 weeks and then weekly for 4 weeks. Diabetes status was assessed by measuring basal and glucagon-stimulated C-peptide, average daily insulin requirement (U/kg/day), and glycosylated hemoglobin (HbA1c). After treatment with $DAB_{486}IL$-2, patients were allowed to enter a maintenance protocol in which a low dose of cyclosporine A (CsA) (5 mg/kg/day) was administered. Response was assessed by decreased insulin requirement (complete ≤0.10 U/kg/day, partial >0.10 and <0.25 U/kg/day) in the presence of good metabolic control (HbA1c≤7.5%). Patients were followed monthly for the duration of response.

Forty-three patients were enrolled: 20 females and 23 males, aged 25 ± 7 years (range 15–41 years). Daily insulin requirement was 0.74 ± 29 U/kg/day at baseline.

1. Safety

$DAB_{486}IL$-2 was well tolerated at all doses with occasional fever, rash, or lower extremity edema (non-dose-related) and/or dose-related hepatic transaminase elevation (2–3 × nl at the highest dose level). There was a transient decrease in the in vitro lymphocyte proliferation to DT antigen during the period of agent administration. During the CsA administration period, plasma creatinine increased mildly from 78.5 ± 11.5 μmol/L at baseline to 85.5 ± 11.3 μmol/L at 12 months.

2. Biological Effects

Twenty-four of 34 patients (71%) experienced a partial remission, including 14 (41%) who experienced a complete remission during the study. At 12 months, 12 of 34 patients (34%) remain in good control with reduced insulin requirements.

According to the cumulative dose of $DAB_{486}IL$-2, the frequency of remissions was higher among the patients who received "low" doses (46% at 12 months) than in those who received "high" doses (10%).

Patients with partial or complete response exhibited better glycemic control and preservation of the endogenous residual insulin reserve (stable or increased C-peptide) than those with no response. In nine of 12 patients in remission at month 12, the response was evident before CsA was introduced.

Thirty-four percent of patients remained in remission at 12 months, suggesting that an induction/maintenance strategy with IL-2 fusion toxin and CsA may be feasible in type 1 diabetes with good tolerability and enhanced metabolic control. A placebo-controlled trial of $DAB_{389}IL$-2 is presently underway.

D. Phase I/II Evaluation of $DAB_{389}IL$-2 in Recalcitrant Psoriasis

A cohort dose-escalation study is underway to evaluate the safety, tolerability, and pharmacokinetics of $DAB_{389}IL$-2 in patients with recalcitrant psoriasis and to assess the pathogenic role of activated T lymphocytes in this disease. In vitro experiments using mitogen-activated human lymphocytes and rapidly proliferating human epidermal keratinocytes established that lymphocytes could be effectively inhibited by 10^{-11} M $DAB_{389}IL$-2, whereas keratinocytes were not inhibited by $DAB_{389}IL$-2 in concentrations as high as 10^{-8} M. Ten patients (five per group) have been evaluated.

$DAB_{389}IL$-2 was administered intravenously to patients at two dose levels (100 kU/kg/day and 200 kU/kg/day). A treatment course consisted of daily infusion of $DAB_{389}IL$-2 for 5 days with follow-up observation and laboratory studies during the subsequent 3 weeks (each treatment cycle was 28 days). Each patient received at least two courses of $DAB_{389}IL$-2 infusions. The response of psoriasis to $DAB_{389}IL$-2 was followed clinically (serial psoriasis area and severity scores) and by histological assessment of lesional biopsies taken before treatment and on days 9 and 25 of each treatment cycle.

1. Safety

The most frequent adverse experiences reported were fever (55%), dizziness (45%), itching (36%), and transaminase elevations (27%). All were transient and most were considered mild.

2. Biological Effects

Histological assessment of psoriatic pathology before treatment showed acanthotic epidermis with psoriasiform rete elongation, a sparse to absent granular layer, frequent parakeratosis, neutrophil presence in the epidermis, and a chronic immune infiltrate in the dermis. The epidermis was hyperproliferative as assessed with antibodies to Ki67 and showed "alternate" or "regenerative" differentiation as detected by keratin 16 expression and altered distribution of filagrin in the epidermis. In each case, CD8+ lymphocytes predominated in the epidermis and CD4+ lymphocytes predominated in the dermis. Activated T lymphocytes expressing the IL-2R alpha chain (CD25+) were present in pretreatment biopsies from all patients.

Eight of 10 patients showed a decrease in acanthotic epidermis in psoriatic lesions following $DAB_{389}IL$-2 administration. Before treatment, mean epidermal thickness was 389 µm and at the end of the second treatment cycle, the mean was 219 µm (44% reduction, approximately 60% reduction in pathological acanthosis, $p < .0005$). CD8+ lymphocytes in psoriatic lesions were also reduced in the same patients. Before treatment, a fixed length of psoriatic epidermis (725 µm) contained an average of 93 intraepidermal CD8+ lymphocytes; at the end of the second treatment cycle, an average of 365 CD8+ lymphocytes were present (mean reduction 60%, $p < .01$). Histochemical analysis of treated psoriatic lesions at the end of the second treatment cycle using antibodies to the Ki67 protein (cell proliferation) and keratin 16 and filaggrin (keratinocyte differentiation) showed complete elimination of psoriatic epidermal pathology in four patients. The elimination of psoriatic epidermal pathology corresponded with marked reductions in CD8 intraepidermal lymphocytes following $DAB_{389}IL$-2 administration (average of 94% reduction in tissue infiltrating lymphocytes in these cases). Conversely, there were no examples in which tissue-infiltrating CD8+ lymphocytes were significantly reduced without proportional reductions in epidermal acanthosis. These pathological improvements were accompanied by clinical improvement, with four patients experiencing almost complete clearing of their psoriasis.

This study is continuing to accrue patients at higher dose levels.

III. SUMMARY

This early clinical experience provides a basis for evaluation of $DAB_{389}IL$-2 in randomized, placebo-controlled clinical trials in patients with early, remediable disease where activated lymphocytes and a proliferative immune response are more likely to predominate. In diseases such as psoriasis, in which the immunopathology of active lesions demonstrates the presence of activated cells, the stage and duration of disease may be less critical.

The results of studies in RA provide preliminary evidence of $DAB_{389}IL$-2 safety and suggest that a subset of patients with long-duration, refractory RA may still improve with specific immunological intervention. Recognizing our limited understanding of the immunopathogenesis of this disease and its underlying pathophysiological mechanisms, it is likely more appropriate to evaluate the efficacy of $DAB_{389}IL$-2-targeted therapy in patients with earlier, more remediable, and less refractory disease.

$DAB_{389}IL$-2 represents a specific immunoregulatory agent with a targeted mechanism of action, offering an opportunity to assess the immunopathogenesis of various autoimmune diseases. No generalized or persistent immunosuppression has been evident in IL-2 fusion toxin clinical trials. Sensitive, specific tests of immune function that measure change in relation to drug treatment will enhance evaluation in dose/schedule-finding studies. Clinical studies in psoriasis, where active lesion immunohistochemistry can be correlated with peripheral blood immunological evaluation, may offer an opportunity to characterize such laboratory tests, potentially enhancing our ability to maximize clinical efficacy for individual patients in phase III clinical trials.

REFERENCES

1. Williams DP, Parker K, Bacha P, Borowski M, Genbauffe F, Strom TB, Murphy JR. Diphtheria toxin receptor binding domain substitution with interleukin-2: genetic construction and properties of a diphtheria toxin–related interleukin-2 fusion protein. Protein Eng 1987; 1:493–498.
2. LeMaistre CF, Meneghetti C, Rosenblum M, Reuben JH, Parker K, Shaw J, Deisseroth A, Woodworth T, Parkinson DR. Phase I trial of an interleukin-2 (IL-2) fusion toxin ($DAB_{486}IL$-2) in hematologic malignancies expressing the IL-2 receptor. Blood 1992; 79:2547–2554.
3. Hesketh P, Caguioa P, Koh H, Dewey H, Facada A, McCaffrey R, Parker K, Nylen P, Woodworth T. Clinical activity of a cytotoxic fusion protein in the treatment of cutaneous T-cell lymphoma. J Clin Oncol 1993; 11:1682–1690.
4. Tepler I, Schwartz G, Parker K, Charette J, Kadin M, Woodworth T, Schnipper L. Phase I trial of an interleukin-2 fusion toxin ($DAB_{486}IL$-2) in hematologic malignancies: complete response in a patient with Hodgkin's disease refractory to chemotherapy. Cancer 1994; 73:1276–1285.
5. LeMaistre CF, Craig FE, Meneghetti C, McMullin Parker K, Reuben JB, Boldt DH, Rosenblum M, Banks P, Woodworth T. Phase I trial of a 90-minute infusion of the fusion toxin $DAB_{486}IL$-2 in hematological cancers. Cancer Res 1993; 53:3930–3934.
6. Kuzel TM, Rosen ST, Gordon LI, Winter J, Samuelson E, Kaul K, Roenigk HH, Nylen P, Woodworth T. Phase I trial of the diphtheria toxin/interleukin-2 fusion protein $DAB_{486}IL$-2: efficacy in mycosis fungoides and other non-Hodgkin's lymphomas. Leukemia Lymphoma 1993; 11:369–377.
7. Woodworth TG, LeMaistre CF, McCaffrey R, Schnipper L, Rosen S, Ratain M. Phase I/II clinical studies of $DAB_{486}IL$-2 fusion toxin in patients with refractory IL-2 receptor expressing malignancies. Proceedings of the American Society of Hematology, 34th Annual Meeting, Anaheim, CA, 1992.
8. Sewell KL, Parker KC, Woodworth TG, Reuben J, Swartz W, Trentham DE. $DAB_{486}IL$-2 fusion toxin in refractory rheumatoid arthritis. Arthritis Rheum 1993; 36: 1223–1233.
9. Moreland LW, Sewell KL, Sullivan WF, Shmerling RH, Parker KC, Swartz WG, Woodworth TG, Koopman WJ, Trentham DE. Treatment of refractory rheumatoid

arthritis with $DAB_{486}IL-2$ fusion toxin: double-blind placebo-controlled study. Proceedings of the ACR, 57th Annual Meeting, November 1993.
10. Timsit, J, Assan R, Mogenet A, Debussche X, Kaloustian E, Attali JR, Chanson P, Chatenoud L, Bach JF, Boitard C, Woodworth TG. Improvement in endogenous insulin secretion in patients with recent-onset type 1 (insulin-dependent) diabetes mellitus, (IDDM) receiving $DAB_{486}IL-2$ fusion toxin in a phase I/II trial. Proceedings of the Clinical Immunology Society, 7th Annual Conference on Clinical Immunology, 1992.
11. Sewell KL, Moreland LW, Cush JJ, Furst DE, Woodworth TG, Meehan RT. Phase I/II double-blind, dose-response trial of a second fusion toxin, $DAB_{389}IL-2$, in rheumatoid arthritis. Proceedings of the ACR, 57th Annual Scientific Meeting, November 1993.
12. Kremer JM, Petrillo GF, Rigby WFC, Plehn SJ, Woodworth TG, Parker KC, Taintor GS. Phase I/II open-label trial of $DAB_{389}IL-2$ administered to patients with active rheumatoid arthritis (RA) receiving treatment with methotrexate (MTX). Proceedings of the ACR, 58th Annual Meeting, October 1994.
13. Gottleib SL, Gilleaudeau P, Johnson R, Estes L, Woodworth TG, Gottleib AB, Krueger JG. Response of psoriasis to a lymphocyte-selective toxin ($DAB_{389}IL-2$) suggests a primary immune, but not keratinocyte pathogenic basis. Nature Med 1995; 1:442–447.
14. Pacheo-Silva A, Bastos MG, Muggia RA, Pankewycz O, Nichols J, Murphy JR, Strom TB, Rubin-Kelley VE. Interleukin-2 receptor targeted fusion toxin ($DAB_{389}IL-2$) treatment blocks diabetogenic autoimmunity in non-obese diabetic mice. Eur J Immunol 1992; 22:697–702

4
Chimeric Anti-CD4 Antibody as a Potential Therapeutic Agent for Rheumatoid Arthritis

LARRY W. MORELAND and WILLIAM J. KOOPMAN
University of Alabama at Birmingham, Birmingham, Alabama

I. INTRODUCTION

Rheumatoid arthritis (RA) is an immunologically mediated, chronic inflammatory disease characterized by synovial cell proliferation, inflammation, and destruction of adjacent articular tissue (1). Although considerable evidence supports the importance of immune processes in the pathogenesis of RA (1,2), the underlying cause is unknown. The prevailing view is that RA is mediated by antigen-activated T cells that infiltrate the synovial membrane and initiate a series of inflammatory processes, resulting in vascular and synovial cell proliferation with resorption of cartilage and bone (3). Nonimmunological pathways also probably contribute to tissue injury and destruction in established RA (4). Degradation of articular extracellular matrix components is a hallmark of RA and is largely mediated by neutral proteinases such as collagenase, stromelysin, elastase, and cathepsin G. A schematic diagram of cellular and humoral pathways implicated in the pathogenesis of RA is presented in Figure 1. The model suggests several opportunities for interruption of these pathways that could be exploited in treating autoimmune disorders such as RA.

T cells are divided into two subpopulations based on surface expression of the CD4 or CD8 antigen. CD4+ T cells are referred to as "helper-inducer" cells, because these cells promote humoral or cell-mediated immunity, whereas CD8+ T cells (cytotoxic/suppressor) mediate cytotoxicity or suppress immunity. The

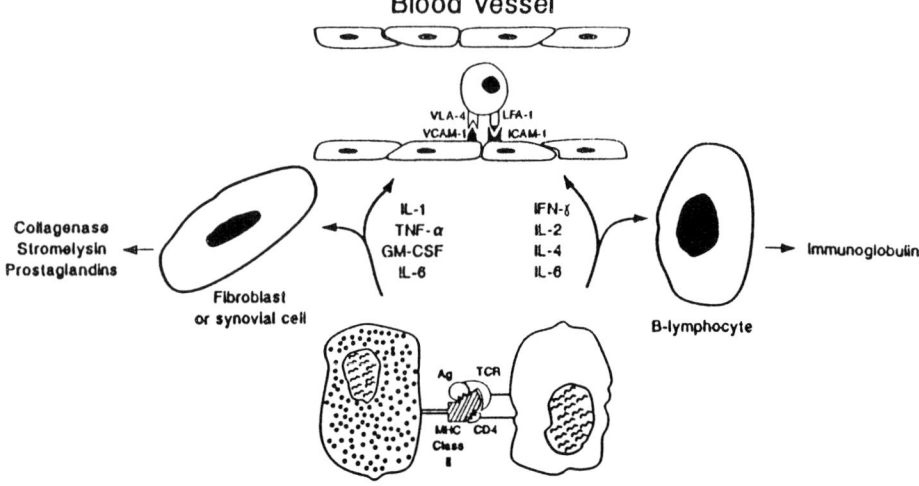

Figure 1 Schematic illustrating the cellular and humoral pathways implicated in the pathogenesis of rheumatoid arthritis. IL = interleukin; TNF-α = tumor necrosis factor-alpha; GM-CSF = granulocyte-macrophage-colony stimulating factor; IFN-γ = interferon-gamma; Ag = antigen; TCR = T-cell receptor; MHC = major histocompatibility complex; VLA-4 = very late antigen-4; VCAM-1 = vascular adhesion molecule-1; LFA-1 = lymphocyte-function-associated antigen-1; ICAM-1 = intercellular adhesion molecule-1. (From the *American Journal of Medical Sciences* 1993; 305:40–51, with permission.)

development of humoral and cellular immunity requires CD4+ T-cell recognition of processed antigen (peptides) complexed with self class II major histocompatibility complex (MHC) molecules on the surface of antigen-presenting cells (APC). This complex is recognized by antigen-specific MHC class II–restricted T-cell antigen receptors. The avidity of this interaction is augmented by the T-cell accessory molecule CD4, which binds to a highly conserved segment of cell surface MHC class II molecules (5). Activation of helper T cells is modulated by immunoregulatory cytokines such as interleukin-1 (IL-1), IL-2, IL-4, IL-6, tumor necrosis factor-α (TNF-α), and interferon gamma (IFN-γ). Several of these cytokines also probably contribute to local inflammation in RA.

Although the etiology of RA remains unclear (1), considerable evidence supports a role for CD4+ T cells in the pathogenesis of the disease, including: (1) predominance of Ia+ CD4+ T cells in the synovial mononuclear infiltrates of patients with RA (6,7); (2) increased numbers of Ia+ T cells in the peripheral blood of RA patients (7); (3) evidence that certain MHC class II alleles

(HLA-DR1 and HLA-DR4, subtypes Dw4, Dw14, and Dw15) are associated with susceptibility to RA (8–11); (4) improvement of preexisting RA in patients developing acquired immunodeficiency syndrome (AIDS) (12); and (5) improvement of refractory RA following thoracic duct drainage (13), total lymphoid irradiation (14), and administration of cyclosporine (15).

The central role of CD4+ T cells in the pathogenesis of several animal models of human autoimmune disease has become increasingly apparent. Efficacy of anti-CD4 MAb therapy in several murine models, including collagen arthritis (16), systemic lupus erythematosus (17), myasthenia gravis (18), experimental allergic encephalomyelitis (19), diabetes mellitus (20), thyroiditis (21), and uveitis (22), strongly supports this view. Pilot studies using several murine MAbs directed against human CD4 in refractory RA patients have yielded promising results (23–28). Although the results varied, some studies have noted a significant reduction in inflammatory markers such as C-reactive protein, erythrocyte sedimentation rate (ESR), and serum rheumatoid factor (26). A source of concern in some of these early trials has been the development of host antibody responses directed against the therapeutic MAb (29,30).

The apparent efficacy of anti-CD4 MAbs in several animal models of autoimmune diseases may reflect several mechanisms. Treatment with the MAb may cause immunosuppression by depleting CD4+ T cells, either by direct killing of the cells (complement-mediated or by inducing apoptosis) or by inducing reticuloendothelial clearance of antibody-coated CD4+ T cells (by complement and Fc receptor-mediated mechanisms). Antibody-dependent cellular cytotoxicity also may contribute to CD4+ T cell depletion. Alternatively, binding of the anti-CD4 MAb to helper T cells may inhibit MHC class II–dependent T-cell responses by blocking the interaction of CD4+ T cells with APC. Finally anti-CD4 MAb therapy may also transmit an inhibitory signal to the CD4+ cells (31,32). When anti-CD4 MAbs are administered in conjunction with a nominal antigen, they induce tolerance to the antigen in mice (33,34). However, it is unknown if such tolerance can be achieved in an already established immune-mediated disease such as RA. Indeed, evidence for tolerance induction by anti-CD4 antibodies in humans is lacking.

There are potential problems with murine MAbs in the treatment of autoimmune diseases. First, the recipient's humoral response to the murine MAb may result in toxicity or neutralize the efficacy of the MAb (35). To overcome this problem, chimeric MAbs have been developed that exhibit lower immunogenicity compared to the parent murine MAb (36). A chimeric MAb consists of human immunoglobulin heavy (H)- and light (L)-chain constant regions fused to the variable H- and L-chain regions of the murine MAb such that the resulting molecule is approximately 75% human and 25% murine in origin. A second potential problem with available MAbs is suppression of normal immune function. In mice, treatment with anti-L3T4 results in depletion of CD4+ T cells, blocks primary and secondary humoral immune responses, and inhibits the generation of cellular immune responses (37). These effects are reversible, but the return of CD4+ T cells is gradual. Prolonged immune suppression is also observed after treatment with F(ab')2 anti-L3T4 that does not deplete the targeted T cells (38). This latter phenomenon may reflect transmission of a negative signal to L3T4+ T cells (38) or inhibition of the interaction between L3T4+ T cells and APC (39).

II. EXPERIENCE WITH CHIMERIC ANTI-CD4 ANTIBODY FOR TREATMENT OF RA

We have used a chimeric MAb (cM-T412) in phase I and phase II trials involving patients with refractory RA (40–42). The objectives of the phase I study were to evaluate the safety, immunogenicity, and biological effects of cM-T412 (40). Twenty-five patients with active refractory RA (all taking methotrexate concomitantly) were treated with incremental doses (10–700 mg) of cM-T412 in an open-label, escalating-dose trial. Levels of circulating CD4+ T cells decreased rapidly postinfusion (Fig. 2) and remained significantly depressed even at 18 months and 30 months following treatment (Fig. 3). Repopulation of CD4+ T cells consisting of increased CD45RA+ (naïve) and CD45RO+ (memory) CD4+ T cells was observed in approximately one-third of the patients between day 14 and 6 months postinfusion (42). Proliferative responses of peripheral blood lymphocytes to mitogens and recall antigens were generally diminished following cM-T412 infusion, with mitogen responses normalizing more rapidly than responses to recall antigens. Adverse events during the first 6 months of follow-up included fever, often associated with myalgias, malaise, and asymptomatic hypotension; these symptoms were self-limited and appeared to correlate with transient elevations of IL-6 (40). One patient, who was also receiving methotrexate and prednisone, died 18 months after receiving 100 mg of cM-T412 (41). This patient developed pneumonia (*Staphylococcus* and *Pneumocystis*) and succumbed to an intracranial bleed. The patient's CD4+ T-cell count had varied between 126 and 566/mm^3 in the 30–240 days following treatment. It is unclear whether cM-T412 alone was responsible for this patient's death or whether other risk factors (e.g., methotrexate, prednisone, malnutrition, and long-standing RA) may have also contributed to his death (41).

Clinical assessments of disease activity were made at baseline and at weekly posttreatment visits for 2 months and then monthly for 4 additional months. Nonblinded assessment of painful joint counts indicated that 43% of patients exhibited ≥50% improvement in tender joint counts at 5 weeks and 33% at 6 months postinfusion. Similar improvements were noted in the swollen joint counts

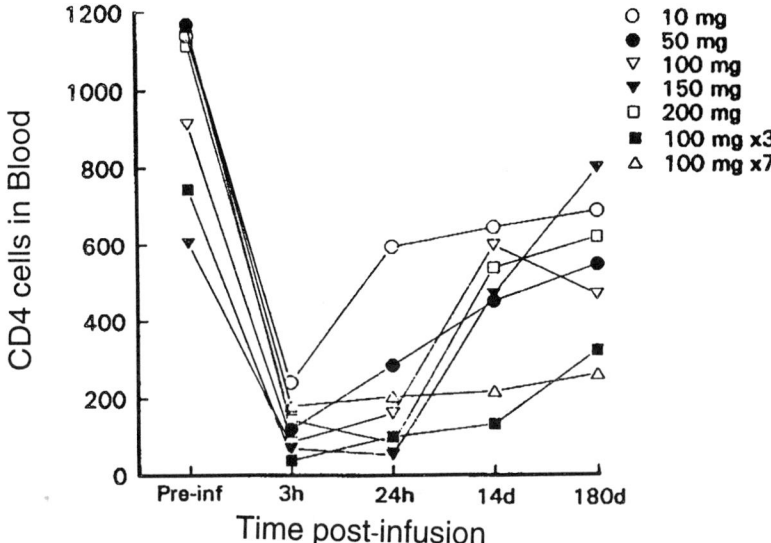

Figure 2 Comparison of peripheral CD4+ T cell counts (cells/μl) over 6 months of follow-up in patients receiving incremental doses of cM-T412 in phase I trial (40). Patient 24 received only 100 mg of the scheduled 700-mg regimen, and those data are not included. inf = infusion. (From Ref. 40, with permission.)

(4). No significant changes were observed in the Westergren ESR, grip strength, duration of morning stiffness, or pain scores.

Based on the results of the phase I study, three phase II double-blind, placebo-controlled trials were subsequently initiated. These studies involved three distinct RA patient populations; (1) refractory to multiple DMARDS (*and* taking concurrent methotrexate); (2) intermediate disease duration (and not on concurrent DMARDS); and (3) early RA (chimeric anti-CD4 was the first antirheumatic drug). Our center participated in a multicenter phase II trial directed toward treatment of refractory RA patients (43).

Sixty-four patients with refractory RA on stable doses of methotrexate were randomized into this phase II double-blind, placebo-controlled trial to receive either monthly (×3) treatments with placebo, or 5, 10, or 50 mg intravenous cM-T412 (43). After the 3-month double-blind period, patients initially receiving placebo were treated with three monthly cycles of 10 mg cM-T412/cycle. Enrollment criteria included active disease as determined by ≥6 swollen joints plus at least two of three additional criteria: an ESR ≥28 mm/hr, morning stiffness ≥45 min, and ≥9 painful joints on palpation. Patients were required to be on a

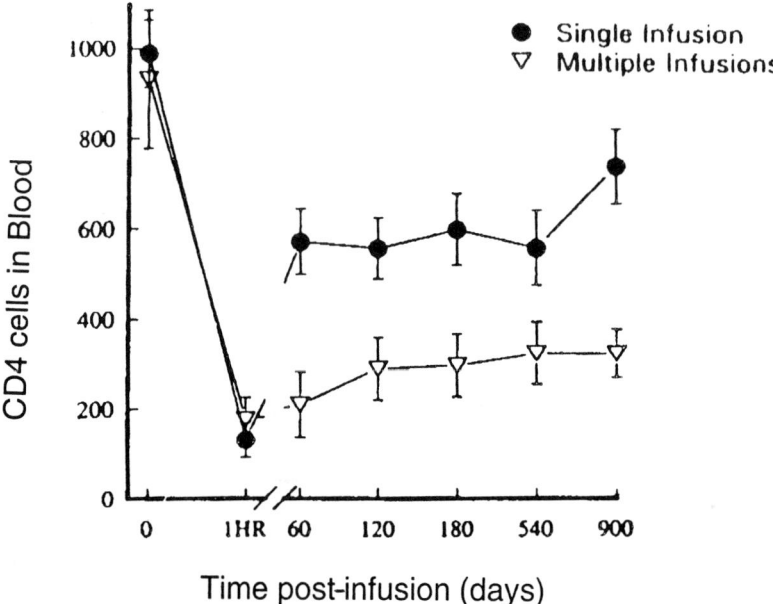

Figure 3 Levels of circulating CD4+ T cells in 23 rheumatoid arthritis patients treated with chimeric anti-CD4 monoclonal antibody cM-T412 in a phase I trial (41). Single infusions were either 10, 50, 150, or 200 mg in 13 patients. Multiple infusions were either 300 mg (100 mg/day × 3 over 1 week) or 700 mg (100 mg/day × 7 days) in 10 patients. Values are the mean ± SEM. inf = infusion. (From Ref. 41, with permission.)

stable dose (at least 3 months) of weekly methotrexate (≤15 mg/week). Concomitant treatment with stable doses of NSAID and/or prednisone ≤10 mg/day) was allowed. Patients had active disease (mean swollen joint count of 22 and mean tender joint count of 31) despite treatment with ≤10 mg daily prednisone (7.6 mg in 11 patients), and methotrexate for at least 3 months (23 ± 33 weeks). The mean weekly dose of methotrexate was 11.7 mg ± 3.7 (range 5–17.5 mg/week). Mean age of patients (42 female, 22 male) was 49.7 years with a mean disease duration of 8.7 years. Using 50% improvement in swollen joint counts as a measure of clinical efficacy, at 3 months (end of placebo-controlled portion of trial), 13%, 13%, 18%, and 13% of patients receiving 50 mg, 10 mg, 5 mg, or placebo, respectively, exhibited a clinical response. Using 50% improvement in tender joint counts as a measure of clinical efficacy, at 3 months 19%, 13%, 12%, and 6% of the patients receiving 50 mg, 10 mg, 5 mg, or placebo, respectively, responded. The CD4+ T-cell counts in the treatment groups pretherapy and up to 6 months follow-up are presented in Table 1. All patients with a CD4+ T-cell

Table 1 Pre- and Posttreatment Levels of Circulating CD4+ T Cells in 64 Patients with Refractory RA Treated with cM-T412 in a Phase II Placebo-Controlled Trial (43)

Treatment group[a]	CD4+ T-cells/mm^3 of blood				
	Preinfusion	1 month	2 months	3 months	6 months
Placebo	922	779	778	747	706
5 mg	891	722	704	612	699
10 mg	1035	805	594	490	595
50 mg	856	470	363	373	353

[a]See text for definition of treatment regimen.

count of <400 have been followed every 3 months until their CD4 T-cell counts are ≥400.

In this phase II trial "flu-like" symptoms (fever, chills, rigor) within 24 hr of the infusion occurred more frequently in the groups receiving 50 mg (29%) and 10 mg (31%) doses than in those receiving 5 mg (12%) or placebo (13%). Other adverse events were infrequent. All patients have been followed for 12 months after the final treatment; no significant adverse events, in particular infectious complications, have occurred.

Thus, treatment with cM-T412 in this cohort of RA patients receiving methotrexate concomitantly was not associated with clinical efficacy. In addition, in a recently completed placebo-controlled trial in Europe where cM-T412 was used in early RA, no clinical benefit was demonstrated (44).

Other therapeutic trials using cM-T412 in treating RA have yielded mixed results (45–48). In a study of nine RA patients, four received 50 mg of cM-T412 intravenously once, three received 50 mg of cM-T412 weekly for 4 consecutive weeks, and two received placebo infusions (45). There was no significant clinical benefit in any of these patients despite a marked reduction in circulating CD4+ T cells. These same investigators used higher doses of cM-T412 (50 mg intravenously daily for 5 days) in 12 patients with refractory RA in an open-label study (46,47). Significant clinical improvement was seen in these refractory RA patients. In six of these patients paired peripheral blood and synovial fluid samples were obtained at pre- and postinfusion on days 1 and 5. The percentage of MAb-coated lymphocytes in synovial fluid after 5 days correlated with the degree of clinical improvement (46).

In another phase I open-label trial in Europe, 32 RA patients received daily doses of 10, 50, or 100 mg of cM-T412 intravenously for 7 days (48). DMARDS were discontinued at least 6 weeks prior to the study. All patients exhibited sustained decreases in circulating CD4+ T cells. Those patients receiving 50-mg ($n = 13$) and 100-mg doses ($n = 3$) experienced significant reductions in disease activity (Ritchie articular index) at 2 and 4 weeks after treatment. These positive clinical responses correlated with serum levels of unbound cM-T412 and not to the peripheral blood counts (49). As with our phase I trial, there was no correlation between the decrease in number of circulating CD4+ T cells and the clinical response. Similarly, sustained decreases in CD4+ T cells were noted at 12 months posttreatment (60% of baseline values) (48).

The human immune response to cM-T412 in these trials varied between 20 and 40% (with one exception) and was generally of low titer (50). In trials where patients received concomitant immunosuppressive agents with cM-T412, the human antichimeric antibody (HACA) responses were lower. For example, in our phase I trial only two of 25 patients developed transient low levels of HACA (40). In the phase I trial reported by van der Lubbe et al., 75% of patients treated with cM-T412 developed HACA responses (48). With this noted exception, the HACA response to cM-T412 was lower than that reported in RA patients receiving murine anti-C4 MAbs.

III. DISCUSSION

Our experience and that of others with chimeric anti-CD4 MAb (cM-T412) indicates that this antibody induces a dose-dependent sustained depletion of circulating CD4+ T cells. This property constrains dosing and frequency of administration of this antibody. The degree of depletion of circulating CD4+ T cells in these trials did not correlate with clinical response to treatment with cM-T412. These data are consistent with evidence in animal models that the efficacy of anti-CD4 antibodies is not dependent on physical depletion of T cells (51–54). These data are consistent with the views that modulation of T-cell function (e.g., tolerance induction) is more likely responsible for the effectiveness of anti-CD4 antibodies in these animal models. For example, in thymectomized NZB/NZW F1 (B/W) mice treated with anti-CD4 MAbs, significant depletion of CD4+ T cells was not sufficient to suppress autoimmunity (51). Suppression of autoimmunity in these mice required sustained functional inhibition of residual CD4+ T cells by means of chronic MAb therapy. Thus, the few CD4+ T cells that were not depleted by anti-CD4 were enough to facilitate the full expression of autoimmunity. Suppression of autoimmunity in NZB/NZW mice can also be activated by administration of nondepleting (F(ab')$_2$ fragments of anti-CD4 MAb to L3T4 (52,53). Of interest, treatment of C57BL/6 and Balb/c mice with F(ab')$_2$ anti-L3T4 has been shown to induce tolerance (54).

The etiology of the prolonged CD4+ T cell depletion resulting from cM-T412 treatment is not known. This effect is enhanced by the administration of multiple doses. This sustained depletion of circulating CD4+ T cells observed in the RA clinical trials contrasts with the transient depression of CD4+ T cells following infusion of murine anti-CD4 MAb (26,27). This difference most likely reflects the biological effects of the human IgG1 Fc domain present in cM-T412. Although the concurrent use of methotrexate potentially could affect reconstitution of CD4+ T cells (55,56), this probably does not have a significant influence on circulating CD4+ T cells as similar prolonged profound depletion of circulating CD4+ T cells has been noted in patients with multiple sclerosis treated with cM-T412 (50).

In our phase I study using cM-T412, there was partial repopulation of CD4+ T cells (≥50% increase between 14 days and 6 months postinfusion) in approximately 40% of patients at 6 months (40,42). This repopulation involved combined increases in both CD45RO+ (memory) and CD45RA+ (naïve) subpopulations of CD4+ T cells. The total CD3+ T-cell frequency was also decreased with no compensatory increase in CD8+ T cells. A similar profound prolonged depletion of lymphocytes (CD4 and CD8) has been associated with CAMPATH-1, a humanized MAb to CDw52, an antigen common to all lymphocytes (57).

In a recent report of CD4+ T-lymphocyte reconstitution following intensive chemotherapy, there was an inverse relation between the age of patients and the CD4+ T-lymphocyte count 6 months after therapy (58). There was a higher proportion of CD45RA+ CD4+ T-lymphocytes in patients with thymic enlargement after chemotherapy than in patients without such thymic enlargement. These results suggest that rapid T-cell regeneration requires residual thymic function. This age-related decrease in thymic functioning and resulting decreased cellular immunity in elderly subjects (59) has significant impact on the type (depleting/nondepleting) of MAbs that may be appropriate for use as therapeutic agents in RA. A similar age-related recovery of CD4+ T-cells following treatment with anti-CD4 MAb has been observed in mice (60).

The failure of cMT-412 to exert significant beneficial clinical effects in our studies might be explained by the inability to administer adequate quantities of the MAb to achieve modulation of T-cell function and/or deplete T cells at local sites of inflammation (46,47). Alternatively, the role of T cells in established RA may be of minimal clinical importance. Studies in progress with nondepleting anti-CD4 antibodies should be helpful in resolving these important issues (61).

REFERENCES

1. Harris ED. Rheumatoid arthritis. Pathophysiology and implications for therapy. N Engl J Med 1990; 322:1277–1289.
2. Winchester RF, Gregersen PK. The molecular basis of susceptibility to rheumatoid arthritis. Springer Semin Immunopathol 1988; 10:119–139.

3. Panayi GS, Lanchbury JS, Kingsley GH. The importance of the T-cell in initiating and maintaining the chronic synovitis of rheumatoid arthritis. Arthritis Rheum 1992; 35:729-735.
4. Koopman WJ, Gay S. Do nonimmunologically mediated pathways play a role in the pathogenesis of rheumatoid arthritis? Rheum Dis Clin North Am 1993; 19:107–122.
5. Weiss A. Structure and function of the T-cell antigen receptor. J Clin Invest 1990; 86:1015–1022.
6. Van Boxel JA, Paget SA. Predominantly T-cell infiltrate in rheumatoid synovial membrane. N Engl J Med 1975; 293:517–520.
7. Burmester GR, Yu DTY, Irani AM, Kunkel HG, Winchester RJ. Ia+T-cells in synovial fluid and tissue of patients with rheumatoid arthritis. Arthritis Rheum 1986; 24:1370–1376.
8. Stastny P. Association of the B-cell alloantigen DRw4 with rheumatoid arthritis. N Engl J Med 1978; 298:869–871.
9. Schiff B, Mizrachi Y, Orgad S, Yaron M, Gazit E. Association of HLA-Aw31 and HLA-DR1 with adult rheumatoid arthritis. Ann Rheum Dis 1982; 41:403–404.
10. Gregersen PK, Silver J, Winchester RJ. The shared epitope hypothesis: an approach to understanding the molecular genetics of susceptibility to rheumatoid arthritis. Arthritis Rheum 1987; 30:1205–1213.
11. Nepom GT, Byers P, Seyfried C, Healey LA, Wilske KR, Stage D, Nepom BS. HLA genes associated with rheumatoid arthritis. Arthritis Rheum 1989; 32:14–21.
12. Calabrese LH, Wilske WS, Parkins AD, Tubbs RR. Rheumatoid arthritis complicated by infection with human immunodeficiency virus and the development of Sjögren's syndrome. Arthritis Rheum 1989; 32:1453–1457.
13. Paulus HE, Machleder HI, Levine S, Yu DTY, MacDonald NS. Lymphocyte involvement in rheumatoid arthritis: studies during thoracic duct drainage. Arthritis Rheum 1977; 20:1249–1262.
14. Strober S, Tanay A, Field E, Hoppe RT, Calin A, Engleman EG, Kotzin B, Brown BW, Kaplan HS. Efficacy of total lymphoid irradiation in intractable rheumatoid arthritis: a double-blind, randomized trial. Ann Intern Med 1985; 102:441–449.
15. Tugwell P, Bombardier C, Gent M, Bennett K, Bensen W, Carette S, Chalmers A, Esdaile J, Klinkhorf A, Kragg G, Ludwin D, Roberts R. Low-dose cyclosporine versus placebo in patients with rheumatoid arthritis. Lancet 1990; 335:1051–1055.
16. Ranges GE, Sriram S, Cooper SM. Prevention of type II collagen-induced arthritis by in vivo treatment with anti-L3T4. J Exp Med 1985; 162:1105–1110.
17. Wofsy D, Seaman WE. Reversal of advanced murine lupus in NZB/NZW F1 mice by treatment with monoclonal antibody to L3T4. J Immunol 1987; 138:3247–3253.
18. Christadoss P, Dauphinee MJ. Immunotherapy for myasthenia gravis: a murine model. J Immunol 1986; 136:2437–2440.
19. Waldor MK, Sriram S, Hardy R, Herzenberg LA, Lanier L, Lim M, Steinman L. Reversal of experimental allergic encephalomyelitis with monoclonal antibody to a T-cell subset marker. Science 1985; 22:415–417.
20. Koike T, Itoh Y, Ishii T, Ito I, Takabayashi K, Maruyama N, Tomioka H, Yoshida S. Preventive effect of monoclonal anti-L3T4 antibody on development of diabetes in NOD mice. Diabetes 1987; 36:539–541.

21. Nabozny GH, Cobbold SP, Waldmann H, Kong YM. Suppression in murine experimental autoimmune thyroiditis: in vivo inhibition of CD4+ T-cell-mediated resistance by a nondepleting rat CD4 monoclonal antibody. Cell Immuno 1991; 138:185–196.
22. Atalla L, Linker-Israeli M, Steinman L, Rao NA. Inhibition of autoimmune uveitis by anti-CD4 antibody. Invest Ophthalmol Visual Sci 1990; 31:1264–1270.
23. Hertzog C, Walker C, Pickler W, Aeschlimann A, Wassmer P, Stockinger H, Knapp W, Rieber P, Müller W. Monoclonal anti-CD4 in arthritis. Lancet 1987; 2: 1461–1462.
24. Wendling D, Wijdenes J, Racadot E, Morel-Fourrier B: Therapeutic use of monoclonal anti-CD4 antibody in rheumatoid arthritis. J Rheum 1991; 18:325–327.
25. Rieter C, Kakavand B, Rieber EP, Schattenkirchner M, Riethmüller G, Krüger K. Treatment of rheumatoid arthritis with monoclonal CD4 antibody M-T151: clinical results and immunopharmacologic effects in an open study, including repeated administration. Arthritis Rheum 1991; 34:525–536.
26. Horneff G, Burmester GR, Emmrich F, Kalden JR. Treatment of rheumatoid arthritis with an anti-CD4 monoclonal antibody. Arthritis Rheum 1991; 34:129–140.
27. Herzog C, Walker C, Müller W, Rieber P, Reiter C, Riethmüeller G, Wassmer P, Stockinger H, Madic O, Pichler WJK. Anti-CD4 treatment of patients with rheumatoid arthritis: I. Effect on clinical course and circulating T-cells. J Autoimmun 1989; 2:627–642.
28. Wendling D, Racadot E, Morel-Fourrier B, Wijdenes J. Treatment of rheumatoid arthritis with anti-CD4 monoclonal antibody: open study of 25 patients with the B-F5 clone. Clin Rheumatol. 1992; 11:542–547.
29. Horneff G, Winkler T, Kalden JR, Emmrich F, Burmester GR. Human anti-mouse antibody response induced by anti-CD4 monoclonal antibody therapy in patients with rheumatoid arthritis. Clin Immunol Immunopathol 1991; 59:89–103.
30. Goldberg D, Morel P, Chatenoud L, Boitard C, Menkes CJ, Bertoye P-H, Revillard J-P, Bach J-F. Immunological effects of high dose administration of anti-CD4 antibody in rheumatoid arthritis patients. J Autoimmun 1991; 4:617–630.
31. Carteron NL, Schimenta CL, Wofsy D. Treatment of murine lupus with $F(ab')_2$ fragments of monoclonal antibody to L3T4. J Immunol 1989; 142:1470–1475.
32. Newell MK, Haugh LJ, Maroun CR, Julius MH. Death of mature T-cells by separate ligation of CD4 and the T-cell receptor for antigen. Nature 1990; 346:286–289.
33. Benjamin RJ, Qin S, Wise MP, Cobbold SP, Waldmann H. Mechanism of monoclonal antibody facilitated tolerance induction: a possible role for the CD4 (L3T4) and CD11a (LFA-1) molecule in self-non-self discrimination. Eur J Immunol 1988; 18:1079–1088.
34. Waldmann H, Cobbold SP. Is tolerance therapy a plausible possibility? Br J Rheum 1991; 30(Suppl):75.
35. Villemain F, Jonker M, Bach JF, Chatenoud L. Fine specificity of antibodies produced in rhesus monkey following in vivo treatment with anti-T cell murine monoclonal antibodies. Eur J Immunol 1986; 16:945–949.
36. LoBuglio AF, Wheeler RH, Trang J, Haynes A, Rogers K, Harvey EB, Sun L, Ghrayeb J, Khazaeli MB. Mouse/human chimeric monoclonal antibody in man: kinetics and immune response. Proc Natl Acad Sci USA 1989; 86:4220–4224.

37. Wofsy D, Maves DC, Woodcok J, Seaman WE. Inhibition of humoral immunity in vivo by monoclonal antibody to L3T4: studies with soluble antigens in intact mice. J Immunol 1985; 135:1698–1701.
38. Wassmer PJ, Chan C, Lögdberg L, Shevach EM. Role of the L3T4 antigen in T-cell activation. II. Inhibition of T-cell activation by monoclonal anti-L3T4 antibodies in the absence of accessory cells. J Immunol 1985; 135:2237–2242.
39. Wilde DB, Marrack P, Kappier J, Dialynas D, Fitch FW. Evidence implicating L3T4 in class II MHC antigen reactivity: monoclonal antibody GK 1.5 (anti-L3T4a) blocks class II MHC antigen-specific proliferation, release of lymphokines, and binding by cloned murine helper T lymphocyte lines. J Immunol. 1993; 131:2178–2183.
40. Moreland LW, Bucy RP, Tilden A, Pratt PW, LoBuglio AF, Khazaeli M, Everson MP, Daddona P, Ghrayeb J, Kilgarriff C, Sanders ME, Koopman WJ. Use of a chimeric anti-CD4 antibody in patients with refractory rheumatoid arthritis. Arthritis Rheum 1993; 36:307–318.
41. Moreland LW, Pratt PW, Bucy RP, Jackson BS, Feldman JW, Koopman WJ. Treatment of refractory rheumatoid arthritis with a chimeric anti-CD4 monoclonal antibody: longterm follow-up of CD4+ T-cell counts. Arthritis Rheum 1994; 37:834–838.
42. Pratt PW, Bucy RP, Moreland LW, Koopman WJ. Thymic processing appears to contribute to repopulation of CD4 lymphocytes after depletion with chimeric CD4 antibody in refractory rheumatoid arthritis. Arthritis Rheum 1992; 35(Suppl): S105.
43. Moreland LW, Pratt PW, Mayes M, Postlethwaite A, Weisman M, Schnitzer T, Lightfoot R, Calabrese L, Zelinger DJ, Woody JN, Koopman WJ. Double-blind, placebo controlled multicenter trial using chimeric monoclonal anti-CD4 antibody in rheumatoid arthritis patients receiving concomitant methotrexate. Arthritis Rheum 1995; 38:1581–1588.
44. van der Lubbe PA, Dijkmans BAC, Markusse HM, Nüssander U, Breedveld FC. Lack of clinical effects of CD4 monoclonal antibody therapy in early rheumatoid arthritis in a placebo controlled trial. Arthritis Rheum 1995; 38:1097–1106.
45. Choy EHS, Chikanza IC, Kingsley GH, Corrigall V, Panayi GS. Treatment of rheumatoid arthritis with single dose or weekly pulses of chimaeric anti-CD4 monoclonal antibody. Scand J Immunol 1992; 36:291–298.
46. Choy EHS, Pitzalis C, Bijl JA, Kingsley GH, Panayi GS. The amount of anti-CD4 monoclonal antibody (MAb) entering the rheumatoid (RA) joint may determine clinical efficacy. Arthritis Rheum 1993; 36(Suppl):S39.
47. Choy EHS, Pitzalis C, Bijl JA, Kingsley GH, Panayi GS. The importance of dosing and dosing regimen of anti-CD4 monoclonal antibody (MAb) in the treatment of rheumatoid arthritis (RA). Arthritis Rheum 1993; 36(Suppl):S129.
48. van der Lubbe PA, Reiter C, Breedweld FC, Krüger K, Schattenkirchner M, Sanders ME, Riethmüller G. Chimeric CD4 monoclonal antibody cM-T412 as a therapeutic approach to rheumatoid arthritis. Arthritis Rheum 1993; 36:1375–1379.
49. van der Lubbe PA, Reiter C, Miltenburg AM, Krüger K, de Ruyter AN, Rieber EP, Bijl JA, Riethmüller G, Breedveld FC. Treatment of rheumatoid arthritis with a chimeric CD4 monoclonal antibody (cM-T412): immunopharmacological aspects and mechanisms of action. Scand J Immunol 1994; 39:286–294.

50. Lindsey JW, Hodgkinson S, Mehta R, Siegel RC, Mitchell D, Lim M, Piercy C, Enzmann D, Steinman L. Phase I clinical trial of repeated treatment with chimeric monoclonal anti-CD4 antibody in multiple sclerosis. Neurology 1994; 44:413–419.
51. Connolly K, Roubinian JR, Wofsy D: Development of murine lupus in CD4 depleted NZB/NZW mice: sustained inhibition of residual CD4+ T cells is required to suppress autoimmunity. J Immunol 1992; 149:3083–3088.
52. Gutstein NL, Wofsy D. Administration of F(ab')$_2$ fragments of monoclonal antibody to L3T4 inhibits humoral immunity in mice without depleting L3T4+ cells. J Immunol 1986; 137:3414–3419.
53. Carteron NL, Schimenti CL, Wofsy D. Treatment of murine lupus with F(ab')$_2$ fragments of monoclonal antibody to L3T4: suppression of autoimmunity does not depend on T helper cell depletion. J Immunol 1989; 142:1470–1475.
54. Carteron NL, Wofsy D, Seaman WE. Induction of murine tolerance during administration of monoclonal antibody to L3T4 does not depend on depletion of L3T4+ cells. J Immunol 1988; 140:713–716.
55. Calabrese LH, Taylor JW, Wilke WS, Segal AM, Valenzuela R, Clough JD. Response of immunoregulatory lymphocyte subsets to methotrexate in rheumatoid arthritis. Cleve Clin J Med 1990; 57:232–241.
56. Rosenthal GJ, Germolec DR, Lamon KR, Ackerman MF, Luster MI. Comparative effects on the immune system of methotrexate and trimetrexate. Int J Immunopharmacol 1987; 9:793–801.
57. Isaacs JD, Watts RA, Hazleman BL, Hale G, Keogan MT, Cobbold SP, Waldmann H. Humanized monoclonal antibody therapy for rheumatoid arthritis. Lancet 1992; 340:748–752.
58. Mackall CL, Fleisher TA, Brown MR, Andrich MP, Chen CC, Feuerstein IM, Horowitz ME, Magrath IT, Shad AT, Steinberg SM, Wexler LH, Gress RE: Age, thymopoiesis, and CD4+ T-lymphocyte regeneration after intensive chemotherapy. N Engl J Med 1995; 332:143–149.
59. Roberts-Thomson IC, Whittingham S, Youngchaiyud U, Mackay IR. Aging, immune response, and mortality. Lancet 1974; 2:368–370.
60. Fleming AL, Field EH, Tolaymat N, Cowdery JS. Age influences recovery of systemic and mucosal immune responses following acute depletion of CD4 T-cells. Clin Immunol Immunopathol 1993; 69:285–291.
61. Moreland LW, Bucy RP, Knowles RW, Wacholtz MC, Haverty TP, Koopman WJ. Treating rheumatoid arthritis with a non-depleting anti-CD4 monoclonal antibody (MAb). J Invest Med 1995; 43(Suppl):362A.

5
CD4 Monoclonal Antibody Therapy in Rheumatoid Arthritis

F. C. BREEDVELD
Leiden University Hospital, Leiden, The Netherlands

I. INTRODUCTION

When the hybridoma technique of Köhler and Milstein was introduced in 1975, its potential for clinical use was immediately perceived (1). Within a decade, therapeutic monoclonal antibodies (MAb) directed against the pan-T-cell antigen CD3 became established in clinical use (2). It took longer to exploit the obvious advantage of a subclass-specific antibody for clinical purposes. Though the T-helper-specific human T4 antigen had already been described in 1979, a systemic study of its usefulness as target for immunotherapies became possible only when a MAb against its murine homolog L3T4 was made available (3). Since then, CD4 antibodies have become a major analytical and therapeutic tool to dissect and influence the immune response in vivo. Many studies have revealed the central role of CD4+ T cells in the majority of experimental autoimmune diseases, which could be prevented by early treatment with CD4 MAb. Moreover, already established experimental autoimmune diseases were susceptible to treatment (4,5). The effort to introduce CD4 MAb into clinical use was mainly nourished by the expectation of a more selective immunosuppression sparing the T-helper-cell-independent arms of the immune response.

II. RATIONALES FOR CD4 Mab AS ANTIARTHRITIC AGENTS

The CD4 glycoprotein is a typical member of the immunoglobulin gene family with an array of four extracellular domains, a transmembrane anchor, and a short intracytoplasmic tail. Though it is also expressed on cells other than T cells, e.g., monocytes and eosinophils, its main function appears to be to guide T lymphocytes toward MHC class II recognition. As coreceptor of the antigen receptor on T cells, CD4 is involved in T-cell differentiation as well as in the process of activation or inactivation, i.e., induction of anergy. Antibodies that crosslink CD4 and the CD3-T-cell receptor (TCR) complex were shown to be strong positive signals for activation, whereas crosslinking of CD4 molecules with antibodies prior to engagement of the T-cell receptor may poise the cells for induction of apoptosis (6,7). Thus CD4 is shown to have a prominent role as coreceptor being responsible for the class II–restricted stimulation of T-helper cells, as well as for modulating T-cell activation in a positive or negative direction.

The use of MAb as therapeutic agents has been studied most extensively in the area of cancer treatment and organ transplantation. These studies have contributed to the understanding of the mechanisms of antibody therapy and possible adverse reactions to antibody application. Most remarkable was the observation that induction of long-lasting tolerance could be achieved with CD4 MAb in transplantation models (8). Tolerance to soluble protein could be achieved if the protein was administered under the umbrella of short-term anti-CD4 treatment. These observations have generated hope that reprogramming of the immune system should be possible even in adult individuals, which may create therapeutic possibilities for autoimmune diseases including rheumatoid arthritis (RA).

Several lines of evidence suggest that such a T-cell activation has a central role in the pathogenesis of RA. The evidence is summarized below.

1. Immunohistochemical studies demonstrated the perivascular accumulation of predominantly activated CD4+ T cells in close contact with B cells and macrophages (9,10).
2. Studies on antigen specificity of synovial tissue–derived T cells revealed reactivity against antigens that predominantly occur in the joints (11,12).
3. T cells may use a restricted number of T-cell-receptor genes (13,14).
4. Immunogenetic studies revealed linkage between MHC and the susceptibility to RA (15,16).
5. Experimental animal models of arthritis have shown that a small number of antigen-specific T cells can converge to joints and initiate a form of arthritis histologically similar to RA (17).

6. Clinical trials aimed at the elimination of T cells have proven to have a beneficial effect on RA (18).

These observations imply that inhibition of the interaction of the TCR/CD4 complex with the potential autoantigen presented by antigen-presenting cells may prove to be a specific and efficient immunotherapy for RA.

III. SELECTION OF CD4 MAb

The in vivo effect of MAb therapy depends first on the epitope to which the MAb binds but also on the interaction of the bound MAb with the natural defense systems. This interaction may lead to cytotoxicity by activating complement or activation of killer cells. MAb bound to T-cell surface molecules may also lead to a decreased expression of the molecule. After injection of anti-CD4 MAb, a considerable percentage of the circulating T cells do not express the TCR/CD4 complex any longer and are therefore immunologically "blind." Variations in the possible target molecule, the different epitopes on one particular target molecule, the binding characteristics, and the Fc portion of the molecule, including the linkage to toxins, make the number of possibilities for T-cell-directed MAb therapy extremely large. Studies that compare the in vitro characteristics of MAb with the in vivo effects are sparse. Nevertheless, the criteria used for selection of CD4 MAb for clinical studies generally include epitope specificity, affinity, isotype, low modulating activity, and maximal capacity to inhibit antigen-specific stimulation of T lymphocytes. Secondary considerations are the amount of antibody produced by individual hybridomas and the growth characteristics or stability of the clones. Because of the immunogenicity of murine MAb chimerization or humanization is found to be mandatory, particularly since repeated administration for the treatment of RA is foreseeable.

IV. IMMUNOPHARMACOLOGY OF CD4 MAb

The serum level of circulating CD4 MAb after injection is a result of the applied dose and the infusion rate. Unbound circulating CD4 MAb were detected only after injections of more than 10 mg (19). These levels peaked to 1–2 mg/L after the infusion of 20 mg and became undetectable 24 hr later (20–22). With murine MAb a half-life of 3 hr was measured whereas a four- to six-fold longer half-life was reported for chimeric MAb (21,22). Following infusion of 10 mg or more of CD4 MAb, all circulating CD4+ cells will be coated with the antibody (19,22). This observation explains the most impressive immediate effect seen after anti-CD4 MAb administration, which is the clearance of CD4+ lymphocytes from the circulation (Fig. 1). After administration of several murine CD4 MAb, the number

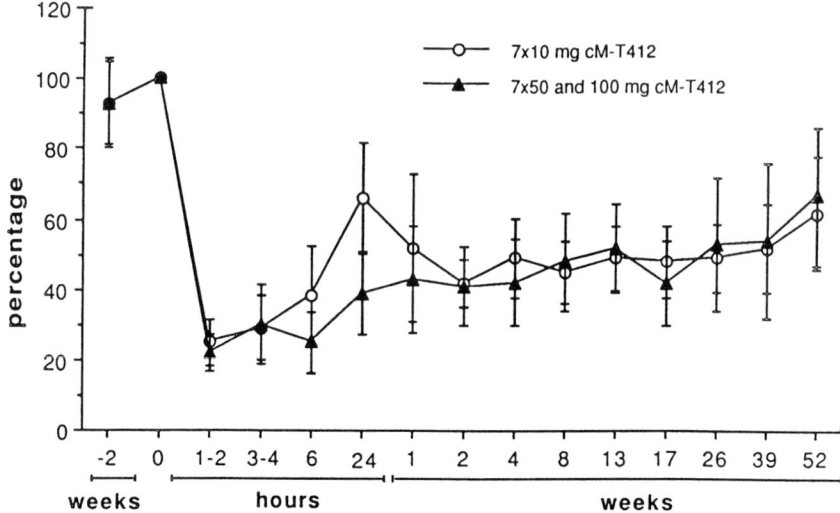

Figure 1 Prolonged depletion of CD4+ cells following injection of CD4 MAb on 7 consecutive days.

of CD4+ lymphocytes reached pretreatment levels within 24 hr but others, such as the MAX 16H5 (22) and in particular the chimeric MAb, induced a prolonged depletion of CD4+ lymphocytes lasting in single cases for more than a year (19). Binding of the MAb to cell surface antigens may lead to a down-regulation of CD4 surface molecules. However, such a modulation was observed only upon administration of the MAX 16H5 murine MAb (22) but not with other anti-CD4 antibodies investigated.

The body distribution of injected CD4 MAb has not been studied in great detail. Injected immunoglobulins accumulate nonspecifically at inflammatory sites (23). Besides nonspecific accumulation, Kinne et al. demonstrated a specific accumulation of anti-CD4 MAb in organs containing high numbers of CD4+ cells (24). In patients with RA, Choy et al. demonstrated that single injections of the cMT412 CD4 MAb led to complete coating of circulating CD4+ T cells but that there was little coating of CD4+ T cells in the joints (25). These authors reported that only after five daily 50-mg infusions did increased coating of synovial fluid cells occur with a reduction in synovial fluid CD4+ T-cell numbers. This observation is in line with those of Tak et al., who showed depletion of T cells from the synovial tissue of RA patients only with dosages of 25 mg or higher for 5 consecutive days (26).

V. EFFECT ON THE IMMUNE SYSTEM

The injection of CD4 MAb is followed by an immediate depression of the number of circulating CD4+ cells lasting from a few days in the case of murine antibodies but persisting for up to many months in the case of chimerized antibodies of the IgG1 class. Differences in the cell-depleting effect of immune antibodies were attributed to the isotype and antigen specificity. To date, the mechanism responsible for the prolonged depletion of circulating CD4+ T cells has not been elucidated. The most obvious explanation is that the depletion is due to trapping of antibody-coated cells in the mononuclear phagocyte system. Recently chimerized CD4 MAb of the IgG4 isotype were studied in arthritis patients. These MAb did not deplete the number of circulating CD4+ cells. Despite the low number of circulating T-helper cells following treatment, no overt clinical signs of immune suppression have been observed (27–29). In fact, the only side effect observed is a flu-like syndrome during the first day of treatment, which could be ascribed to a lymphokine release syndrome. The majority of individuals treated with murine and chimeric CD4 MAb develop human anti-mouse antibodies. Interestingly, these antibodies generally did not induce allergic reactions upon repeated treatment. The levels of CD4 MAb-induced anti-mouse antibodies seemed to be lower than the CD3 MAb-induced titers, which may be explained by a specific immunosuppression caused by CD4 MAb therapy.

VI. CD4 MONOCLONAL ANTIBODIES AS TREATMENT OF RA

CD4 MAb have been investigated for the treatment of RA in a limited number of patients, who were in an advanced stage of disease. Without exception all the published uncontrolled phase I and II studies applying multiple-dosage regimens of CD4 MAb in patients with RA reported a direct amelioration of arthritis activity lasting for several months (20,22,30–34). Open-label clinical trials utilizing a variety of murine CD4 MAb have been reported in approximately 75 RA patients. Clinical results following administration of two murine IgG1 MAb (MAX 16H5 and BF-5) and three IgG2a MAb (BL4, MT151, and VIT4) were remarkably similar. Clinical responses as judged by a significant reduction in the Ritchie index were observed in approximately 60–75% of the patients, but all evaluations were unblinded. Generally 3 months of benefit were reported, with a range of 1–12 months. Despite the strong effect of number of circulating CD4+ cells, there has been no consistent improvement in the traditional laboratory measures of disease activity such as acute-phase proteins.

To date, only the 16H5 and BF-5 produced a significant decrease in the serum C-reactive protein levels. Three of these studies involved the cM-T412 CD4 MAb, which is a chimeric antibody with an IgG1 Fc region. One study reported a significant clinical response in patients treated with 50 mg in contrast to those treated with 10 mg cM-T412 daily for 7 days (19). Another study reported beneficial clinical effects of 100 mg cM-T412 daily for 7 days as well as of single-dose infusions with 10, 50, 100, 150, or 200 mg CD4 MAb in patients who were concomitantly treated with a stable dose of methotrexate (MTX) (33). The third study reported significant improvement of arthritis activity in RA patients during the first 6 weeks following treatment with 50 mg CD4 MAb for 5 consecutive days (34). These antibodies did not induce decreased levels of acute-phase proteins. Two placebo-controlled studies were performed on single-dose administration of CD4 MAb, both with cM-T412. In one study single-dose infusions with placebo or 5, 10, or 50 mg CD4 MAb were given for 3 consecutive months to 64 MTX-treated RA patients (35). In the other controlled study two patients were randomized to placebo and seven to single-dose infusions with 50 mg cM-T412 (36). No significant clinical benefit of cM-T412 administration was shown in these two studies. A recent study has investigated the efficacy of multiple-dosage regimens of CD4 MAb treatment of RA in a placebo-controlled, randomized trial. The patients studied were in an early phase of the disease and had received no disease-modifying antirheumatic drug previously (37). The main conclusion of this study is that treatment with neither 10, 25, or 50 mg of cM-T412 CD4 MAb on 5 consecutive days induced an improvement of arthritis activity. cM-T412 induced a sustained depression of circulating CD4+ cells, which was similar for all three dosages and in the same range as reported previously. A direct relationship between the clinical effect and the decrease in the number of circulating CD4+ cells was not found in any of the trials discussed. The results of this study are in clear contrast with the therapeutic effect of multiple-dosage regimens of CD4 MAb treatment observed in previous open studies on RA patients.

Several factors may account for the discrepancies, the most obvious being the use of a placebo-controlled design. Placebo effects in the treatment of RA have long been recognized (38). The discrepancy between the improvement in clinical parameters of disease and not in the objective laboratory parameters, such as ESR and serum CRP levels, is compatible with a placebo effect in the open studies. The type of CD4 MAb employed may underlie the difference in results with some studies. The CD4 MAb investigated so far in the treatment of RA differ with respect to epitope specificity, affinity, and Fc isotype, in vitro effects on T-cell functions, modulatory effect on the CD4 surface molecule, and in the capacity to induce a decrease in the number of circulating CD4+ cells in vivo. Since the relationship between these antibody characteristics and clinical efficacy is not known, further insight into possible differences in clinical efficacy between different CD4 MAb has to come from direct clinical comparison.

VII. CONCLUDING REMARKS

It can only be speculated why CD4 MAb treatment is clinically not effective in the randomized controlled trials. Despite the profound depletion of circulating CD4+ T cells, the effect on T cells in the synovial tissue may have been insufficient. Support for this hypothesis was recently presented by Tak et al. (39). Immunohistochemical analysis of synovial biopsies obtained from RA patients directly before and 4 weeks after treatment with CD4 MAb did not reveal statistically significant differences between the mean scores for infiltration with CD4+ cells. Some decrease in synovial tissue CD4+ cells, both the CD45RA and CD45RO+, was found only in patients (e.g., clinical nonresponders) treated with more than 125 mg CD4 MAb. In addition, it was found that proinflammatory cytokines such as IL-1β or TNFα could still be detected in the synovial tissue after treatment. It is therefore conceivable that even a reduced number of CD4+ cells found in the synovium may have maintained enough inflammation to explain the lack of clinical effect in these patients. Accumulation of unbound cM-T412 in the peripheral blood only occurs with relatively high dosages (28) and penetration of unbound CD4 MAb into synovial tissue might be essential to induce a more profound blockage of CD4+ cells in the synovial tissue. Therefore, the clinical effect of higher dosages of CD4 MAb is of interest. This would also be in line with the higher dosages studied in animal models. The use of higher dosages with depleting chimeric antibodies studied so far was blocked by the fear of side effects following the induction of prolonged and severe lymphocytopenia. Recently studied nondepleting primalized IDEC-CE 9.1 CD4 MAb may solve this problem (40).

It may also be worth considering the combined use of various MAb. Combining two different MAb to distinct CD4 epitopes permitted a reduction in dose to achieve the same immune suppression, and the combination of CD4 MAb and anti-TNF MAb was effective in the treatment of an arthritis model (41,42). Since only a limited number of MAb can be evaluated clinically, studies should be focused on the relationship between antiarthritic and immunological effects. Observation from this will support the development of this therapeutic approach and will further supply important information about the pathogenesis of the disease.

In conclusion, at present there is not firm evidence that treatment with MAb directed against the CD4 molecule has therapeutic efficacy in RA. Whether this lack of efficacy should be ascribed to the selection of an antibody with low efficacy in the controlled trials or to the design of an inappropriate treatment regimen needs additional study.

REFERENCES

1. Köhler G, Milstein C. Continuous cultures of fused cells secreting antibody of predefined specificity. Nature 1975; 256:495–497.

2. Ortho Multicentre Transplant Study Group. A randomized clinical trial of OKT3 monoclonal antibody for acute rejection of cadaveric renal transplants. N Engl J Med 1985; 313:337–342.
3. Dialynas DP, Quan ZS, Wall KA, Pierres A, Guintas J, Loken MR, Pierres FR, Fitch FW. Characterization of the murine T cell surface molecule designated L3T4, identified by monoclonal antibody GK 1.5; similarity of L3T4 to the human Leu-3/T4 molecule. J Immunol 1983; 131:2445–2451.
4. Wofsy D, Seaman WE. Reversal of advanced murine lupus with NZB/NZW F1 mice by treatment with monoclonal antibody to L3T4. J Immunol 1987; 138:3247–3253.
5. Billingham MEJ, Hicks CA, Carney SL. Monoclonal antibodies and arthritis. Agents Actions 1990; 29:77–87.
6. Bierer BE, Sleckman BP, Ratnofsky SE, Burakoff SJ. The biologic role of CD2, CD4 and CD8 in T cell activation. Annu Rev Immunol 1989; 7:579–599.
7. Newell MK, Haughn LJ, Maroun CR, Julius MH. Death of mature T cells by separate ligation of CD4 and the T cell receptor for antigen. Nature 1990; 347:286–289.
8. Waldmann H. Manipulation of T cell responses with monoclonal antibodies. Annu Rev Immunol 1989; 7:407–444.
9. Pitzalis C, Kingsley G, Lanchbury JSS, Murphy J, Panayi GS. Expression of HLA-DR, DQ and DP antigens and interleukin-2 receptor on synovial fluid T lymphocyte subsets in rheumatoid arthritis: evidence for "frustrated" activation. J Rheumatol 1987; 14: 662–666.
10. Duke O, Panayi GS, Janossy G, Poulter LW. An immunohistological analysis of lymphocyte subpopulations and their microenvironment in the synovial membranes of patients with rheumatoid arthritis using monoclonal antibodies. Clin Exp Immunol 1982; 49:22–30.
11. Res PCM, Schaar CG, Breedveld FC, et al. Synovial fluid T cell reactivity against 65 kD heat shock protein of mycobacteria in early chronic arthritis. Lancet 1988; 478–480.
12. Klareskog L, Forsum U, Scheynius A, Kabelitz D, Wigzell H. Evidence in support of a self-perpetuating HLA-DR-dependent delayed-type cell reaction in rheumatoid arthritis. Med Sci 1982; 79: 3632–3636.
13. Paliard X, West SG, Lafferty JA, et al. Evidence for the effects of a superantigen in rheumatoid arthritis. Science 1992; 253:325–329.
14. van Laar JM, Miltenburg AMM, Verdonk MJA, et al. T cell receptor B-chain gene rearrangements of T cell populations expanded from multiple sites of synovial tissue obtained from a patient with rheumatoid arthritis. Scand J Immunol 1992; 35:187–194.
15. Teyton L, Lotteau V, Turmel P, et al. HLA DR, DQ, and DP antigen expression in rheumatoid synovial cells: a biochemical and quantitative study. J Immunol 1987; 138:1730–1738.
16. Burmester GR, Jahn B, Gramatzki M, Zacher J, Kalden JR. Activated T cells in vivo and in vitro: divergence in expression of Tac and Ia antigens in the nonblastoid small T cells of inflammation and normal T cells activated in vitro. J Immunol 1984; 133:1230–1234.
17. Holoshitz J, Matitiau A, Cohen IR. Arthritis induced in rats by cloned T lymphocytes responsive to mycobacteria but not to collagen type II. J Clin Invest 1984; 73:211–215.

18. Breedveld FC, de Vries RRP. Anti-CD4 antibodies in rheumatoid arthritis. Clin Exp Rheumatol 1992; 10:325–326.
19. van der Lubbe PA, Reiter C, Breedveld FC, Krüger K, Schattenkirchner M, Sanders ME, et al. Chimeric CD4 monoclonal antibody cM-T412 as a therapeutic approach to rheumatoid arthritis. Arthritis Rheum 1993; 36:1375–1379.
20. Reiter C, Kakawand B, Rieber EP, Schattenkirchner M, Reithmüller G, Krüger K. Treatment of rheumatoid arthritis with monoclonal CD4 antibody M-T151. Arthritis Rheum 1991; 34:525–536.
21. LoBuglio AF, Wheeler RH, Trang J, Hanes A, Rogers K, Harvey EB, Sun L, Grayeb J, Khazaeli MB. Mouse/human chimeric monoclonal antibody in man: kinetics and immune response. Proc Natl Acad Sci USA 1989; 86:4220–4224.
22. Horneff G, Burmester GR, Emmerich F, Kalden JR. Treatment of rheumatoid arthritis with an anti-CD4 monoclonal antibody. Arthritis Rheum 1991; 34:129–140.
23. de Bois MHW, Arndt JW, Tak PP, Kluin PM, van der Velde EA, Pauwels EKJ, Breedveld FC. $^{99}Tc^m$-labelled polyclonal human immunoglobulin G scintigraphy before and after intra-articular knee injection of triamcinolone hexacetonide in patients with rheumatoid arthritis. Nucl Med Commun 1993; 14:883–887.
24. Kinne RW, Becker W, Simon G, Paganelli G, Palombo-Kinne E, Wolski A, Block S, Schwarz A, Wolf F, Emmerich F. Joint uptake and distribution of a technetium-99m-labeled anti-rat CD4 monoclonal antibody in rat adjuvant arthritis. J Nucl Med 1993; 34:92–98.
25. Choy EHS, Pitzales C, Bijl JA, Kingsley GH, Panayi GS. The amount of anti-CD4 of monoclonal antibody entering the rheumatoid joint may determine clinical efficacy. Arthritis Rheum 1993; 36:S39.
26. Tak PP, van der Lubbe PA, Daha MR, Kluin PhM, Meinders AE, Breedveld FC. Reduction of synovial inflammation without clinical improvement after CD4 monoclonal antibody treatment in early rheumatoid arthritis. Clin Rheumatol 1994; 13:188.
27. Moreland LW, Pratt PW, Bucy RP, Jackson BS, Feldman JW, Koopman WJ. Treatment of refractory rheumatoid arthritis with a chimeric anti-CD4 monoclonal antibody: long-term followup of CD4+ T cell counts. Arthritis Rheum 1994; 37:834–838.
28. van der Lubbe PA, Reiter C, Miltenburg AM, Krüger K, de Ruyter AN, Rieber EP, Bijl JA, Riethmüller G, Breedveld FC. Treatment of rheumatoid arthritis with a chimeric CD4 monoclonal antibody (cM-T412): immunopharmacological aspects and mechanisms of action. Scand J Immunol 1994; 39:286–294.
29. Riethmüller G, Rieber EP, Kiefersauer S, Prinz J, van der Lubbe P, Meiser B, Breedveld F, Eisenburg J, Krüger K, Deusch K, Sanders M, Reiter C. From antilymphocyte serum to therapeutic monoclonal antibodies: first experiences with a chimeric CD4 antibody in the treatment of autoimmune disease. Immunol Rev 1992; 129:81–104.
30. Herzog C, Walker C, Müller W, Rieber P, Reiter C, Riethmüller G, et al. Anti-CD4 antibody treatment of patients with rheumatoid arthritis: I. Effect on clinical course and circulating T cells. J Autoimmun 1989; 2:627–642.
31. Goldberg D, Morel P, Chatenoud L, Boitard C, Menkes CJ, Bertoye PH, et al. Immunological effects of high dose administration of anti-CD4 antibody in rheumatoid arthritis patients. J Autoimmun 1991; 4:617–630.

32. Wendling D, Racadot E, Morel-Fourrier B, Wijdenes J. Treatment of rheumatoid arthritis with anti-CD4 monoclonal antibody: open study of 25 patients with the B-F5 clone. Clin Rheumatol 1992; 11:542–547.
33. Moreland LW, Bucy RP, Tilden A, Pratt PW, LoBuglio AF, Khazaeli M, et al. Use of a chimeric monoclonal anti-CD4 antibody in patients with refractory rheumatoid arthritis. Arthritis Rheum 1993; 36:307–318.
34. Choy EHS, Pitzalis C, Bijl JA, Kingsley GH, Panayi GS. The importance of dose and dosing regimen of anti-CD4 monoclonal antibody in the treatment of rheumatoid arthritis. Arthritis Rheum 1993; 36:S129 (abstract).
35. Moreland L, Pratt P, Mayes M, Postlethwaite A, Weisman M, Schnitzer T, et al. Minimal efficacy of a depleting chimeric anti-CD4 (cM-T412) in treatment of patients with refractory rheumatoid arthritis receiving concomitant methotrexate. Arthritis Rheum 193; 36:S39 (abstract).
36. Choy EHS, Chikanza IC, Kingsley Gh, Corrigall V, Panayi GS. Treatment of rheumatoid arthritis with single dose or weekly pulses of chimeric anti-CD4 monoclonal antibody. Scand J Immunol 1992; 36:291–298.
37. van der Lubbe PA, Dijkmans BAC, Markusse HM, Nässander U, Breedveld FC. Lack of clinical effect of CD4 monoclonal antibody therapy in early rheumatoid arthritis: a placebo controlled trial. Arthritis Rheum 1994; 37:S807.
38. Paulus HE, Egger MJ, Ward JR, Williams HJ. Cooperate Systematic Studies of Rheumatic Diseases Group. Analysis of improvement in individual rheumatoid arthritis patients treated with disease-modifying antirheumatic drugs, based on the findings in patients treated with placebo. Arthritis Rheum 1990; 33:477–484.
39. Tak PP, van der Lubbe PA, Daha MR, Smeets TJM, Kluin PM, Meinders EA, Bijl H, Breedveld FC. Reduction of synovial inflammation after anti-CD4 monoclonal antibody treatment in early rheumatoid arthritis. Arthritis Rheum 1994; 37(Suppl):S337.
40. Solinger Am, Yocum DE, Tesser J, Gluck O, O'Sullivan E, Henkel C, Cornett M, Lipani J. Clinical activity in an early phase I trial of primatized IDEC-CE9.1-, an anti-CD4 monoclonal antibody in rheumatoid arthritis. Arthritis Rheum 1994; 37(Suppl):S1060.
41. Qin SX, Cobbold S, Tighe H, Benjamin R, Waldmann H. CD4 monoclonal antibody pairs for immune suppression and tolerance induction. Eur J Immunol 1987; 17:1159–1165.
42. Williams RO, Mason LJ, Feldman M, Maini RN. Synergy between anti-CD4 and anti-TNF in the amelioration of established collagen induced arthritis. Proc Natl Acad Sci USA (in press).

6

The Use of CE9.1, a Primatized Monoclonal Anti-CD4, in the Treatment of Rheumatoid Arthritis

DAVID E. YOCUM
University of Arizona College of Medicine, Tucson, Arizona

ALAN M. SOLINGER
IDEC Pharmaceuticals Corporation, and University of California, San Diego, California

JOHN A. LIPANI
SmithKline Beecham Pharmaceuticals, Collegeville, Pennsylvania

I. INTRODUCTION

Murine monoclonal antibodies (MAb) have been used effectively for immunotherapy (1,2). However, patients quickly develop human anti-mouse antibodies (HAMA response) that render the treatment ineffective. The chimeric antibodies appear biologically superior to the murine antibodies since they are part human and part murine (3,4). However, a HAMA response still develops with these antibodies. Even fully humanized MAb have triggered production of anti-idiotypic antibodies (5). PRIMATIZED antibodies are part human and part monkey with the variable regions showing a high degree of homology to analogous human variable regions and a high affinity to human antigens. These antibodies may be as effective as the murine or chimeric antibodies and also may be less immunogenic. In addition, grafting the human constant region allows the choice of effector functions, i.e., a molecule with depleting or nondepleting activities, complement fixation, and antibody-dependent cellular cytotoxicity.

The primate/human anti-CD4 antibody IDEC-CE9.1 is an IgG1 lambda antibody containing macaque light- and heavy-chain variable region domains and human gamma 1 heavy-chain and lambda light-chain constant regions. IDEC-CE9.1 reacts with the CD4 antigen found on the surface of human and chimpanzee

CD4-positive (CD4+) T cells and on established CD4+ T-cell lines. No antibody reactivity exists with CD4+ T cells from several other species (6). IDEC-CE9.1 antibody may be a candidate for long-term immunotherapy of chronic diseases such as rheumatoid arthritis (RA).

II. ROLE OF CD4 CELLS IN RA

The CD4 molecule is a member of the immunoglobulin supergene family. It is a type I surface glycoprotein with a large (≈370-Kd residue) extracellular segment composed of four immunoglobulin-like domains, a single transmembrane domain, and a short cytoplasmic tail (38 residues). It has a relative molecular mass of 55,000 Daltons and contains two glycosylation sites on domains 3 and 4 (7). The triggering of T cells that results from an immune challenge involves presentation of the antigen, in association with the class II molecule, to the T-cell receptor. It has been suggested that regions of the CD4 molecule contact nonpolymorphic regions of the class II MHC molecule on the antigen-presenting cell. Such an engaged CD4 molecule probably transmits an intracellular signal, most likely through the association of its cytoplasmic tail with the cytoplasmic tyrosine kinase $p56^{lck}$.

In the pathogenesis of certain autoimmune diseases, when tolerance to self-antigens breaks down, CD4+ T cells foster and aggravate inflammatory conditions. This is accomplished through recruitment of inflammatory cells of the hematopoietic lineage, production of antibodies and cytokines, and activation of killer cells (8). The CD4+ T cells have been implicated in several autoimmune diseases such as RA, psoriasis, insulin-dependent diabetes mellitus, systemic lupus erythematosus, inflammatory bowel diseases, and cutaneous T-cell lymphoma.

Several lines of evidence suggest that early RA is an autoimmune disease driven by dysfunctional CD4+ T lymphocytes. Activated CD4+ T cells predominate in the initial synovial lesion (8,9). Certain MHC class II genes in the DR1 and DR4 loci confer an increased risk for developing RA (10). A purely T-cell-mediated arthritis can be induced in Lewis rats by complete Freund adjuvant and transferred by T cells to naïve animals (11). In addition, T-cell-depleting maneuvers such as thoracic duct drainage, total lymphoid irradiation, or lymphapheresis have been associated with disease improvement in some patients (12–14). More recently, anecdotal evidence of resolution of RA in patients developing AIDS (15) and transient improvement of disease symptoms with treatments using agents designed to interrupt T-cell function add to this body of evidence (16). However, current strategies are focusing on more selective agents that block unwanted immune responses without causing solid-organ toxicity or other major side effects. One such treatment strategy is the use of anti-CD4 antibodies for the selective removal or inactivation of T cells responsible for disease progression.

In animal models of autoimmunity and transplantation, anti-CD4 antibodies have been shown to arrest or reverse disease progression. Early results from trials with murine MAb in the treatment of RA and other autoimmune disorders have been promising (1,3,4,17).

Although clinical improvement was noted during these trials, it was not always correlated with T-cell depletion (3). In one high-dose trial of a murine antibody, significant but transient clinical improvement occurred with few adverse effects; saturation of binding sites on circulating CD4+ cells was achieved (1). However, the use of murine antibodies in immunocompetent hosts was limited by the human anti-mouse antibody (HAMA) response.

III. CE9.1 ANTI-CD4

The chimeric primate/human anti-CD4 MAb IDEC-CE9.1 is a human gamma 1 lambda antibody with primate (cynomolgus) variable regions isolated from a primate anti-CD4 MAb designated CE9.1. The chimeric antibody, which is secreted by the Chinese hamster ovary (CHO) transfectoma clone AB6-10E10-100A12, binds with high affinity to CD4+ T cells and mediates human effector functions in vitro. The transfectoma was initially cultured in hollow-fiber bioreactors to produce the antibody for clinical use; subsequent lots of antibody will be produced in suspension culture.

In a direct binding assay, radiolabeled chimeric anti-CD4 antibody IDEC-CE9.1 bound to the CD4+ human T-cell line SupT-1 with an apparent affinity of 3.2×10^{-11} M, which is similar to the affinity for the IDEC-CE9.1 monkey antibody (2.1×10^{-11} M) and for the murine anti-CD4 antibody, IF3. In a competitive binding assay, the IC_{50} values determined for the IDEC-CE9.1 antibody and for the IDEC-CE9.1 and IF3 antibodies were all roughly 0.2 nM. Results of previous work demonstrate that 1F3 competes effectively with Leu3a antibody, is highly effective in blocking CD4/gp120 binding, and has an apparent affinity for the CD4 antigen of approximately 3.1×10^{-11} M. The inhibitory effect of IDEC-CE9.1 antibody on human T-cell proliferative responses in vitro was shown by its inhibition of mixed lymphocyte reactions (Fig. 1) and by its inhibition of tetanus toxoid–specific secondary responses (Fig. 2); the respective IC_{50} values were 3.5 ng/ml and 70 ng/ml.

IV. ANIMAL STUDIES

Chimpanzee studies were performed to determine the effect of the IDEC-CE9.1 antibody on depletion and recovery of CD4+ T cells. A preliminary non-GLP dose

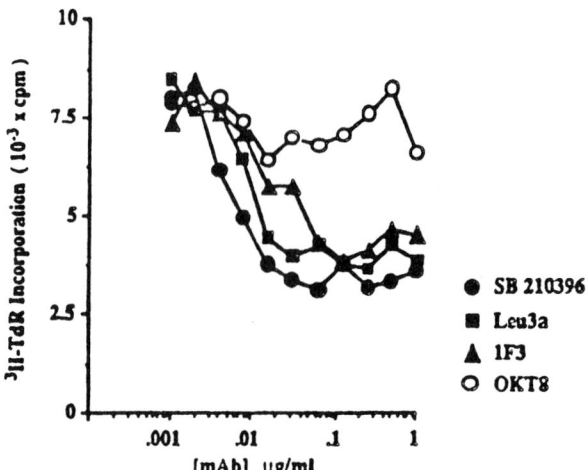

Figure 1 Effect of IDEC-CE9.1 on a human one-way mixed lymphocyte reaction.

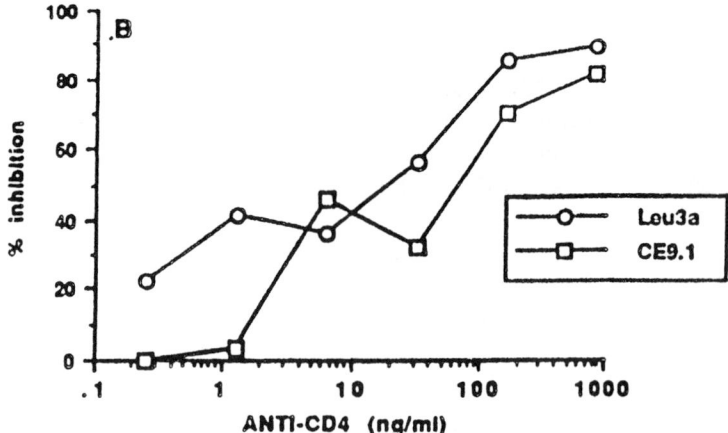

Figure 2 Effect of IDEC-CE9.1 on human T-cell proliferation induced by the recall antigen, tetanus toxoid (30 µg/ml).

escalation study in one chimpanzee is summarized briefly. A dose-ranging GLP study in one chimpanzee and a high-dose pharmacology/toxicology GLP study involving six chimpanzees were also completed.

In the preliminary study, results indicate that the minimally effective dose that reduced CD4+ cells was 1.0 mg/kg. Results from the dose-ranging study indicate that antibody doses up to and including 0.1 mg/kg do not decrease the CD4+ T-cell population. Further, while the 0.3 mg/kg dose was minimally effective, the 1.0 mg/kg dose of IDEC-CE9.1 antibody caused a marked but transient reduction in CD4+ T cells.

In the high dose study, initial exposure to 10.0 mg/kg IV of IDEC-CE9.1 antibody infusions depleted 40–50% of CD4+ T cells from the animals' circulation and modulated the CD4 antigen from the remaining 50–60% of cells. Subsequently, modulated cells recovered and expressed surface CD4 molecules within 2–4 weeks, even after repeated antibody administration. However, replacement of cells actually removed from the circulation was variable and required longer periods of time after multiple high-dose injections.

V. PHASE I SINGLE-DOSE, DOSE-FINDING TRIAL IN HUMAN RA

A phase I, open-label, single-dose, dose-escalation study was carried out in 25 patients with seropositive RA (Table 1). The doses of IDEC-CE9.1 ranged from 0.03 to 4.0 mg/kg given as a single IV infusion over 4 hr.

All patients were functional class II or III RA, with a mean age of 57.2 years and an average of 14.7 years of disease. All patients had failed four or fewer DMARDS as a result of inefficacy (Table 2).

The effect of IDEC-CE9.1 was monitored for 2 weeks after which patients could go on another slow-acting antirheumatic drug. At the end of 2 weeks, eight and seven of 25 evaluable patients had a 20% or greater reduction in swollen and tender joints, respectively (Table 3). Five patients had a greater than 30% response in both measures.

Unlike previously tested anti-CD4 MAb, no infusion-related events occurred (i.e., hypotension, fever, tachycardia, etc.). Eight patients experienced at least one adverse experience and four had more than one (Table 4). None of the adverse events were considered related to the MAb. In addition, one patient had a postvenule capillary vasculitis secondary to his underlying RA.

VI. EFFECTS ON CD4 CELLS AND OTHER IMMUNE FUNCTIONS

The CD4 counts were reduced in 23 of 25 patients, with the nadir occurring within 1 day and recovery within 7 days (Fig. 3). One patient (4.0 mg/kg) had a CD4 count less than 250 cells/mm^3 at 3 months. However, at week 4 following

Table 1 Patient Characteristics at Baseline (IDEC Study 7001)

Patient characteristics	CE9.1 dose (mg/kg)								
	0.03	0.06	0.12	0.25	0.50	1.0	2.0	4.0	Total
Sex									
Female	3	2	0	3	0	2	2	3	15
Male	0	1	3	0	3	1	1	1	10
Race									
Caucasian	1	3	3	3	2	3	3	2	20
Hispanic	1	0	0	0	1	0	0	2	4
Native American	1	0	0	0	0	0	0	0	1
Anatomical stage									
II	0	1	1	0	2	2	1	2	9
III	3	2	2	3	1	1	2	1	15
No X-rays	0	0	1	0	0	0	0	0	1
Functional class									
II	2	2	3	3	2	3	0	2	17
III	1	1	0	0	1	0	3	2	8
Age									
Mean	66.7	56.3	55.5	44.5	56.2	54.7	68.4	55.6	57.2
Range	55–74	45–60	55–56	36–42	50–69	42–67	64–75	44–74	36–75
Disease duration									
Mean years	14.4	10.2	13.2	11.1	15.8	13.0	23.1	16.3	14.7
Range	6–20	7–13	6–19	5–18	4–26	12–15	12–36	7–30	4–36

Table 2 Previous DMARD Use (IDEC Study 7001)

Previous DMARD treatment	CE9.1 dose (mg/kg)								
	0.03	0.06	0.12	0.25	0.5	1.0	2.0	4.0	Total
Methotrexate	3	3	3	3	2	3	2	4	23
Azathioprine	0	1	1	2	1	3	2	4	14
Penicillamine	0	3	0	1	2	0	0	2	8
Hydroxychloroquine	1	2	0	2	1	3	2	4	15
IM gold	3	3	2	1	3	2	3	3	20
Oral gold	0	2	0	1	0	0	0	0	3
Sulfasalazine	0	1	0	1	1	0	1	1	5
Cyclosporine	1	1	0	1	1	0	0	1	5
Chlorambucil	0	0	0	0	0	0	0	1	1

IM = intramuscular.

Table 3 Effect of CE9.1 on the Number of Swollen and Tender Joints at Week 2 Postinfusion (IDEC Protocol 7001)

CE9.1 (mg/kg)	% change in number of swollen joints				
	↑ or 0	1–9%	10–19%	20–29%	≥30%
0.03	1	2			
0.06		1		2	
0.12		1			2
0.25	1	1			1
0.5	1				2
1.0	1		1	1	
2.0	3				
4.0	1	2	1		
All doses	8	7	2	3	5
	% change in number of tender joints				
	↑ or 0	1–9%	10–19%	20–29%	≥30%
0.03	1	1		1	
0.06	1		2		
0.12	1				2
0.25	1		2		
0.5			1		2
1.0	1	1			1
2.0	1		1	1	
4.0	4				
All doses	10	2	6	2	5

Table 4 Possible/Unknown Adverse Experiences Reported During the Study or Within 30 Days of the Last Follow-Up Visit

Adverse experience	Severity	Corrective treatment	Attribution
Rash, breast (*Candida*-like)	Mild	Lamisil qid prn	Unknown
Diarrhea (5 days)	Mild	?? Other treatment	Possibly related
Chills	Moderate	None	Possibly related
Upper respiratory infection	Moderate	Augmentin; phenergan	Possibly related
Arthralgias	Moderate	?? Medication	Possibly related

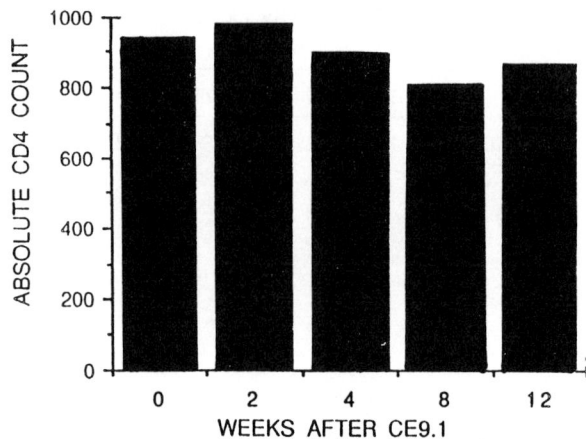

Figure 3 Average absolute CD4 counts for all ($n = 25$) RA patients treated with IDEC-CE9.1 over the 12 weeks of the single-dose treatment study.

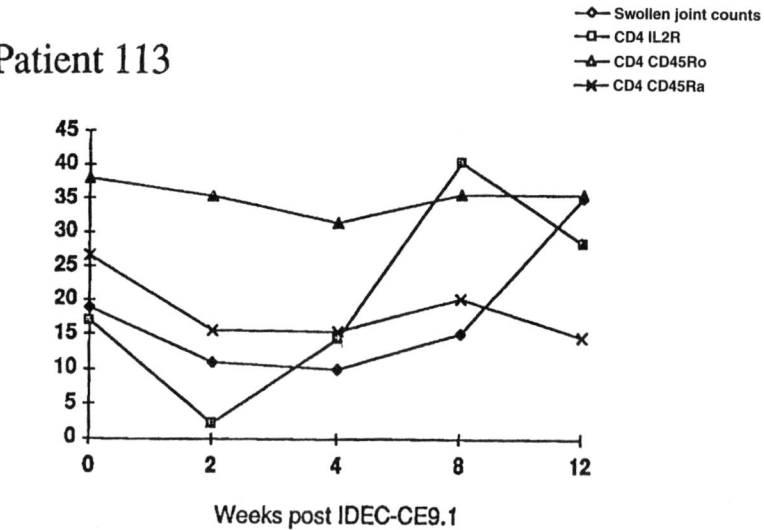

Figure 4 Effect of IDEC-CE9.1 on percentages of various CD4 subpopulations (CD25/IL2r, CD45Ro, and CD45Ra) correlated with the swollen joint count in patient 113, a patient who had a good clinical response.

infusion, the count was 1101, suggesting artifactual data. Follow-up data are being collected.

While IDEC-CE9.1 did not deplete CD4+ cells, CD25 appears to be down-modulated off CD4+ cells (Fig. 4), and in some patients, this was associated with a positive clinical response. Similarly, CD4 was also down-modulated. Polymorphic Dr expression slowly decreased after infusion unrelated to clinical response. Mitogen- and antigen-induced proliferation of peripheral mononuclear cells was suppressed in most patients for up to 1 month. The duration and intensity of this suppression were not dose-related. This type of immunological data may provide insight into the mechanism of action of this product.

VII. SUMMARY

IDEC-CE9.1, a PRIMATIZED anti-CD4 MAb, has been infused in single doses up to 573 mg in 25 patients with RA without infusion-related adverse experiences. Overall, CD4 suppression appeared minimal and 30% of patients experienced 20% or greater clinical response at 2 weeks. Further study with multiple dosing is needed to fully evaluate the safety and potential efficacy of this product.

REFERENCES

1. Horneff G, Burmester GR, Emmrich F, Kalden Jr. Treatment of rheumatoid arthritis with an anti-CD4 monoclonal antibody. Arthritis Rheum 1991; 34:129–140.
2. Stand V, Lipsky P, Cannon G, Calabrese L, Wiesnhutter C, Cohen SB, Olsen N, Lee M, Lorenz TJ, Nelson B, and the CD5 Plus Rheumatoid Arthritis Investigator's Group. Effects of administration of an anti-CD5 plus immunoconjugate in rheumatoid arthritis: results of two phase II studies. Arthritis Rheum 1993; 36:620–630.
3. Moreland LW, Bucy RP, Tilden A, Pratt PW, Lobuglio AF, Khazael M, Everson MP, Daddona P, Ghrayeb J, Kilgaffiff C, Sanders ME, Koopman WJ. Use of a chimeric monoclonal anti-CD4 antibody in patients with refractory rheumatoid arthritis. Arthritis Rheum 1993; 36:307–318.
4. Choy EHS, Chikanza IC, Kingsley GH, Corrigal V, Panayi GS. Treatment of rheumatoid arthritis with single dose or weekly pulses of chimeric anti-CD4 monoclonal antibody. Scand J Immunol 1992; 36:291–298.
5. Isaacs JD, Watts RA, Hazleman BL, Hale G, Keogan MT, Cobbold SP, Waldmann H. Humanized monoclonal therapy for rheumatoid arthritis. Lancet 1992; 340:748–752.
6. Newman R, Alberts J, Anderson D, Carner K, Heard C, Norton F, Raab R, Reff M, Shuey S, Hanna N. "Primatization" of recombinant antibodies for immunotherapy of human diseases: a macaque/human chimeric antibody against human CD4. Biotechnology 1992; 10:1455–1460.
7. Shevach EM. Intercellular interactions in the immune response. In: Oppenheim JJ, Shevach EM, eds. Immunophysiology: The Role of Cells and Cytokines in Immunity and Inflammation. New York: Oxford University Press, 1990:104–128.

8. Panayi GS, Lanchbury JS, Kingsley GH. The importance of the T cell in initiating and maintaining the chronic synovitis of rheumatoid arthritis. Arthritis Rheum 1992; 35:729–735.
9. Malone DG, Wahl SM, Tsokus M, et al. Immune function in severe, active rheumatoid arthritis: a relationship between peripheral blood mononuclear cell proliferation to soluble antigens and synovial tissue immunohistologic characteristics. J Clin Invest 1984; 74:1173–1185.
10. Harris ED. Rheumatoid arthritis: pathophysiology and implications for therapy. N Engl J Med 1990; 322:1277–1289.
11. Allen JB, Malone DG, Wahl SM, et al. Role of the thymus in streptococcal cell wall induced arthritis and granuloma formation. J Clin Invest 1985; 76:1042–1056.
12. Kotzin BL, Strober S, Engleman EG, Calin A, Hoppe RT, Kansas GS, Terrell CP, Kaplan HS. Treatment of intractable rheumatoid arthritis with total lymphoid irradiation. N Engl J Med 1981; 305:969–976.
13. Paulus HE, Machleder HI, Levine S, Yu DT, MacDonald NS. Lymphocyte involvement in rheumatoid arthritis: studies during thoracic duct drainage. Arthritis Rheum 1977; 20:1249–1262.
14. Kursh J, Wright DG, Klippel JH. Lymphocyte depletion by continuous flow cell centrifugation in rheumatoid arthritis: clinical effects. Arthritis Rheum 1979; 22: 1055–1059.
15. Calabrese LH, Wilske WAS, Parkins, et al. Rheumatoid arthritis complicated by infection with human immuno deficiency virus and the development of Sjögrens syndrome. Arthritis Rheum 1989; 32:1453–1457.
16. Yocum DE, Klippel JH, Wilder RL, Gerber NL, Austin HA III, Wahl SM, Les GL, Minor JR, Preuss, Yarburo C. Cyclosporin A in severe treatment-refractory rheumatoid arthritis: a randomized study. Ann Intern Med 1988; 109:863–869.
17. Wendling D, Racadot E, Morel-Fourrier B, Wijdenes J. Treatment of rheumatoid arthritis with anti-CD4 monoclonal antibody: open study of 25 patients with the B-F5 clone. Clin Rheum 1992; 11:542–547.

7
CAMPATH-1H Therapy in Autoimmune Diseases

RICHARD A. WATTS and JOHN D. ISAACS
Addenbrooke's Hospital, Cambridge, England

I. INTRODUCTION

Autoimmune diseases such as rheumatoid arthritis (RA) are considered to be T-cell-mediated and might therefore be expected to respond to therapies directed against the T lymphocyte. Previous therapies with lymphapheresis, total nodal irradiation, and thoracic duct drainage all showed clinical benefit, but responses were not maintained and the therapies were too toxic for routine use. More recently, cyclosporine (a fungal metabolite with a specific action against T cells) has been shown to be an effective therapy in autoimmune disease.

Antilymphocyte monoclonal antibodies (MAbs) have potent immunosuppressive properties and are capable of inducing antigen-specific tolerance (1). Data from animal experiments suggest that MAbs with a broad spectrum of specificities may be required to induce disease control (2). CD4 MAbs can induce tolerance in situations where it is anticipated that there are small numbers of autoreactive T cells, but CD8 MAbs are required in addition when there is a greater number of autoreactive T cells. Attempts to "reprogram" the immune system suggest that it may be necessary to "debulk" lymphocyte numbers using a depleting MAb and then modulate function of the remaining cells with blocking MAbs. Using such regimens it is possible to tolerize mice actively rejecting an allogeneic skin graft, a process akin to established autoimmune disease (3).

Rodent antibodies usually induce an antiglobulin response in humans (4). This does not necessarily prevent repeated infusions, but these may be less effective and potentially dangerous. Genetic engineering techniques have led to the development of recombinant antibodies, either chimeric (rodent variable region in association with a human constant region), primatized (macaque variable and human constant region), or fully humanized in which the complementarity-determining regions are all that remain of the "parent" rodent MAb. The aim of these manipulations is to reduce immunogenicity by minimizing the nonhuman component of the antibody. While antiallotype and anti-idiotype responses are still possible, chimeric and humanized antibodies appear to be less immunogenic than rodent antibodies (5,6).

CAMPATH-1H is a humanized MAb directed against the glycoprotein CD52. This surface molecule is present on all human B and T lymphocytes at a density of around 400,000 copies per cell and at a lower density on human monocytes/macrophages (7). The CD52 molecule is not present on other myeloid cell types or CD34+ pluripotent stem cells. Extensive immunohistochemical studies of human tissue cross-reactivity have shown specific binding only to peripheral and resident lymphoid cells and mature (but not testicular) spermatozoa (7,8).

CAMPATH-1H has a human IgG1 constant region and is efficient at complement- and cell-mediated lysis of human lymphocytes (7). Natural killer (NK) cells and monocytes are bound but spared killing by CAMPATH-1H. The function of CD52 is at present under investigation. CAMPATH-1H was derived by humanization of the rodent antibody CAMPATH-1G (9).

II. CLINICAL STUDIES USING CAMPATH-1H

CAMPATH-1H has been used to treat patients with RA, systemic vasculitis, and multiple sclerosis (MS).

A. Biological Activity

CAMPATH-1H has been administered by the intravenous and subcutaneous routes and has potent biological activity. A single intravenous infusion of CAMPATH-1H results in a rapid fall in peripheral blood lymphocyte (PBL) count in all patients, with the nadir count occurring within a few hours (10). The suppression of PBL count lasts for at least 20 months following a single infusion (11) and for at least 32 months after multiple intravenous infusion (12). CD8+ T cells appear to return to the peripheral circulation more quickly than CD4+ T cells. Peripheral blood B-cell (CD19+) numbers approach baseline values by 60 days after treatment. NK cells in peripheral blood are not affected by CAMPATH-1H, an observation consistent with in vitro data using the rat IgM

MAb CAMPATH-1H. Monocyte (CD14+) numbers are transiently reduced after CAMPATH-1H but return to baseline values within a few weeks of treatment. Subcutaneous administration also results in profound lymphopenia; however, following a single dose the nadir count is reached more slowly at around day 4 after injection (13). Suppression is, however, similarly prolonged.

B. Clinical Studies in RA

The initial investigation of CAMPATH-1H in RA was conducted in Cambridge in a single-center open study (6). In this study eight patients received a total dose of 60 mg intravenously over 10 days. There was significant clinical improvement in 7/8 patients for up to 100 days with a maximal 53% and 71% fall in Ritchie index and joint score, respectively. Use of an increased dose (40 mg daily for 5 days) resulted in no greater clinical response but was associated with a greater frequency of side effects (see below) (15). There was no correlation between total peripheral blood lymphocyte count and magnitude or duration of clinical response. C-reactive protein levels fell modestly but there was no change in the erythrocyte sedimentation rate.

A series of multicenter trials of CAMPATH-1H in RA have been performed sponsored by the Wellcome Foundation (10). All have been open, uncontrolled, multicenter phase I studies. Studies of single or multiple intravenous and single- or multiple-dose subcutaneous therapy have been conducted. All patients had refractory active RA of long duration and had received a number of disease-modifying agents.

In a single-dose intravenous study 33 patients received 1.0–100 mg of CAMPATH-1H (10). Overall 20 patients reached a Paulus response, but duration of response was extremely short (median less than 2 weeks). The Paulus response was reached within 3 days of infusion. There was no clear correlation between dose and response; however, a longer duration of response was seen in patients receiving more than 30 mg CAMPATH-1H (10).

In a multicenter study of multiple intravenous dosing, 41 patients received total doses of 100, 250, or 400 mg of CAMPATH-1H over either 5 or 10 days (14). Significant improvements in painful joint score and swollen joint score were seen 130 days after treatment. Nearly 50% of patients achieved a Paulus (50%) response at 1 month with 20% of patients still responding at 6 months. There was a trend for longer response duration at higher doses.

CAMPATH-1H has been administered subcutaneously to 24 patients with RA, in doses of 0.3–10 mg/day for 10 days (16). A Paulus response was obtained in 56% of patients but took up to 4 weeks to occur. The median duration of response was 1 month and no relationship to dosage was observed. Eight patients responded for 2 months.

Single-dose subcutaneous CAMPATH-1H (30 mg) has been used in six patients with RA (13). There was a significant reduction in joint pain and swelling, which lasted for at least 1 month, but the response was not sustained.

Synovial tissue has been obtained from two patients undergoing joint arthroplasty at days 110 and 203 after treatment (17). Both patients had received single-dose CAMPATH-1H and had reduced peripheral blood CD3+ and CD4+ T-cell numbers. There was active synovitis at the time of surgery. In one patient there was hypertrophy of the synovial lining layer, with a diffuse mononuclear cell infiltrate with CD4+, CD8+, and CD19+ cells together with lymphoid aggregates of CD3+ and CD4+CD45RO+ cells with few CD8+ cells. In the second patient there was only an infiltrate with CD3+ and CD4+CD45RO+ cells. Thus there was a prominent synovial lymphocyte infiltrate following CAMPATH-1H at a time when there was a peripheral blood lymphopenia.

Overall these open-label trials of CAMPATH-1H therapy in RA suggest that there is an immediate clinical response with improvement in joint score and pain. These improvements are not, however, maintained. There was no correlation between peripheral blood lymphocyte count and duration of response, although a fall in PBL count was associated with improvement. Peripheral blood lymphocyte depletion may be necessary for induction of remission but not for its maintenance.

C. Systemic Vasculitis

The systemic vasculitides are believed to be immune-mediated diseases, and experimental evidence suggests that T cells may have a direct role in causing vessel damage (18).

CAMPATH-1H has been used to treat patients with systemic vasculitis refractory to conventional treatment with corticosteroids and cytotoxic drugs. Four patients with T-cell-associated vasculitis (microscopic polyarteritis two, Behçet's syndrome one, Sjögren's syndrome one) received up to a total dose of 40 mg CAMPATH-1H (19). Three patients also received a CD4 MAb. The duration of response was 3–54 months. One patient responded to repeated infusions but developed an antiglobulin response (see below). Response to treatment was rapid within 72 hr, suggesting that cell-mediated rather than antibody-mediated mechanisms were predominantly affected (19).

Six additional patients with ANCA-associated vasculitis (Wegener's granulomatosis four, microscopic polyarteritis two) received similar doses of CAMPATH-1H (20). In four of these patients CAMPATH-1H therapy was combined with a CD4 MAb. A good clinical response was seen in 5/6, with remissions of 6 months. However, two patients remained dialysis-dependent. One patient died at day 7 following a cerebrovascular accident. Clinical improvement was associated with a fall in ANCA titers, improvement in chest radiograph appearance, and clearing of skin lesions.

Thus CAMPATH-1H appears to be a promising agent in the treatment of systemic vasculitis, but further controlled clinical studies are necessary.

D. Multiple Sclerosis

The T lymphocyte is believed to play a pivotal role in the pathogenesis of the inflammatory demyelinating lesion of multiple sclerosis (21,22). The T cell may therefore be an appropriate target for immunotherapy with MAbs. CAMPATH-1H has been given in an open-label trial to seven patients with MS (23). These patients were assessed using repeated magnetic resonance imaging (MRI) of the central nervous system to monitor the development of neurological lesions. CAMPATH-1H was given intravenously, 120 mg over 10 days. There was a decrease in the rate of appearance of new lesions as determined by MRI scanning. CAMPATH-1H was not detectable in cerebrospinal fluid from five patients who consented to repeat lumbar puncture.

III. SIDE EFFECTS OF CAMPATH-1H THERAPY

CAMPATH-1H is associated with both immediate and long-term side effects. The first infusion of CAMPATH-1H results in a rapid fall in peripheral blood lymphocyte count, which is accompanied by fever up to 40°C, rigors, urticarial skin rash, and hypotension. This is similar to that seen after OKT3 treatment (24), in which the first dose syndrome follows release of cytokines from lymphocytes and monocytes into the circulation. Elevated levels of tumor necrosis factor-alpha, interferon-gamma, and interleukin-6 have been demonstrated following CAMPATH-1H infusion (23). These side effects are dose-dependent and occur more severely at higher doses (40–100 mg). Subsequent infusions are accompanied by minimal side effects. Cytokine antagonists should be able to reduce or abolish these symptoms. Subcutaneous administration of CAMPATH-1H is accompanied by a similar side effect profile but at equivalent doses it is less severe, presumably reflecting the slower rate of lympholysis after subcutaneous administration than after intravenous administration.

The mechanism underlying the prolonged suppression of peripheral blood CD4+ T cells is not understood. It is unclear whether recirculation has been interrupted or whether there is a failure to produce new CD4+ T cells. Certainly the infection risk is low relative to the T-cell numbers in peripheral blood. There is evidence that the repopulating CD4+ T-cell population after CAMPATH-1H is depleted of naïve T cells and that those remaining carry the memory phenotype (indicative of proliferating T cells) (25). In this study, furthermore, analysis of

TCR β chains revealed a limited spectrum and clonal expression of T-cell specificities (26), although this has not been confirmed by other groups.

The presence of CD4 with a memory phenotype together with intact neutrophil and NK-cell function, and after a few weeks normal monocyte numbers should provide effective host defenses against most infectious pathogens. Prophylaxis against infection has not been used routinely. Around 10% of patients develop infections, mostly minor (localized mucocutaneous herpes and *Candida* infections). In many cases contributing factors have been identified (e.g., salmonellosis in a food handler, pseudomembranous colitis in a patient on antibiotics, enterohemorrhagic *Escherichia coli* enteritis after raw beef consumption) (10). One patient developed disseminated coccidioidomycosis after CAMPATH-1H therapy followed immediately by methotrexate and prednisolone. A second patient developed pulmonary infiltrates 10 weeks after CAMPATH-1H therapy; these were shown to be caused by a nontuberculous mycobacterial infection. These infections occurred when the lymphocyte count was less than 0.5×10^9/L (10).

CAMPATH-1H is a humanized antibody and as such would be expected to be less immunogenic than rodent or chimeric antibodies; however, there remains the possibility of development of anti-idiotypic or antiallotypic responses. Such antiglobulin responses might result in loss of efficacy of antibody and potentially result in anaphylactic reaction. In our initial study 3/4 patients who were retreated developed an antiglobulin response (6). In a patient with systemic vasculitis, development of an antiglobulin response resulted in loss of efficacy. The antiglobulin was successfully removed by plasma exchange but reappeared after treatment with CAMPATH-1H (19).

Induction of malignancy, particularly lymphomas, is a potential long-term consequence of CAMPATH-1H therapy, as with other immunosuppressive therapy.

Two types of injection site reaction occur after subcutaneous CAMPATH-1H. Severe burning is experienced by all patients immediately upon injection; this lasts for only a few minutes and appears to be related to the injection vehicle (13). A milder delayed skin reaction characterized by erythema and swelling occurs in 50% of patients; it lasts for up to 10 days and appears to be related to the biological activity of the MAb. This reaction seems to be less severe with higher doses of CAMPATH-1H.

IV. CONCLUSIONS

CAMPATH-1H has a promising effect in the treatment of autoimmune disease. The aim of immunotherapy in the autoimmune diseases is to induce tolerance to the provoking antigen. It may be necessary as in animals to use broad-spectrum

MAbs to achieve this aim rather than MAbs with a narrow target specificity, e.g., CD4, IL-2 receptor. This may be particularly true in late disease where there are large numbers of autoreactive T cells. A logical approach to late disease may be to use a low dose of CAMPATH-1H to debulk the autoaggressive load and to follow this debulking with a nondepleting CD4 MAb.

REFERENCES

1. Waldmann H. Manipulation of T cell responses with monoclonal antibodies. Annu Rev Immunol 1989; 7:407–444.
2. Cobbold SP, Qin S, Leong LYW, Martin G, Waldmann H. Reprogramming the immune system for peripheral tolerance with CD4 and CD8 monoclonal antibodies. Immunol Rev 1992; 129:165–201.
3. Cobbold SP, Martin G. Waldmann H. The induction of skin graft tolerance in major histocompatibility-mismatched or primed recipients: primed T cells can be tolerised in the periphery with anti-CD4 and anti-CD8 antibodies. Eur J Immunol 1990; 20:2747–2755.
4. Isaacs JD. The antiglobulin response to therapeutic antibodies. Semin Immunol 1990; 2:449–456.
5. Kirkham BW, Thien F, Pelton BK, et al. Chimeric CD7 monoclonal antibody therapy in rheumatoid arthritis. J Rheumatol 1992; 19:1348–1352.
6. Isaacs JD, Watts RA, Hazleman BL, et al. Humanised monoclonal antibody therapy for rheumatoid arthritis. Lancet 1992; 340:748–752.
7. Hale G, Xia M-Q, Tighe HP, Dyer MJS, Waldmann H. The CAMPATH-1H antigen (CDw52). Tissue Antigens 1990; 35:118–127.
8. Hirsch T, Havemann K, Krouse W, Ziegler A, Uchanska-Ziegler B. Use of human spermatozoa and small cell lung cancer lines to characterise MAb directed against NK and non-lineage antigens. In: Krapp W, ed. Leucocyte Typing IV: White Cell Differentiation Antigens. Oxford: Oxford University Press, 1989:667.
9. Riechmann L, Clark M. Waldmann H, Winter G. Reshaping human antibodies for therapy. Nature 1988; 332:323–327.
10. Johnston JM, Speer WR. Treatment of rheumatoid arthritis with humanised monoclonal antibody CAMPATH-1H. In: Proceedings: Early Decisions in DMARD Therapy. III. Biologic Agents in Autoimmune Diseases. Arthritis Foundation, 1994:155–165.
11. Weinblatt ME, Coblyn J, Maier A, et al. Continued lymphocyte suppression following single dose monoclonal antibody therapy with CAMPATH-1H—a 20 month follow up. Arthritis Rheum 1994; 37(Suppl):S420.
12. Watts RA, Isaacs JD, Hale G, Hazleman BL, Waldmann H. Peripheral blood lymphocytes subsets after CAMPATH-1H therapy for RA—a 3 year follow up. Arthritis Rheum 1994; 37(Suppl):S338.
13. Watts RA, Manna V, Hazleman BL. Single dose subcutaneous CAMPATH-1H in the treatment of rheumatoid arthritis. Br J Rheum 1993; 32(Suppl 2):21 (abstract).
14. Isaacs JD, Manna VK, Hazleman BL, et al. CAMPATH-1H in RA—a study of multiple IV dosing. Arthritis Rheum 1993; 36(Suppl):40.

15. Watts RA, Isaacs JD, Hale G, Hazleman BL, Waldmann H. High dose CAMPATH-1H therapy in rheumatoid arthritis. Br J Rheum 1993; 32(Suppl 1):54 (abstract).
16. Johnston JM, Hays AE, Heitman CK, et al. Treatment of rheumatoid arthritis patients by subcutaneous injection of CAMPATH-1H. Arthritis Rheum 1992; 35:S105.
17. Ruderman EM, Weinblatt ME, Thurmond LM, Pinkus GS, Gravallese EM. Synovial tissue response to treatment with CAMPATH-1H. Arthritis Rheum 1994; 37(Suppl):S420.
18. Wofsy D, Seaman WE. Reversal of advanced lupus in NZB/NZW F1 mice by treatment with monoclonal antibody to L3T4. J Immunol 1987; 138:3247–3253.
19. Lockwood CM, Thiru S, Isaacs JD, Hale G, Waldmann H. Long term remission of intractable systemic vasculitis with monoclonal antibody therapy. Lancet 1993; 341:1620–1622.
20. Lockwood CM. Treatment of systemic vasculitis by monoclonal antibody. Eular Bull 1994; 23:80–85.
21. Raine CS. Multiple sclerosis: a pivotal role for the T cell in lesion development. Neuropathol Appl Neurobiol 1991; 17:265–274.
22. ffrench-Constant C. The pathogenesis of multiple sclerosis. Lancet 1994; 343:271–275.
23. Moreau T, Thorpe J, Miller D, et al. Preliminary evidence from magnetic resonance imaging for reduction in disease activity after lymphocyte depletion in multiple sclerosis. Lancet 1994; 344:298–301.
24. Chatenoud L, Bach J-F. OKT3 in allogenic transplantation: clinical efficacy, mode of action, and side effects. In: NBurlingham WJ, ed. A Critical Analysis of Monoclonal Antibody Therapy in Transplantation. Boca Raton, FL: CRC Press, 1992.
25. Ganten T-M, Jendro MC, Fulbright JW, et al. restricted diversity of the T cell repertoire after therapeutic T cell depletion with monoclonal antibody CAMPATH-1H. Arthritis Rheum 1993; 36(Suppl):40.
26. Sundey JS, Denning SM, Jacobs MR, St. Clair EW. The effect of CAMPATH-1H on the T cell repertoire (TCR) of patients with rheumatoid arthritis. Arthritis Rheum 1993; 36(Suppl):40.

8
CAMPATH-1H in Rheumatoid Arthritis
United States Experience

DAVID E. YOCUM
University of Arizona College of Medicine, Tucson, Arizona

JEFFREY M. JOHNSTON
Glaxo Wellcome, Inc., Research Triangle Park, North Carolina

I. INTRODUCTION

Rheumatoid arthritis (RA) is a chronic inflammatory disease of unknown etiology that is associated with several immunological abnormalities (1–4). While a key feature of this disease is the heterogeneity of the synovial immunohistology, the tissues of many patients exhibit intense infiltration of lymphocytes (5). These lymphocytes, activated by an unknown antigenic stimulus, are felt to play a central role in the pathogeneses of RA (6,7).

Antilymphocyte therapy including total nodal irradiation (8), lymphopheresis (9), cyclosporine (10), and monoclonal antibody (MAb) therapy (11–14) to lymphocyte antigens (CD4, CD5, CD7) have been used in the therapy of RA with variable success. CAMPATH*-1H, a humanized MAb CD52(15), an antigen of unknown function on a majority of mononuclear cells, has recently been used in a variety of protocols in the treatment of RA. The data from these studies will be discussed.

---

* CAMPATH is a trademark of Wellcome group companies and is registered in the U.S. Patent and Trademark Office.

II. CAMPATH-1 ANTIGEN

The CAMPATH-1 antigen, designated CD52, is a heavily glycosylated, non-modulating glycoprotein of molecular weight 21–28 kDa, which, in humans, is predominantly expressed on peripheral blood lymphocytes and monocyte/macrophage cells (15,16). CAMPATH-1 antigen presence has also been detected on a small proportion of neutrophils and is present in human sperm (17).

The antigen is not present on platelets, erythroid and myeloid bone marrow cells, or hematopoietic stem cells. The function of CAMPATH-1 in the human immune system is unknown, although limited data suggest the antigen may play a role in T-cell activation and/or cell signaling (15). Antibodies raised against the CAMPATH-1 antigen recognize greater than 95% of peripheral blood lymphocytes, tonsil cells, and thymocytes as well as blood monocytes. While such antibodies efficiently lyse lymphocytes, they are not lytic of monocytes but rather appear to induce transient margination.

III. HUMANIZED CAMPATH-1H ANTIBODY

Several MAbs directed against the same surface antigen (CD52) were isolated from a fusion of a rat myeloma cell line with spleen cells of a rat immunized with human T lymphocytes (18). An IgM from this fusion (CAMPATH-1M) has been used extensively ex vivo to purge T lymphocytes from bone marrow harvests prior to transplantation to prevent graft-versus-host disease (GVHD) (19). CAMPATH-1G, an IgG2b active in cell-mediated cytotoxicity as well as complement lysis, has been used in vivo for salvage treatment of lymphoid malignancies (20,21), immunosuppressive prophylaxis, and treatment of GVHD and renal transplant rejection (22). Although CAMPATH-1G was well tolerated and efficacious, an anti-rat immunoglobulin response developed rapidly in a majority of patients tested and bronchospasm was reported in two such patients.

To minimize this problem and to optimize effector function, the antibody was humanized using a human IgG1 genomic framework (23). The reshaped antibody, CAMPATH-1H, is highly active in vitro and has shown early in vivo efficacy in lymphoma (24), and in autoimmune disease (25,26), patients treated at Cambridge University.

IV. CAMPATH-1H PRECLINICAL DATA (27)

Studies in cynomolgus monkeys using single-dose CAMPATH-1H demonstrated lymphocyte depletion at intravenous (iv) or subcutaneous (sc) doses of 1 mg/kg or higher. Lymphocyte recovery occurred within 2–6 weeks and anti-CAMPATH-1H responses developed within 7–10 days following injection. No clinical, laboratory, or pathological adverse events were noted. A multi-dose study of CAMPATH-1H administered iv or sc to monkeys for 14 or 30 consecutive days also demonstrated lymphocyte depletion and similar CAMPATH-1H serum concentrations at the same iv and sc dose. Anti-CAMPATH-1H responses developed in at least six of the eight treated animals. Unexplained neutropenia was observed in animals treated for 30 days; however, no other laboratory abnormalities were seen, and there was no clinical or pathological evidence of toxicity. Another study, investigating cardiovascular and respiratory effects of single iv infusions at doses of 3, 10, or 30 mg/kg, demonstrated dose-related hypotension persisting 1–3 hr postinfusion. One animal given 30 mg/kg exhibited a more marked and progressive hypotension and tachycardia, which culminated in death 6.5 hr after dosing.

CAMPATH-1H binding, both antigen-specific [detected by use of labeled $F(ab')_2$ fragment] and nonspecific (e.g., Fc reactivity), has been demonstrated in a number of human tissues. As expected, specific binding was documented to lymphocytes, monocytes, macrophages, and bone marrow lymphoid cells. In addition, an estimated 1–5% of neutrophils were found to bind $F(ab')_2$. Small mononuclear cells of lymphoid origin in all nonhemopoietic tissues tested showed specific binding, as did mature spermatozoa and epithelial cells within the epididymis and seminal vesicle. The antigen was not detected on spermatogenic cells or spermatozoa within the testis. These studies concluded that in vivo CAMPATH-1H-specific binding would result in a substantial reduction of lymphoid cell numbers and could possibly impair male fertility. Nonspecific Fc binding was observed in the following tissues: adrenal, carotid body, cerebellum, cerebrum, dorsal root ganglia, liver, peripheral nerve, pituitary, prostate, spinal cord, sympathetic nerve/nerve ganglia, testis, and vagus nerve. This nonspecific binding was thought to be of little clinical significance due to (1) the limited access of antibody to many of the implicated tissues, (2) the presence of large amounts of endogenous IgG (extremely effective in blocking in vitro CAMPATH-1H Fc binding), and (3) the lack of activation of effector function when antibody is nonspecifically bound.

V. CAMPATH-1H CLINICAL EXPERIENCE IN RA

Three pilot studies in RA have been conducted in the United States. These phase I/II protocols were designed as parallel, multicenter, open-label, dose-escalation, cohort studies. The major difference between protocols was the dose regimen/ route of administration (28–31). The first study investigated the effects of CAMPATH-1H at escalating single iv doses of 1, 3, 10, 30, 60, 100 mg, while the second used a 10-day sc regimen with daily doses of 0.3, 1, 3, 10, and 30 mg. The third study evaluated clinical activity of 100, 250, and 400 mg, cumulative iv doses given over 5 or 10 days. Potential efficacy was assessed using the Paulus criteria (32) with a single modification: *either* CRP or ESR improvement was acceptable as the acute phase reactant component. These studies along with several adjunctive immunological studies have been published only in abstract form.

The iv study was a collaboration between study centers in Canada, The Netherlands, the United Kingdom, and the United States (28). A total of 40 patients were enrolled in the single-dose, dose-escalation iv study. A total of 33 patients were evaluable for disease response (completed assigned dose and at least one postdose assessment). All patients were evaluated for safety. The overall response rate in this trial according to the modified Paulus criteria was 61% (20 of 33) among evaluable patients. This response did not show a strict dose-response relationship, although significantly more patients experienced swelling score improvements in dose cohorts of >30 mg/day.

The responses were achieved rapidly (less than 3 days postinfusion) but the duration of response was transient and response lasted only a median of 2 weeks across all cohorts. All patients exhibited a rapid fall in absolute lymphocyte counts to a nadir level as low as 8% of the predose lymphocyte (Fig. 1). These counts

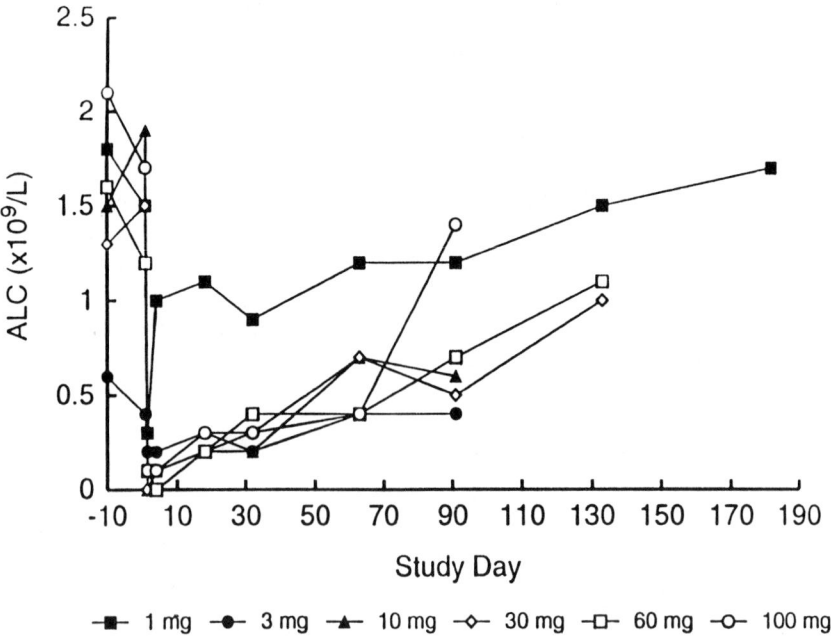

Figure 1 CAMPATH-1H: rheumatoid arthritis studies median absolute lymphocyte count in single-dose iv study.

remained at or near these nadir levels throughout the first month for all but the 1-mg cohort. Weinblatt et al. reported on 13 such patients treated at Brigham and Women's Hospital in Boston (33). While there was a rapid drop in absolute lymphocyte count in all patients, only seven patients developed improvement as measured by Paulus criteria. Six of 13 patients received a second infusion and one patient received a total of four separate infusions. Significant reduction of the absolute lymphocyte count, including CD3+, CD4+, and CD8+ cells, was noted to continue after a mean of 232 (±54) days. The mean absolute lymphocyte count had dropped from 1.293 cells × 10^9/L to 0.478 cells × 10^9/L, with the CD4+ cells falling from 0.733 cells × 10^9/L to 0.119 cells × 10^9/L. No significant change was noted in the total leukocyte count, number of monocytes, number of B cells, or number of natural killer cells. No infections or other serious adverse events were noted. However, despite the sustained peripheral lymphopenia, all patients experienced return of RA activity.

The second study, done only within the United States, enrolled a total of 30 patients for treatment with a 10-day sc regimen of CAMPATH-1H (29). Of these 30, 27 were evaluable for disease response. As noted in the single-dose iv study, there was a dramatic fall in absolute lymphocyte counts to <5% of the predose level. Lymphocyte counts recovered slowly, with only the lowest-dose cohort (0.3 mg/day for 10 days) exceeding 50% of the baseline absolute lymphocyte count at 1 month posttreatment (Fig. 2). In vitro proliferation studies demonstrated that lymphocyte responses to both mitogens and soluble recall antigens were suppressed (34). Clinical improvement was noted by the modified Paulus criteria in 56% of patients. This response lasted a mean of 1 month. A dose-response relationship was observed for swelling scores but not for Paulus response (Fig. 3).

Follow-up evaluation of the lymphocyte phenotypic profile on some of these patients was done at Duke University and Mayo Clinic. Ganten et al. evaluated lymphocytes of 11 patients and noted that reconstitution of the T-cell compartment was slow with the CD4+ cells as low as 0.1 cells × 10^9/L after 48 weeks (35). It was noted that a majority (8 of 11) of these patients had less than 25% memory ($CD45+Rc^+$ T cells) prior to treatment; however, following treatment this population exceeded 80% in four of 11 patients. V_s-specific antibody probes demonstrated a limited spectrum and clonal expansion of T-cell specificity. This study concluded that treatment with CAMPATH-1H may lead to long-term reduction in the size of the naïve CD4 population (CD45+ Rc^+ T cells). Simile studies of nine patients were done by Sundy et al. at Duke University, using similar techniques (36). These investigators noted low numbers of absolute CD4+ T cells in the recovery stage following CAMPATH-1H treatment. A more extensive antibody profile (anti-γδ anti-Vα2, and anti-Vβ MAbs) demonstrated greater than 50% posttreatment change in at least one subset. However, no statistically significant difference was noted in these changes. The group concluded that there was no

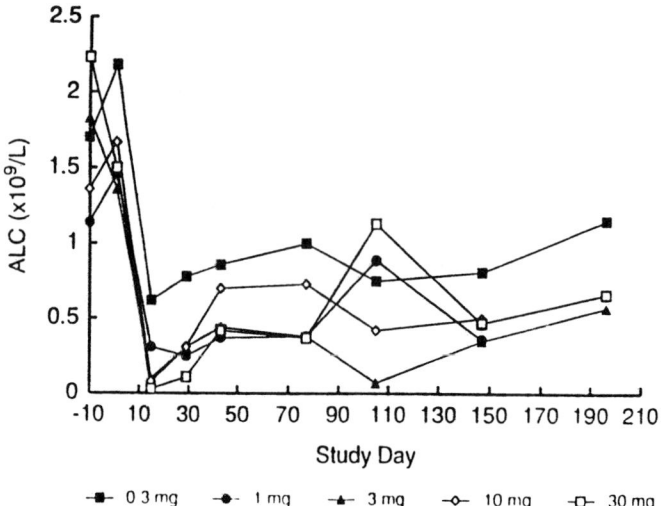

Figure 2 CAMPATH-1H: rheumatoid arthritis studies median absolute lymphocyte count in 10-day SC study.

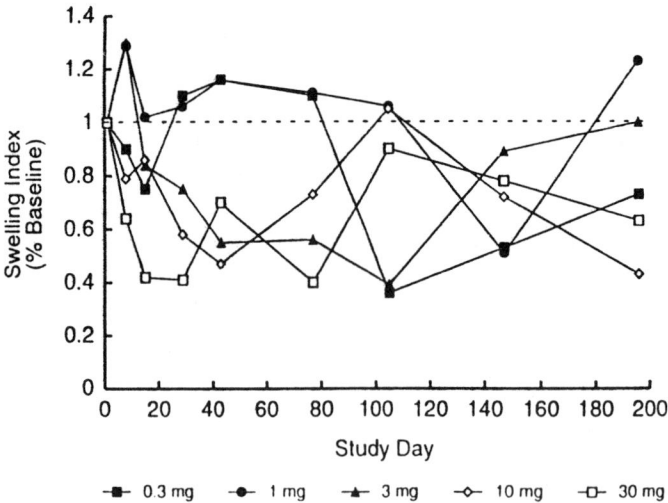

Figure 3 CAMPATH-1H: rheumatoid arthritis studies median swelling index in 10-day SC study.

consistent pattern in lymphocyte reconstitution following T-cell depletion in either the Vα or Vβ subset.

In the third study, done in both the United States and England (31), 41 patients were treated in one of six cohorts receiving repetitive dosing of a total of 100, 250, or 400 mg of CAMPATH-1H over 5 or 10 days, as before, lymphopenia, particularly CD4+ and CD8+ cytopenia, was rapid, near-absolute, and prolonged (Fig. 4). The absolute lymphocyte count, as well as monocyte and NK-cell counts, was significantly lower in the 400-mg cohort. Eighty-six percent of eligible patients experienced significant clinical improvement based on the Paulus criteria and 20% of patients maintained a 50% Paulus response at 6 months. The patients with the best clinical response were all in the 250- and 400-mg dose cohorts. Nine of 31 patients tested developed antiglobulins to CAMPATH-1H; however, sensitization did not correlate with the presence or duration of a therapeutic response.

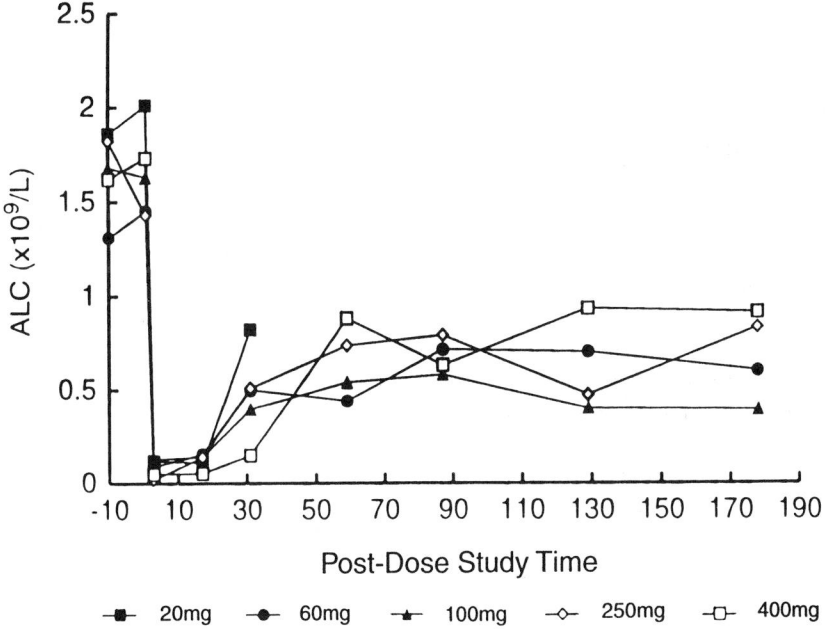

Figure 4 CAMPATH-1H: rheumatoid arthritis studies median absolute lymphocyte count in 5- and 10-day iv study.

VI. CLINICAL SIDE EFFECTS OF CAMPATH-1H IN RA PATIENTS

In the above studies, a significant side effect profile was noted, which was acute and included headache, fever, chills, nausea, and vomiting in the majority of patients. These systemic reactions occurred during or shortly after infusion and were usually resolved in 24 hr. However, some patients maintained a low-grade fever beyond 24 hr. In addition, asymptomatic hypotension, often delayed in onset, was noted less frequently. Following repetitive dosing, reactions were noted much less frequently on days 2–10. The marked decrease in lymphocytes during the first 24 hr suggests that these side effects were due to cytokine release and not allergic reactions. It is likely that the significant drop in lymphocytes during these first 24 hr left few cells to cause such a significant reaction following later injections. In addition, significant delayed skin reactions at injection sites in the sc study were noted although the severity was usually mild and not related to dose. An increased incidence of infection, predominantly of the herpesvirus group, has accompanied the early lymphopenia; however, the frequency of severe infection appears similar to that of the refractory RA population in general. Nonetheless, two infection-related deaths have been noted in CAMPATH-1H trials for RA.

SUMMARY

CAMPATH-1H is the first fully humanized MAb to be studied for therapeutic use in humans; it is a biologically potent antilymphocyte agent, which exhibits rapid and prolonged lymphocyte depletion in patients treated for RA. Although NK and monocyte depletion was transient and B-cell count returned to the normal range within a few months, T cells (both CD4+ and CD8+ subsets) remained significantly depressed beyond the 6-month study period. Clinical response, as defined by modified Paulus criteria, occurred rapidly in the majority of patients; however, the response was of limited duration. In general, responses were more pronounced and of longer duration in patients receiving higher doses; in addition, single doses were much less effective clinically than daily dosing for 5 or 10 days despite similar degrees of induced lymphopenia. Subcutaneous dosing was biologically and clinically as effective as iv infusions and appeared to be less toxic. Nonetheless, relapse of disease activity was apparent in most patients despite persistence of CD4+ lymphocytopenia.

A dose-related acute side effect profile consistent with a lymphokine release syndrome was evident after the first few doses of CAMPATH-1H. In addition, a delayed local cutaneous reaction of brief duration developed in most patients treated with sc MAb and did not appear to be dose-related. Although several infections were observed during the period of absolute lymphopenia following dosing, most were self-limited and responded to treatment; however, two fatalities related to infection have been reported in study patients. Many patients developed low-titer anti-idiotypic antibody responses to CAMPATH-1H, but there did not seem to be any relationship between such response and clinical effect or toxicity despite retreatment in a number of patients.

Improved clinical effects without the dramatic lymphopenia and acute toxicity might be achieved with smaller repetitive sc doses given over a longer period. In addition, combination studies incorporating CAMPATH-1H and other biological response modifiers may allow lower, less immunosuppressive doses with additive or even synergistic clinical effects.

REFERENCES

1. Decker JL, Malone DG, Haraoul B, Wahl SM, Schrieber L, Klippel JH, Steinberg AD, Wilder RL. Rheumatoid arthritis: evolving concepts of pathogenesis and treatment. Ann Intern Med 1984; 101:810–824.
2. Alpert SD, Koide J, Takada S, Engleman EG. T cell regulatory disturbances in rheumatic diseases. Rheum Dis Clin North Am 1987; 13:431–455.
3. Haraoul B, Wilder RL, Malone DG, Allen JB, Katone IM, Wahl SM. Immune function in severe, active rheumatoid arthritis: a relationship between peripheral blood mononuclear cell proliferation to soluble antigens and mononuclear cell subset profiles. J Immunol 1984; 133:697–701.
4. Boling EP, Ohishi T, Wahl SM, Misiti J, Wistar R Jr, Wilder RL. Humoral immune function in severe, active rheumatoid arthritis. Clin Immunol Immunopathol 1987; 43:185–194.
5. Malone DG, Wahl SM, Tsokos M, Cattel H, Decker JL, Wilder RL. Immune function in severe, active rheumatoid arthritis: a relationship between peripheral blood mononuclear cell proliferation to soluble antigens and synovial tissue immunohistologic characteristics. J Clin Invest 1984; 74:1173–1186.
6. Van Boxel JA, Paget SA. Predominantly T-cell infiltrate in rheumatoid synovial membranes. N Engl J Med 1975; 283:517–520.
7. Smiley JD, Sachs C, Ziff M. In vitro synthesis of immunoglobulin by rheumatoid synovial membrane. J Clin Invest 1968; 47:624–632.
8. Trantham DE, Belil JA, Anderson RJ, et al. Clinical and immunologic effects of fractionated total lymphoid irradiation in refractory rheumatoid arthritis. N Engl J Med 1981; 305:976–982.
9. Paulus HE, Machieder HI, Levine S, et al. Lymphocyte involvement in rheumatoid arthritis: studies during thoracic duct drainage. Arthritis Rheum 1977; 20:1248–1262.
10. Yocum DE, Klippel JH, Wilder RL, Garger NL, Austin HA, Wahl SM, Lesko L, Minor JR, Preuss HG, Yarboro C, Berkabile C, Dougherty S. Cyclosporin A in severe, treatment-refractory rheumatoid arthritis: a randomized study. Ann Intern Med 1988; 109:863–869.
11. Horneff G, Burmester GR, Emmrich F, Kalden Jr. Treatment of rheumatoid arthritis with an anti-CD4 monoclonal antibody. Arthritis Rheum 1991; 34:129–140.

12. Moreland, LW, Bucy RP, Tilden A. Use of a chimeric monoclonal anti-CD4 antibody in patients with refractory rheumatoid arthritis. Arthritis Rheum 1993; 36:307–318.
13. Strand V, Lipsky PE, et al. Effects of administration of anti-CD5 plus immunoconjugate in rheumatoid arthritis. Arthritis Rheum 1993; 36:620–630.
14. Kirkham B, Chikanza I, Pitzalis C, et al. Response to monoclonal CD7 antibody in rheumatoid arthritis. Lancet 1988; 1:589.
15. Hale G, Xia M-Q, Tighe HP, et al. The CAMPATH-1 antigen (CDw52). Tissue Antigens 1990; 35:119–127.
16. Schoissman SF, Boumsell L, Gilds W, et al. CD antigens 1993: update. J Immunol 1994; 152(1):1–2.
17. Hale G, Rye PD, Warford A, et al. The glycosyl phosphatidylinositol-anchored lymphocyte antigen CDw53 is associated with the epididymal maturation of human spermatozoa. J Reprod Immunol 1993; 23:189–205.
18. Hale G, Bright S, Chumbley G, et al. Removal of T-cells from bone marrow for transplantation: a monoclonal antilymphocyte antibody that fixes human complement. Blood 1983; 62:873–882.
19. Hale G, Waldmann H. CAMPATH-1 monoclonal antibodies in bone marrow transplantation. J Hematother 1994; 3:15–31.
20. Dyer MJS, Hale G, Hayhoe FGH, Waldmann H. Effects of CAMPATH-1 antibodies in vivo in patients with lymphoid malignancies: influence of antibody isotype. Blood 1989; 73:1431–1439.
21. Dyer MJS, Hale G, Marcus R, Waldmann H. Remission induction in patients with lymphoid malignancies using unconjugated CAMPATH-1 monoclonal antibodies. Leukaemia Lymphoma 1990; 2:179–193.
22. Hale G, Waldmann H. CAMPATH-1 for prevention of graft-versus-host disease and graft rejection: summary of results from a multi-center study. BM Trans 1988; 3(Suppl 1):11–14.
23. Reichmann L, Clark M. Waldmann H, Winter G. Reshaping human antibodies for therapy. Nature 1988; 322–327.
24. Hale G, Dyer MJS, Clark MR, et al. Remission induction in non-Hodgkin's lymphoma with reshaped human monoclonal antibody CAMPATH-1H. Lancet 1988; 2:1394–1399.
25. Mathieson PW, Cobbold SP, Hale G, et al. Monoclonal-antibody therapy in systemic vasculitis. N Engl J Med 1990; 323:250–254.
26. Isaacs J, Watts R, Hazleman B, et al. Humanized monoclonal antibody therapy for rheumatoid arthritis. Lancet 1992; 340:748–752.
27. Johnston, JM. Personal communication.
28. Weinblatt ME, Johnston JM, Hazleman BL, Manna VK, and the CAMPATH-1H RA Investigators. Treatment of rheumatoid arthritis with single-dose infusion of CAMPATH-1H. Arthritis Rheum 1992; 35:S105.
29. Johnston JM, St Clair EW, Yocum DE, Matteson EL. Treatment of rheumatoid patients by subcutaneous injection of CAMPATH-1H. Arthritis Rheum 1992; 35:S105.
30. Johnston JM, Spreen WR. Treatment of rheumatoid arthritis with humanized monoclonal antibody, CAMPATH-1H. Proceedings: Early Decisions in DMARD Development, III. Biologic Agents in Autoimmune Diseases, March 4–5, 1993, pp. 55–64.

31. Isaacs JD, Manns VK, Hazleman BL, et al. CAMPATH-1H in RA-A study of multiple IV dosing. Arthritis Rheum 1993; 36:540 (abstract).
32. Paulus HE, Egger MJ, Ward JR, et al. Analysis of improvement in individual rheumatoid arthritis patients treated with disease-modifying antirheumatic drugs, based on the findings in patients treated with placebo. Arthritis Rheum 1990; 33:477–484.
33. Weinblatt ME, Coblyn J, Maler A, Anderson B, Helfgott S, Thurmond L, Spreen W, Johnston JM. Sustained lymphocyte suppression after single dose CAMPATH-1H infusion: an 8 month follow up. Arthritis Rheum 1993; 36:S40.
34. Thakor MS, Seavar NA, Liebler KE, Cornett MK, Johnston JM, Yocum DE. The effects of CAMPATH-1H on lymphocyte cell surface markers and proliferation. Arthritis Rheum 1992; 35:S105.
35. Ganten T-M, Jendro MC, Fulbright JW, Matteson EL, Weyand CM, Gurenzy JJ. Restricted diversity of the T cell receptor repertoire after therapeutic T cell depletion with the monoclonal antibody CAMPATH-1H. Arthritis Rheum 1993; 36:S40.
36. Sundy JS, Dennigan SM, Jacobs MR, St. Clair EW. The effect of CAMPATH-1H monoclonal antibody therapy on the T cell receptor (TCR) repertoire of patients with rheumatoid arthritis. Arthritis Rheum 1993; 36:S40.

III
Cytokine Targeted Therapies

Cytokine networks regulate inflammation and the immune response through a balance of inhibitory and stimulatory effects, including autocrine and paracrine regulatory mechanisms. The presence of cytokine receptors on most cell types and the pleiotropism and redundancy of effects in cytokine families help to explain the diversity of their effects. Whereas IL-1, TNFα and IL-6 are considered pro-inflammatory and are implicated in many of the pathogenic processes underlying autoimmunity, IFNγ, TGFβ, IL-4 and IL-10 may suppress the inflammatory response.

Interferonγ was the first biologic agent studied in autoimmune disease, and has been utilized in the treatment of rheumatoid arthritis (RA) and juvenile RA (Chapter 9). Two forms of interferonβ have received approval for the treatment of multiple sclerosis (MS): interferonß2 (Betaseron) and interferonß1a (Avonex).

Several techniques have been developed to inhibit the effector functions of cytokines. These include receptor antagonist proteins, soluble receptors and monoclonal antibodies (MAbs) to either cytokines or their receptors. Despite initial disappointing results in clinical trials of sepsis, many of these agents appear promising in the treatment of autoimmune disease.

IL-1 has long been implicated in the initiation as well as the perpetuation of the inflammatory process. Clinical trials of the IL-1 receptor antagonist (IL-1ra) continue; a placebo RCT is currently underway in 400 patients in RA (1). The benefit of gene transfer of IL-1ra expression has been demonstrated in several animal models; a pilot protocol is underway in humans (2,3). The type I soluble IL-1 receptor (sIL-1R) has been evaluated in two placebo-controlled trials in RA, after intraarticular and subcutaneous administration (Chapter 12). Their disappointing results may be explained because the type II sIL-1R has a higher affinity for IL-1β and less affinity for IL-1ra than the type I receptor, and may therefore be a more effective inhibitor of IL-1 (4).

III: Cytokine Targeted Therapies

Recently, considerable preclinical and clinical data have accumulated implicating the role played by TNFα in the underlying pathogenesis of RA and other inflammatory conditions (5,6). Placebo RCTs of both the chimeric (cA2) and humanized (CDP571) anti-TNFα MAbs have demonstrated significant clinical responses with concordant reductions in malaise, fatigue, CRP and serum IL-6 levels (Chapters 10 and 11). Benefit was rapid in onset and continued with readministration. Adverse effects have been mild; development of ANA, anti-DNA and anti-cardiolipin antibodies after treatment remain unexplained. Similar salutory effects have recently been reported in patients with Crohn disease.

Several soluble TNFα receptor proteins (sTNF-R) are also undergoing evaluation in patients with RA. Results following administration of the type II (or p75) sTNF-R (Chapter 13) have been less dramatic than with the anti-TNFα MAbs, although a large RCT was recently completed. Treatment with the type I (p55) sTNF-R (or the sTNF binding protein) may theoretically be superior, but data following their administration have not been reported (7).

Based on our recent experience in animal models and the clinic, it is likely that a treatment that effectively inhibits the effects of both IL-1 and TNFα will offer additional significant benefit, of longer duration, and may abrogate the associated loss of body cell mass, bone, and cartilage in diseases such as RA.

REFERENCES

1. Campion G. Recombinant human IL-1 receptor antagonist as therapy for rheumatoid arthritis. In Strand V, Johnson K, Simon L, eds. Biologic Agents in Autoimmune Diseases IV. Arthritis Foundation, 1966:59–66.
2. Evans C, Robbins PD. Prospects for treating arthritis by gene therapy. J Rheumatol 1994; 21:779–782.
3. Makarov SS, Olsen JC, Johnston WN, et al. Suppression of experimental arthritis by gene transfer of IL-1 ra cDNA. Proc Natl Acad Sci 1996; 93:402–406.
4. Colotta F, Dower SK, Sims JE, Mantovani A. The type II "decoy" receptor: a novel regulatory pathway for interleukin 1. Immunol Today 1994; 15:562–566.
5. Roubenoff R, Roubenoff RA, Cannon JG, Kehayias JJ, Zhuang H, Dawson-Hughes B, Dinarello CA, Rosenberg IH. Rheumatoid cachexia: Cytokine driven hypermetabolism accompanying reduced body cell mass in chronic inflammation. J Clin Invest 1994; 93:2379–2386.
6. Cope AP, Londei M, Chu NR, Cohen SBA, Elliott MJ, Brennan FM, Maini RN, Feldmann M. Chronic exposure to TNF in vitro impairs the activation of T cells through the TCR/CD3 complex; reversal in vivo by anti TNFa antibodies in patients with RA. J Clin Invest 1994; 94:749–760.
7. Evans TJ, Moyes D, Carpenter A, et al. Protective effect of 55- but not 75-kd soluble tumor necrosis factor receptor immunoglobulin fusion proteins in an animal model of gram negative sepsis. J Exp Med 1994; 180:2173.

9
Interferon-Gamma in the Treatment of Rheumatoid Arthritis

ERIC M. VEYS, HERMAN MIELANTS, and GUST VERBRUGGEN
Ghent University Hospital, Ghent, Belgium

I. INTRODUCTION

The interferons (IFN) are naturally occurring cytokines with various biological functions; they were originally identified by their ability to protect cells and the host from the cytopathic effect of viruses. These molecules possess also a variety of immunoregulatory and antiproliferative effects. It became clear that several polypeptides had IFN activity. IFN-α, previously called leukocyte IFN, and IFN-β, previously called fibroblast IFN, had primarily antiviral and antiproliferative activity (1). IFN-γ, previously called immune IFN, is produced by activated T cells and has immunomodulatory as well as antiviral and antiproliferative properties (2). As a macrophage-activating factor (3), it increases the biosynthesis rate and the membrane density of class II histocompatibility molecules in cells presenting a basal expression of class II molecules such as Ia^+ monocytes and thymic epithelium (4,5). Human recombinant IFN-γ (hrIFN-γ) may also induce de novo expression of class II antigens on cells that in normal conditions do not express these antigens on their surface, such as vascular endothelium and fibroblasts (6) and also articular chondrocytes (7). By increasing the density of the class II expression on the membranes of the antigen-presenting cells, IFN-γ enhances the activation of specific T-helper lymphocytes and consequently also T-helper-cell-dependent immunity (8). The enhancement of antibody-dependent cellular cytotoxicity and of natural killer-cell activity is also documented. There is evidence that only little IFN-γ is present in the rheumatoid environment (9–16). However, the low representation of IFN-γ in rheumatoid arthritis (RA) is not completely understood.

The multiple effects of IFN-gamma on the metabolism of target cells in RA are ambiguous and not definitely established (17–22). The sequence of the interference of IFN-γ on the cytokine cascade is not unequivocally defined. However, some rationale for the therapeutic use of IFN-γ in RA can be distilled from all these data.

In the present overview, the significance of IFN-γ in the rheumatoid environment will be discussed as well as the effect of IFN-γ on the metabolism of connective tissue target cells (fibroblasts and chondrocytes). The results of the clinical trials performed in RA will be compared and the question will be asked whether the compound has a place in the therapeutic arsenal of RA.

II. IFN-γ IN THE RHEUMATOID SYNOVIAL ENVIRONMENT

In the first studies of cytokines in rheumatoid arthritis, interferon was determined by bioassays and high titers of IFN activity were identified in serum and synovial effusions. However, the biological assays used at that time were subject to a lack of specificity, and other factors were identified in synovial fluid that could give false positive results.

More recently, in vivo observations and in vitro experiments suggest that little or no IFN-γ is present in the rheumatoid synovium. Rheumatoid synovial tissue explants, which actively produce prostaglandins and collagenase, do not produce IFN-γ as measured by immunoassay (9). Immunofluorescence studies of rheumatoid synovial membrane show little IFN-γ (10). Synovial fluid monocytes from RA patients produce much less IFN-γ when stimulated by IL-2 than synovial fluid monocytes from gouty patients (11). In contrast to data obtained with bioassays, the determination of IFN-γ in synovial fluid from RA patients using specific radioimmunoassays and enzyme-linked immunosorbent assays (ELISA) gives clearly lower values than in other inflammatory joint effusions (12). Firestein and Zvaifler have shown that, unlike in normal peripheral blood T lymphocytes, in RA synovial fluid T cells, and in peripheral blood T cells that have been stimulated by IL-2 or by autologous mixed lymphocyte reaction, IFN-γ production is decreased in the presence of physiological concentrations of prostaglandins and is almost completely abolished at pharmacological concentrations (12). The proliferation of isolated lymphocytes infected by Epstein-Barr virus is depressed by autologous T cells and macrophages, due to the effect of IFN-γ and of nonspecific cytotoxic T cells. Finally, infected B cells are eliminated by specific cytotoxic T cells. If the same experiments are performed using RA B cells infected by the Epstein-Barr virus, their proliferation continues out of regulation, due to a failure of cytokine production by rheumatoid T cells, including IFN-γ (13,14). The low representation of IFN-γ in the rheumatoid synovial environment may be due to local overconsumption of this cytokine in the immune-mediated inflammatory reaction or to an intrinsic defect in IFN-γ production in RA, as suggested by some authors (15,16). Pharmacological administration of the cytokine in these instances could succeed in reducing disease activity. Low IFN-γ levels in RA might also be due to a physiological down-regulation, to protect the joint against some deleterious effects of IFN-γ on the local articular target cells. Pharmacological administration of IFN-γ in this instance might consequently induce an exacerbation of disease activity in RA.

In theory, the potential detrimental effect of IFN-γ administration to patients with autoimmune diseases lies in the induction of class II molecule expression by cells that normally to not present these antigens on their surface and are thus incapable of antigen presentation to helper T cells. If the antigen is a viral product expressed on the membrane of the infected cell, the aberrant class II expression induced by IFN-γ can be followed by the activation of a specific helper-T-cell line, resulting in a potential benefit for the host. If the antigen presented by the new class II positive cells is an autologous cell membrane molecule, one that is unique to a particular tissue and is tolerated by the host-helper T cells, the endresult could be an autoimmune attack against the IFN-γ-induced class II positive tissue (4). However, it is noteworthy that expression of class II antigens on the surface membrane is not synonymous with the ability of these cells to act as antigen-presenting cells.

Consequently, it appears from most of the studies that IFN-γ is present only in very small amounts in the RA synovial environment. However, stimulation of synovial tissue cells with a T-cell mitogen induces large amounts of IFN-γ, indicating that cells capable of producing the cytokine are present in the synovium. IFN-γ gene expression in the synovium has given conflicting results; some authors report no detectable IFN-γ RNA transcripts (17), while others appear to find some (18). More recently, polymerase chain reaction and in situ hybridization have given concordant results and have led to the conclusion that only 0.1–0.5% of all T cells in RA synovial membrane expressed IFN-γ mRNA (19).

The relative absence of IFN-γ in the rheumatoid synovium is difficult to explain. Some factors inhibiting cytokine synthesis and T-cell activation, such as transforming growth factor (TGF)-β are found in larger amounts in the synovium. Although it is conceivable that IFN-γ is not the major macrophage-activating factor in RA, we have to keep in mind that, as for MHC class II induction, tumor necrosis factor alpha (TNF-α) synergizes with IFN-γ, and granulocyte macrophage-colony stimulating factor (GM-CSF) is additive with IFN-γ. Consequently, small amounts of IFN-γ delivered in the microenvironment may be sufficient to stimulate target cells. Finally, it is possible that T cells in the RA synovium have not received the appropriate activation signal. Indeed the production of the synovial T-cell lymphokines in RA is defective (IL-2, IL-3, IL-4, IL-6, IFN-γ, TNF-α, and GM-GSF), whereas the production of synovial macrophage cytokines is elevated (IL-1, IL-6, IL-8, TNF-α, GM-CSF, and TGF-β).

Other possible deterious effects of insufficient IFN-γ production in RA have been recently suggested in a number of studies. IFN-γ has been shown to downregulate the autocrine IL-1-induced IL-1 production by human peripheral blood monocytes (20). Insufficient IFN-γ production in the RA synovium may be responsible for prolonged IL-1 release by macrophages in the synovial fluid and synovial membrane in this disease (21). IFN-γ should have the same regulatory function on IL-1-induced neutrophil-activating peptide (NAP)/IL-8 release in the RA environment (22). A decreased natural killer cell activity has been demonstrated in RA patients who had been good responders to levamisole treatment (23). Like levamisole, IFN-γ has been shown to enhance the natural killer cell activity in different models (24), and an increase in the serum IFN-γ level has been observed in RA patients treated with levamisole (25).

III. EFFECTS OF IFN-γ ON CONNECTIVE TISSUE TARGET CELLS IN RA

A. Effects on the Metabolism of Synovial Fibroblasts

Some authors report a stimulatory effect of IFN-γ on proliferation of synovial fibroblasts (26,27) while others describe an inhibitory effect (28). According to these authors, IFN-γ and TNF-α seem to have opposite actions on cultured synovial fibroblasts. IFN-γ inhibits TNF-α-induced synoviocyte proliferation, collagenase production, and GM-CSF secretion, while TNF-α in turn inhibits the IFN-γ-mediated HLA-DR induction on synovial fibroblasts. Therefore, in RA, the relative absence of IFN-γ could explain the unlimited stimulation of synovial fibroblasts by RNF-α. IFN-γ has no effect on the hyaluronan production and the sulfated glycosaminoglycan synthesis of the fibroblast-like synoviocytes (29), whereas several authors claimed its inhibitory effect on the collagen synthesis by these cells (30–32). IFN-γ also abrogates IL-1-induced collagenase release (28) and IL-1 and TNFα-induced prostaglandin E_2 release (27,28).

B. Effect of IFN-γ on the Metabolism of the Articular Chondrocytes

IFN-γ decreases the proliferation of monolayer-cultured human articular chondrocytes (33). According to some authors, it has no effect on proteoglycan synthesis by these cells (29). Others have shown a decrease of the aggrecan synthesis by human articular chondrocytes in monolayer culture conditions (33) and of the collagen type II synthesis as well (34,35). IFN-γ has no direct effect on enzyme release by chondrocytes (36), but it diminishes the IL-1-induced collagenase production (37) and the TNF-α-induced stromelysin release (36). IFN-γ increases the synthesis of tissue inhibitory metalloprotease (TIMP) by articular chondrocytes (37).

IV. THERAPEUTIC USE OF IFN-γ IN RA

The low representation of IFN-γ in the rheumatoid environment, the mutually antagonistic effect of IFN-γ and TNF-α, the down-regulation of the autocrine IL-1-induced IL-1 production by monocytes by IFN-γ (20), and the inhibitory effect of the IL-1-induced collagenase release by fibroblastic synoviocytes by IFN-γ and its inhibitory effect on the proliferation of these cells (28) provided the rationale to use IFN-γ in the treatment of RA.

The efficacy of IFN-γ in the treatment of RA has been suggested in open studies (38–40) and in a short-term, double-blind study in which two dosages of IFN-γ (10 versus 100 µg) are compared (41). In these pilot studies hr-IFN-γ is administered subcutaneously; this mode of administration has in fact been used in all clinical studies conducted in RA until now. Data from open studies suggest that in patients who responded to IFN-γ after 3 or 4 weeks of treatment, it would, be preferable to reduce the schedule, e.g., from 50 µg seven times per week to three or four times per week, to maintain the beneficial effect (42).

To date, four placebo-controlled, double-blind studies have been conducted (43–46); the duration of the trials, the number of patients included, and the dropout rate are presented in Table 1.

A. Dosages and Schedules

The dosages and schedules of IFN-γ were different in these studies, being 50 µg daily for 20 days, then 50 µg every second day up to day 28 in the German study

Table 1 Number of Patients, Duration of the Treatment, and Dropout Rate in the Four Placebo-Controlled Trials.

		Number of patients included			Number of dropouts	
Ref.	Duration	IFN-γ	Placebo	Total	IFN-γ	Placebo
Lemmel et al., 1988 (43)	28 days	40	39	79	3	3
Cannon et al., 1989 (45)	12 weeks	54	51	105	12	9
Veys et al. 1988 (44)	24 weeks	13	13	26	2	2
Veys et al., in press (46)	24 weeks	99	96	195	34	24

(43), 100 μg 5 days/week for 12 weeks in the American study (44), 100 μg daily for the first 5 days and then twice weekly for the following 23 weeks in the Belgian study (45), and 50 μg 7 days/week for 3 weeks, followed by 50 μg 3 times per week for 5 weeks and 50 μg twice weekly for 16 weeks in the international study.

B. Dropouts

Occurrence of untoward effects was the only reason for dropout in the German study (43). In the American study (44) reasons for withdrawal were the occurrence of adverse drug reactions in two patients treated with IFN-γ and in three patients treated with placebo, and disease flare in seven patients in the IFN-γ group and in four patients in the placebo group. In the Belgian study (45), reasons for dropout were consent withdrawal by two patients in the placebo group and in one patient in the IFN-γ group. One patient of the IFN-γ group was withdrawn because of the occurrence of a concomitant bacterial infection after 4 weeks. In the international study (46) withdrawal due to adverse experiences occurred in four patients of both groups; treatment failure was the reason for dropout in 22 patients of the IFN-γ group and in 16 patients of the placebo group. Consent was withdrawn by seven patients in the IFN-γ group and two patients in the placebo group. Two patients of both groups were withdrawn for other reasons.

C. Adverse Reactions

Adverse reactions were reported in nearly 10% of the patients of the three first double-blind studies (43–45) and included fever, chills, nausea, headache, dizziness, vasodilatation, reaction at the injection site, fatigue, and diarrhea. In the international study (46), 50% of the patients of both groups, placebo and IFN-γ, reported side effects, but here the investigator asked for the whole list of adverse reactions. In fact, the number of serious adverse effects was not higher than in previous studies. Four patients (4%) in each group were withdrawn for adverse reactions).

No significant changes in the safety biological variables were noticed in the four studies.

D. Efficacy

Concerning efficacy, the trials can be summarized as follows:

1. *In the German study* (43), the statistical analysis concentrated on the number of patients considered to be responders and nonresponders according to the improvement in the Ritchie and/or the Lansbury joint pain index by 30%. In the placebo group, 11 patients responded and 28 did not; in the IFN-γ group, 23

patients were classified as responders and 17 as nonresponders; these differences are statistically significant using the chi-square test ($p = .01$).

2. *In the American study* (44), significant intergroup changes were observed in the IFN-γ-treated group for all the clinical variables, except the grip strength; in the placebo group, only borderline significant improvement was observed in the joint tenderness count and the joint tenderness score. Intergroup evaluation revealed a significant difference in favor of IFN-γ for the Stanford Health Assessment questionnaire (HAQ) (47). The erythrocyte sedimentation rate (ESR) was significantly increased in both groups after 12 weeks of treatment.

3. *In the Belgian study* (45), no significant intergroup changes were observed: the authors found significant improvement in the joint tenderness score in the IFN-γ group and a significant increase in the number of subcutaneous nodules in the placebo-treated group.

4. *in the international trial* (46), IFN-γ induced a significant improvement from baseline to the endpoint visit for each measurement (number of tender joints, number of swollen joints, physician and patient assessment of disease activity). However, the placebo-treated group showed similar improvement. No significant intergroup changes were observed. In both groups, ESR did not vary significantly.

The schedules and dosages of IFN-γ used in these trials suggest it to have mild beneficial effects on the disease course of RA, when compared with placebo. It remains questionable whether these results have clinical relevance. The absence of serious adverse effects combined with a rather weak efficacy was confirmed in prospective, 2-year (48) and 5-year (49) follow-ups of the American cohort of patients included in the aforementioned double-blind trial (44). Seventy patients completing the 12-week double-blind trial were enrolled for the prospective protocol; 43 (61%) patients after 1 year, 24 (34%) after 2 years, 18 (26%) after 3 years, 11 (15%) after 4 years, and 8 (11%) after 5 years continued the treatment with sustained clinical benefit. The main reason for discontinuing the drug over 5 years was the lack of continued benefit. As a matter of fact, no evidence exists to clearly establish that, at the dosages and schedules used, subcutaneous administration of IFN-γ alters any immunological variables in RA. Some immunomodulatory effect was shown after 12 weeks of treatment by a decrease in the number of circulating B cells (50). Unfortunately, the authors did not include the CD5 marker in their evaluation.

IFN-γ administration did not influence ESR or acute-phase proteins in the trials discussed.

In a pilot study of nine patients with severe systemic juvenile RA, a beneficial effect of IFN-γ (0.5 μg/kg body weight, daily for 3 weeks and then three times per week for 3 weeks) has been reported in seven cases (51). However, IFN-γ has been described to induce an exacerbation of systemic lupus erythematodes in one patient (52).

V. CONCLUSIONS

Differing and apparently conflicting data on the role of IFN-γ in the pathogenesis of RA, as to its immunoregulatory effect as well as its actions on connective tissue cells, render the use of this compound in the treatment of the disease questionable. Nonetheless, some investigators found in these data a rationale to look to the effect of subcutaneous administration of IFN-γ on the disease activity.

In the major trials that were performed, no serious adverse reactions were seen. Regarding efficacy, only one fact is clear: the administration of IFN-γ did not induce disease activity exacerbation. Beneficial effects were weak and inconstant; only the German study showed the number of responders to be significantly higher in the IFN-γ-treated group compared to placebo (43). The other three trials (44–46) did not show significant intergroup changes.

The question can be asked whether the weak efficacy apparent from these studies has clinical significance. However, we must realize that the optimal schedule and dosage of this biological compound and also the mode of administration are still unknown. More fundamental research on these issues, also in animal models, is needed before new therapeutic attempts in humans are undertaken.

REFERENCES

1. National Institutes of Health. Interferon nomenclature. Nature 1980; 286:110.
2. Trinchieri G, Perussia B. Immune interferon: a pleiotropic lymphokine with multiple effects. Immunol Today 1985; 6:131–136.
3. Svedersky LP, Benton CV, Berger WH, Rinderknecht E, Harkins RN, Palladino MA. Biological and antigenic similarities of murine interferon-gamma and macrophage-activating factor. J Exp Med 1984; 159:812–817.
4. Cowing C, Frohman MA. Interferon, class II histocompatibility antigens and autoimmunity. In: Pincus SH, Pisetsky DS, Rosenwasser LJ, eds. Biologically Based Immunomodulators in the Therapy of Rheumatic Disease. New York: Elsevier, 1986:349–355.
5. Steeg PS, Moore RN, Johnson HM, Oppenheim JJ. Regulation of murine macrophage Ia antigen expression by a lymphokine with immune interferon activity. J Exp Med 1982; 156:1780–1793.
6. Pober JS, Collins T, Gimbrone MA, et al. Lymphocytes recognize human vascular endothelial and dermal fibroblast Ia antigens induced by recombinant immune interferon. Nature 1983; 305:726–729.
7. Jahn B, Burmester GR, Schmid H, Weseloh G, Rohner P, Kalden JR. Changes in cell surface antigen expression on human articular chondrocytes induced by gamma interferon. Arthritis Rheum 1987; 30:674.
8. Nakamura M, Manser T, Pearson GDN, Daley MJ, Gefter ML. Effect of IFN-gamma on the immune response in vivo and on gene expression in vitro. Nature 1984; 307:381–382.

9. Chin JE, Winterrowd GE, Krzesicki RF, Saunders ME. Role of cytokines in inflammatory synovitis: the coordinate regulation of intercellular adhesion molecule I and HLA class I and class II antigens in rheumatoid synovial fibroblasts. Arthritis Rheum 1990; 33:1776–1786.
10. Husby G, Williams RC. Immunohistochemical studies of interleukin-2 and gamma-interferon in rheumatoid arthritis. Arthritis Rheum 1985; 24:174–181.
11. Combe B, Pope RM, Fishbach LM, et al. Interleukin-2 in rheumatoid arthritis: production of and response to interleukin-2 in rheumatoid synovial fluid, synovial tissue and peripheral blood. Clin Exp Rheumatol 1985; 59:520–528.
12. Firestein GS, Zvaifler NJ. Interferon and rheumatoid arthritis. In: Pincus SH, Pisetsky DS, Rosenwasser LJH, eds. Biologically Based Immunomodulators in the Therapy of Rheumatic Diseases. New York: Elsevier, 1986:369–378.
13. Hasler F, Bluestein EL, Zvaifler NJ, Epstein LB. Analysis of the defects responsible for the impaired regulation of Epstein-Barr virus-induced B cell proliferation by rheumatoid arthritis lymphocytes. I. Diminished gamma-interferon production in response to autologous stimulation. J Exp Med 1983; 157:173–188.
14. Lotz M, Tsoukas CD, Curd JG, Carson DA, Vaughan JH. Effects of recombinant human interferons on rheumatoid arthritis B lymphocytes activated by Epstein-Barr virus. J Rheumatol 1987; 14:42–45.
15. Toubert A, Sadouk M, De La Tour B, Vaquero C, Amor B. Gamma-interferon mRNA expression is decreased upon in vitro T lymphocyte activation in active rheumatoid arthritis patients. Arthritis Rheum 1989; 32(Suppl 4):B47 (abstract).
16. Couret M, Combe B, Reme T, Sany J. Production d'interleukine-2 et d'interferon-gamma dans la polyarthrite rhumatoide au cours d'une étude longitudinale chez des patients traités par méthotrexate. Rev Rhum 1990; 57:641–645.
17. Firestein GS, Alvaro-Gracia JM, Maki R. Quantitative analysis of cytokine gene expression in rheumatoid arthritis. J Immunol 1990; 144:3347–3353.
18. Buchan G, Barrett K, Fujita T, Taniguchi T, Maini R, Feldmann M. Detection of activated T-cell products in the rheumatoid joint using cDNA probes to interleukin-2 (IL-2) receptor and IFN-gamma. Clin Exp Immunol 1988; 71:295–301.
19. Simon K, Seipelt E, Braun J, Sieper J. T-cell cytokine patterns in synovial membrane of reactive arthritis and rheumatoid arthritis. Arthritis Rheum 1993; 36:S156.
20. Ghezzi P, Dinarello PA. Interleukin-1. III. Specific inhibition of interleukin-1 production by gamma-interferon. J Immunol 1988; 140:4238.
21. Ruschen S, Lemm G, Warnatz H. Spontaneous and LPS-stimulated production of intracellular IL-1beta by synovial macrophages in rheumatoid arthritis is inhibited by IFN-gamma. Clin Exp Immunol 1989; 76:246–251.
22. Seitz M, Dewald B, Gerber N, Baggiolini M. Enhanced production of neutrophil-activating peptide-1/interleukin-8 in rheumatoid arthritis. J Clin Invest 1991; 57: 463–469.
23. Barada FA, O'Brien W, Horwitz DA. Defective monocyte cytotoxicity in rheumatoid arthritis: a correlation with disease activity and reversal by levamisole. Arthritis Rheum 1982; 25:10–16.
24. Nakamura M, Manser T, Pearson G, et al. Effect of gamma-interferon on the immune response in vivo and on gene expression in vitro. Nature 1984; 307:381–384.

25. Veys EM, Luyten F, Mielants H, et al. Side-effects of levamisole: recent aspects. In: Rainsford KD, Velo GP, eds. Side-Effects of Anti-inflammatory Drugs. Part 2: Studies in Major Organ Systems. Lancaster: MTP Press, 1987: 235–243.
26. Brinckeroff C, Guyre PM. Increased proliferation of human synovial fibroblasts treated with recombinant immune interferon. J Immunol 1985; 134:3142–3145.
27. Taylor DJ, Feldmann M, Evansson JM, Woolley DE. Comparative and combined effects of transforming growth factors alpha and beta, interleukin-1 and interferon-gamma on rheumatoid synovial cell proliferation, glycolysis and prostaglandin E production. Rheumatol Int 1989; 9:65–70.
28. Nakajima H, Hiyama Y, Tsukada W, et al. Effects of interferon-gamma on cultured synovial cells from patients with rheumatoid arthritis: inhibition of cell growth, prostaglandin E_2 and collagenase release. Ann Rheum Dis 1990; 49:512–516.
29. Yaron I, Meyer FA, Dayer JM, Bleiberg I, Yaron M. Some recombinant human cytokines stimulate glycosaminoglycan synthesis in human synovial fibroblast cultures and inhibit it in human articular cartilage cultures. Arthritis Rheum 1989; 32:173–180.
30. Jimenez SA, Freundlich B, Rosenbloom J. Selective inhibition of human diploid fibroblast collagen synthesis by interferons. J Clin Invest 1984; 74:1112–1116.
31. Amento EP, Bhan AK, McCullagh KG, Krane SM. Influence of gamma interferon on synovial fibroblast-like cells: Ia induction and inhibition of collagen synthesis. J Clin Invest 1985; 76:837–848.
32. Daireaux M, Redini F, Loyau G, Pujol JP. Effects of associated cytokines (IL-1, TNF-alpha, IFN-gamma and TGF-beta) on collagen and glycosaminoglycan production by cultured human synovial cells. Int J Tissue react 1990; 12:21–31.
33. Verbruggen G, Malfait AM, Veys EM, Gyselbrecht L, Lambert J, Almquist KF. Influence of interferon-gamma on isolated chondrocytes from human articular cartilage: dose-dependent inhibition of cell proliferation and proteoglycan synthesis. J Rheumatol 1993; 20:1020-1026.
34. Goldring MD. Control of collagen synthesis in human chondrocyte cultures by immune interferon and IL-1. J Rheumatol 1987; 14:64–66.
35. Goldring MB, Sandell LJ, Stephenson ML, Krane SM. Immune interferon suppresses levels of procollagen mRNA and type II collagen synthesis in cultured human articular and cortal chondrocytes. J Biol Chem 1986; 261:9049-9055.
36. Bunning RAD, Russel RGG. The effect of tumor necrosis factor alpha and gamma-interferon on the resorption of human articular cartilage and on the production of prostaglandin E and of caseinase activity by human articular chondrocytes. Arthritis Rheum 1989; 32:780–784.
37. Andrews HJ, Bunning RA, Plumpton RA, Clark IM, Russell RG, Cawston TE. Inhibition of interleukin-1-induced collagenase production in human articular chondrocytes in vitro by recombinant interferon-gamma. Arthritis Rheum 1990; 33: 1733–1738.
38. Lemmel EM, Botzenhardt U. Chronische Polyarthritis: Immunätiopathogenese. In: Mathies H, ed. Handbuch der Inneren Medizin. Rheumatologie B, Spezieller Teil I, Gelenke. Berlin: Springer, 1984:4–8.
39. Obert HJ, Hofschneider PH. Interferon in chronic polyarthritis: positive effect in clinical evaluation. Dtsch Med Wochenschr 1985; 116:1766–1769.

40. Zilly A, Obert HJ. Therapie der chronischen Polyarthritis. Ergebnisse einer Pilot-Studie mit Lymphokines. München Med Wochenschr 1986; 128:87–89.
41. Wolfe F, Cathey MA, Hawley DJ, Baker JP, Schindler JD. Clinical trial with R-IFN-gamma in rheumatoid arthritis. In: Pincus SH, Pisetsky DS, Rosenwasser LJ, eds. Biologically Based Immunomodulators in the Therapy of Rheumatic Diseases. New York: Elsevier, 1986:379–396.
42. Sprekeler R, Lemmel EM, Obert HJ. Correlation of clinical and serological findings in patients with rheumatoid arthritis treated for one year with interferon-gamma. Z Rheumatol 1990; 49:1–7.
43. Lemmel EM, Bracketz D, Franke M, et al. Results of a multicentre placebo-controlled double-blind randomized phase III clinical study of treatment of rheumatoid arthritis with recombinant interferon-gamma. Rheumatol Int 1988; 8:87–93.
44. Veys EM, Mielants H, Verbruggen G, et al. Interferon-gamma in rheumatoid arthritis: a double-blind study comparing human interferon-gamma with placebo. J Rheumatol 1988; 15:570–574.
45. Cannon GW, Pincus SH, Emkey RD, et al. Double-blind trial of recombinant gamma-interferon versus placebo in the treatment of rheumatoid arthritis. Arthritis Rheum 1989; 32:964–973.
46. Veys EM, Menkes CJ, Emmery P. A clinical study comparing recombinant interferon gamma with placebo for twenty four weeks in the treatment of rheumatoid arthritis (submitted for publication).
47. Fries JF, Spitz P, Kraines RG, Holman HR. Measurement of patient outcome in arthritis. Arthritis Rheum 1980; 23:137–145.
48. Cannon GW, Emkey RD, Denes A, et al. Prospective two-year follow-up of recombinant interferon-gamma in rheumatoid arthritis. J Rheumatol 1990; 17:304–310.
49. Cannon GW, Emkey RD, Denes A, et al. Prospective five-year follow-up of recombinant interferon-gamma in rheumatoid arthritis. Arthritis Rheum 1991; 34(Suppl 9):S91.
50. Pincus SH, Cannon GW, Ward RJ. In vivo administration of interferon-gamma to patients with rheumatoid arthritis decreases number of circulating B cells. J Rheumatol 1990; 17:751–757.
51. Pernice W, Schuckmans L, Dippell J, et al. Therapy for systemic juvenile rheumatoid arthritis with gamma-interferon: a pilot study of nine patients. Arthritis Rheum 1989; 32:643–646.
52. Machold KP, Smolen JS. Interferon-gamma-induced exacerbation of systemic lupus erythematosus. J Rheumatol 1990; 17:831–832.

10

Tumor Necrosis Factor Blockade in Rheumatoid Arthritis

MICHAEL J. ELLIOTT, MARC FELDMANN, and RAVINDER N. MAINI
Kennedy Institute of Rheumatology, London, England

I. INTRODUCTION

Advances in the understanding of basic disease mechanisms in rheumatoid arthritis (RA) are leading to the development of new, rational therapies. In one widely accepted view of disease pathogenesis, RA is initiated and possibly maintained by an immune response involving the recognition by T cells of an autoantigen (yet uncharacterized) in the context of a specific "shared epitope" HLA-DRβ gene product (1), with consequent T-cell activation and the initiation of an inflammatory cascade. Although cytokines may be involved in the initiating immune response, they are particularly important in the subsequent processes of inflammation and it is to this level that anticytokine therapy is primarily directed.

Inflammatory cytokines present in abundance in RA tissues include interleukin-1 (IL-1), interleukin-6 (IL-6), tumor necrosis factor-α (TNF-α), macrophage colony-stimulating factor (M-CSF), granulocyte macrophage colony-stimulating factor (GM-CSF), interleukin-8 (IL-8), and leukemia inhibitory factor (LIF). Although one can make a good case for the involvement of each of these mediators in the disease process in RA, the biological properties of IL-1 and TNF-α make them particularly attractive therapeutic targets. The rationale for the use of IL-1 antagonists in RA is described elsewhere in this book. Here, we present the case for TNF-α blockade in this disease and describe the results of our clinical trials in patients using a specific, TNF-blocking monoclonal antibody (MAb).

II. TNF-α IN RHEUMATOID ARTHRITIS

TNF-α is present in RA synovial fluid (2), is produced in large amounts by unstimulated synovial cells cultured in vitro (3), and is detectable by immunohistology within the synovial lining layer, in interstitial monocyte/macrophage-like cells, and at the cartilage/pannus junction (4,5).

TNF-α achieves biological effects by interaction with specific receptors, the p55 and p75 TNF receptors (TNF-R). TNF-R are expressed in a wide range of cells in RA synovium (6). Following proteolytic cleavage, cell-associated TNF receptors are released in soluble form (sTNF-R) where they act to modulate TNF bioactivity. The levels of p55 and p75 sTNF-R are increased in patients with RA compared with controls, but the degree of up-regulation is insufficient to neutralize all TNF produced, resulting in persistent TNF bioactivity. Higher levels of soluble receptors are found in the synovial fluid compared with serum, suggesting local production within the joint (7).

The biological actions of TNF-α are relevant to the disease process in RA. Exposure to TNF in vitro induces the activation of endothelium, resulting in the expression of adhesion molecules that favor leukocyte migration; the activation of granulocytes, resulting in enhanced phagocytosis, degranulation, and the generation of oxygen radicals; the stimulation of fibroblast growth and cytokine production; and the costimulation with IL-2 of T-cell proliferation and lymphokine production (reviewed in Ref. 8). Importantly, TNF acts (together with IL-1) to promote cartilage and bone destruction in RA, both through the stimulation of metalloproteinase production by fibroblasts and synovial cells and by the suppression of synthesis of matrix components by connective tissue cells (8).

The effects of TNF on articular disease have been studied in vivo in several models of arthritis. Direct intra-articular injection of TNF was shown to result in worsening of clinical disease in rats with type II collagen-induced arthritis (CIA) (9). Overexpression of TNF-α in mice transgenic for a 3'-modified human TNF gene resulted in the development of chronic inflammatory polyarthritis, which could be prevented with MAb to TNF (10). Using a converse approach, the systemic administration of neutralizing MAb to TNF resulted in disease amelioration in murine CIA (11,12). Importantly, this improvement was seen even when treatment with anti-TNF was commenced after the development of clinical arthritis. Histological studies in these mice demonstrated protection by anti-TNF from the joint destruction characteristic of this model (11). Clinical improvement was also seen in murine CIA following treatment with a recombinant human soluble TNF receptor: Fc fusion protein (13).

These experiments support the notion that TNF-α is a cytokine of particular importance in the pathogenesis of inflammatory arthritis and provide a strong rationale for the use of specific TNF-blocking drugs in human RA.

III. CLINICAL TRIALS WITH MONOCLONAL ANTI-TNF (cA2) IN RA

Over the past 3 years, we have tested the properties of a neutralizing chimeric (human/mouse) MAb to human TNF-α (cA2) in the treatment of patients with RA. Our program has had two aims: first, to seek further evidence regarding the role of TNF-α in vivo in patients with inflammatory arthritis, and second, to investigate the potential for cA2 as a therapeutic agent. The rationale behind chimerization (improved pharmacokinetics and reduced immunogenicity), together with the details of the production and characterization of cA2, have been published recently (14). Of particular importance has been the demonstration that the chimerized antibody retains high-affinity binding to TNF and is capable of neutralizing TNF actions in vitro (14). Details of three clinical studies using this antibody are outlined below:

A. Open-Label Trial of cA2

In our first, pilot study with cA2, we selected 20 patients with active RA who had a history of long-standing, refractory disease (15). The patients had failed on average four standard disease-modifying antirheumatic drugs (DMARDS) and had a median disease duration of 10.5 years. The majority of patients were seropositive for rheumatoid factor and all had evidence of erosive disease. The selection of this severe disease group was made for ethical reasons, and such patients might be considered least likely to respond to a new therapeutic intervention. Patients taking DMARDS had these discontinued at least 1 month prior to trial entry, but those taking nonsteroidal anti-inflammatory drugs (NSAIDS) and/or low-dose oral corticosteroids were allowed to continue on stable doses.

The dose of cA2 administered was chosen on the basis of previous experience in the CIA model (11). Patients received a total of 20 mg/kg cA2 in either two or four infusions over a 2-week period, each infusion taking 2 hr.

1. Clinical Response

Response to treatment was assessed on the basis of a composite response index, described by Paulus et al. (16). This index incorporates five separate clinical assessments and one laboratory measure of disease activity (ESR) to determine the presence or absence of a response at a given time. Of the 19 patients who completed the infusion protocol in the open-label study, all showed a response according to this index by week 6, with an average response duration of approximately 3 months. Each of the individual disease activity assessments, including the swollen and tender joint counts, the pain score, the duration of morning stiffness, and the grip strength, showed significant improvement following treatment. Although there was considerable interpatient variability, the average improvement in the tender and swollen joints counts exceeded 70%, indicating a clinically important improvement.

2. Change in Laboratory Indices

Clinical changes were accompanied by trends toward improvement in the white cell and platelet counts and in the level of rheumatoid factor, but the most impressive laboratory changes were in measures of the acute-phase response. Of the 19 patients with elevated CRP at study entry, 17 showed normalization of levels during the trial, with an average improvement at week 6 of 80%. Other acute-phase proteins including serum amyloid A (SAA) and haptoglobin showed significant improvements (Ref. 15 and unpublished data), and these changes were reflected in the ESR, which improved on average by almost 60% at week 6.

IL-6 has been shown to be a major regulator of hepatic acute-phase protein synthesis (17) and to be inducible in human synovial cells by other cytokines including IL-1 and TNF-α (18). Measurement of IL-6 in the sera collected in the open-label trial showed elevated levels at entry in most patients, followed by rapid falls after treatment with cA2. Many patients achieved values in the normal range by day 1. A comparison of the rates of change for serum IL-6 and the individual acute-phase proteins revealed maximal changes for IL-6 preceding those for SAA and CRP (unpublished data), supporting, but not proving, the hypothesis that regulation of acute-phase protein synthesis by cA2 is indirect, mediated by suppression of IL-6 synthesis or secretion.

B. Placebo-Controlled Trial

The promising data obtained in the open-label trial made it mandatory that we proceed to test cA2 in a study of more rigorous design. We performed a randomized, placebo-controlled trial on four sites, recruiting a patient group with similar disease phenotype and duration to that selected for the open-label trial (19). Patients were randomized to one of three different treatment group, receiving either placebo (0.1% human serum albumin) or low- or high-dose cA2 (1 or 10 mg/kg, respectively). Patients received a single, 2-hr infusion in an outpatient setting and were followed for response.

1. Clinical Response

The predetermined, primary outcome measure was the achievement of a Paulus response at week 4. Using this measure, large and highly significant differences were seen between the three treatment groups, with 19 of the 24 patients treated with high-dose cA2 responding, compared with only two of 24 treated with placebo (Fig. 1). Patients treated with low-dose cA2 showed an intermediate response rate, demonstrating a clear dose-response relationship. When the data were analyzed according to a more stringent response index (Paulus 50%), still more than half the high-dose cA2 patients achieved a response. Analysis of the individual disease activity assessments showed highly significant improvements in the cA2 groups compared with placebo, with similar degrees of improvement to those seen in the open-label study (15,19).

2. Change in Laboratory Indices

As in the open-label study, treatment with cA2 was associated with rapid and statistically significant falls in CRP and ESR, with a clear dose-response relationship in evidence (19). Other laboratory parameters that showed significant improvement in cA2-treated groups included the white cell and platelet counts and the hemoglobin. The latter change was of particular interest, since TNF is a known inhibitor of erythropoiesis (20,21) and serum TNF levels in RA have been shown to relate to the degree of anemia (22). In the 4 weeks following treatment with

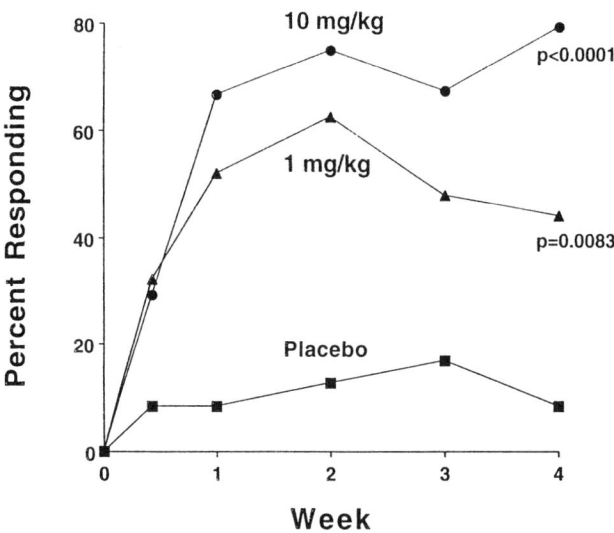

Figure 1 Responses to week 4 in patients treated with placebo (0.1% human serum albumin) or low- or high-dose cA2 (1 or 10 mg/kg respectively). Responses in individual patients were calculated according to Paulus 20% criteria. p values represent significance of changes compared with placebo by Fisher's exact test.

high-dose cA2, the mean hemoglobin rose by 0.4 g/dl, contrasting with a fall of 0.7 g/dl in patients treated with placebo. Although these changes were small, the difference between cA2 and placebo groups was highly significant (Fig. 2). These changes may reflect a direct action of cA2 in the inhibition of TNF-α-mediated suppression of hemopoiesis in the bone marrow.

Overall, the results of the placebo-controlled trial have confirmed the findings from the open-label study and provide good evidence that cA2 is highly effective in the short-term suppression of disease.

The accurate prediction of the clinical utility of cA2 in our open-label pilot study is gratifying and stands in contrast to the recent experience in clinical studies of anti-CD4 therapy in RA. Several small, uncontrolled studies testing different anti-CD4 MAb in RA have demonstrated significant and in some cases impressive improvements in individual patients, but the use of a chimerized anti-CD4 antibody in two well-designed, placebo-controlled studies has not supported the view that anti-CD4 has clinical efficacy (reviewed in Ref. 23).

Three features appeared to distinguish our pilot study of cA2 in RA from the majority of early studies using anti-CD4. First, cA2 induced significant improvement across the whole range of clinical assessments used, while in many cases, anti-CD4 was shown to affect only some of many assessments made. Second, analysis of responses on an individual patient basis revealed a universal response according to composite indices with cA2, whereas most studies testing anti-CD4 have demonstrated improvement in a subset of patients only. Third and perhaps most important, the clinical changes observed with cA2 were accompanied by

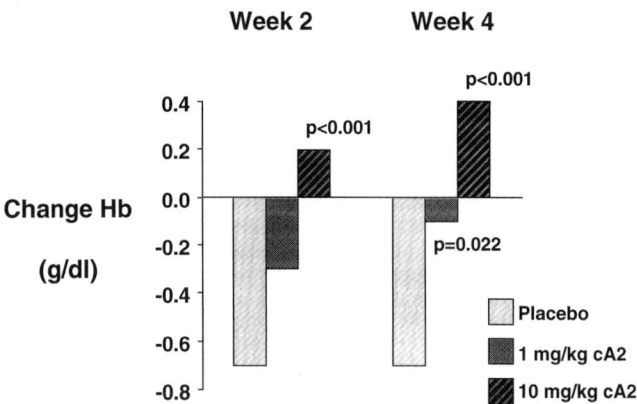

Figure 2 Change in hemoglobin 2 and 4 weeks after a single infusion of placebo (0.1% human serum albumin) or low- or high-dose cA2 (1 or 10 mg/kg, respectively). *p* values represent significance of changes compared to placebo, by analysis of variance.

improvements in laboratory measures of disease activity such as the ESR and CRP, while such changes have been unusual in trials testing T-cell-directed antibodies.

While one must be cautious about drawing conclusions from this limited analysis, the results suggest that consistency of clinical response, both within individual patients and across treatment groups, is important in the accurate prediction of efficacy.

C. Repeated Therapy with cA2

The third study has tested the feasibility of repeated therapy with cA2 in a small group of patients, originally enrolled in the open-label trial (24). This study was designed so that patients would receive treatment with cA2 upon disease relapse, defined as loss of response to the previous cycle. Cycle 1 was defined as the entire two- or four-infusion protocol of the open-label trial (total dose 20 mg/kg). Cycles 2–4 were administered as single infusions, each of 10 mg/kg. The outcome of this study was that each treatment cycle was followed by a clinical response according to the Paulus index, with average maximum improvements in individual disease activity assessments exceeding 80%. Although the magnitude of treatment responses was maintained with successive cycles, a trend toward shortening response duration was noted in some patients. This study provided further evidence for the efficacy of cA2, indicated the feasibility of repeated therapy with this antibody, and suggested that it may be effective in the setting of acute disease flares.

D. Safety Profile of cA2

The administration of cA2 was well tolerated, with little evidence of acute infusion reactions such as those that have complicated treatment with anti-T-cell therapy (reviewed in Ref. 23). Adverse events in general have been infrequent and mild. In view of animal data suggesting that TNF has a role in antibacterial resistance (25,26), an increased incidence or severity of infection might have been predicted with an effective TNF-blocking agent. In any event, we saw only two, relatively mild infective events in the open-label study, both of which responded promptly to treatment. In the placebo-controlled trial, one infective event was seen in the high-dose cA2 group, five in the low-dose cA2 group, and one among patients treated with placebo. These were for the most part trivial upper respiratory tract or skin infections, but one case of pneumonia occurred in the low-dose cA2 group. While the experience is still too limited to reach a firm conclusion on the effect of cA2 on infection rate, the events observed to date would not be limiting for the use of cA2 in most patients.

Regular patient interviews, clinical examination, and laboratory testing have revealed no other consistent pattern of adverse effects associated with cA2. One intriguing observation, however, has been the development in two patients of autoantibodies normally associated with systemic lupus erythematosus (SLE) (15,27). The first patient had long-standing seropositive polyarticular disease with extra-articular features, while the second had seronegative RA limited to the joints. Immunological testing prior to trial entry revealed a weakly positive ANA in both cases but no specific autoantibodies. Six weeks following treatment with cA2, however, both patients developed strongly positive fluorescent antinuclear antibody tests together with significantly elevated levels of antibodies to double-stranded DNA. Additional changes in the first patient included the development of anticardiolipin antibodies, elevation in serum immunoglobulins and circulating immune complexes, and reduction in serum complement components. Despite these changes, no clinical features of SLE developed and the laboratory changes eventually returned to normal.

These findings are of interest in view of experimental data from NZB/NZW F_1 (lupus prone) mice, where TNF protects against the development of autoimmunity and anti-TNF accelerates the development of autoimmune nephritis (reviewed in Ref. 27). Whereas blockade of TNF-α is clearly beneficial in RA, we speculate that it may promote the development of lupus-associated autoimmunity in some patients and is not likely to be a therapy of choice in SLE itself.

IV. MECHANISMS OF ACTION OF cA2

A number of potential mechanisms for the action of cA2 can be proposed, based on the known actions of TNF-α. Many lines of evidence are converging, however, to suggest that regulation of leukocyte emigration is a prime mode of action for this antibody. cA2 induces changes in circulating leukocyte counts within 24 hr of administration, including a rise in circulating lymphocytes consistent with a sudden reduction in their ability to recirculate (unpublished data). Immunohistochemical studies performed on synovial biopsies taken pre- and post-treatment with cA2 have demonstrated a reduction in mononuclear cell infiltrates within the synovial tissue and a down-regulation of endothelial cell adhesion molecules (28). These findings are consistent with the idea that blockade of TNF action allows endothelial deactivation and therefore interference with the disordered leukocyte migration that occurs at the inflammatory site. Since cA2 inhibits IL-6 synthesis, at least as judged by serum IL-6 levels, experiments are underway to determine whether synovial production of other cytokines, including those chemotactic for blood leukocytes, might also be down-regulated following treatment with cA2, providing a second mechanism for reduced leukocyte migration.

Other outcomes resulting from interference with TNF action in the fluid phase might include a reduction in myeloid cell activation and its consequences in terms of inflammatory mediator production and inhibition of fibroblast proliferation and of the processes of cartilage and bone degradation.

Another potential mechanism of action of cA2 involves the direct binding of cA2 to TNF expressed at the cell membrane, with consequent complement-mediated or antibody-dependent cell lysis. Treatment of patients with depleting MAb to T-cell antigens induces rapid lysis of circulating leukocytes and a clinical syndrome thought to be due to release of cytokines and other mediators from damaged cells. Although the clinical syndrome is not seen following treatment with cA2, rapid falls are seen in circulating granulocyte counts (unpublished data), a finding that might be consistent with in vivo cell lysis.

cA2 may also modulate cell function following the binding of antibody to cells expressing Fc receptors. While this mechanism has not been excluded by the controls included in our trials so far, the failure of a similar MAb with the same Fc but different Fv specificity (CD4) to induce clinical improvements in similar patient groups suggests that this mechanism is unlikely to be prominent.

V. CLINICAL UTILITY OF cA2 AND FUTURE STUDIES

The data presented suggest that systemically administered cA2 may find a role in clinical practice in the short-term suppression of synovitis. Clinical situations in which such an approach may be valuable include the induction of remission in a patient at first presentation or the management of a flare during the course of the disease. The use of cA2 in this situation has been supported by our experience with the repeated use of cA2 in patients having disease flares (24).

Short-term disease suppression might also be of use in the management of certain extra-articular features of RA, particularly rheumatoid vasculitis. TNF has important actions on endothelial function that are likely to be relevant to the pathogenesis of vasculitis. Among these actions are the induction of a procoagulant state at the endothelial surface; the induction of endothelial adhesion molecules involved in leukocyte trafficking, including E-selectin, ICAM-1, and VCAM-1; and the induction of leukocyte chemokines such as IL-8 (29,30). More direct evidence for the involvement of TNF in systemic vasculitis was provided by Deguchi et al. (31), who demonstrated enhanced TNF-α gene transcription in peripheral blood mononuclear cells in patients with vasculitis, compared with healthy subjects. Immunohistological studies in temporal arteritis have confirmed the presence of TNF within the inflammatory lesion in this condition (32). Although the data are incomplete, the evidence suggests that TNF is involved in the pathogenesis of systemic vasculitis. Blockade of TNF action with cA2 may therefore provide a new means of treating this condition.

The possibility of injecting cA2 into individual joints, providing short-term local control of synovitis, is also of interest but has not yet been tested in the clinic. Such an approach would need to be shown to be more efficacious, or associated with fewer side effects than intra-articular corticosteroids, in order for it to be a significant clinical advance.

A further use for cA2 may be in the management of secondary amyloidosis. Estimates of the prevalence of secondary (AA) amyloidosis in RA have varied widely, but a recent aspiration biopsy study in hospitalized patients (33) and a separate autopsy study (34) have suggested rates of histological amyloidosis as high as 20%. Clinically significant amyloidosis is less frequent, but this condition contributes to the excess mortality in adult RA (35) and particularly in juvenile chronic arthritis (36). These facts reflect the serious nature of secondary amyloidosis in rheumatic diseases and the limited success with current therapeutic modalities (reviewed in Ref. 37). A major factor in the development of secondary amyloidosis in RA and other inflammatory disorders such as Crohn's disease and familial Mediterranean fever is the presence of a continuing acute-phase response with elevation in SAA levels. In view of the potent effects of cA2 in the regulation of SAA, successful long-term application of this drug might be of benefit in the treatment or prevention of amyloidosis.

Although these areas are of interest, the major challenge for future studies is the attainment of long-term disease suppression with modification of disease outcome. The important actions of TNF in promoting bone and cartilage destruction (reviewed in Ref. 8) make this cytokine a promising target in any rational attempt to alter disease outcome. Further support for this notion is provided by the dramatic response of acute-phase measures to treatment with cA2. The relationship between acute-phase proteins and joint damage in RA has been discussed in a recent review (38). Levels of acute-phase proteins such as CRP reflect disease activity in RA and correlate with and may be predictive of radiographic progression. In patients who respond to disease-modifying drugs, the decrease in CRP levels correlates with the slowing of radiographic progression. On this basis, cA2 could be expected to modify disease if successful long-term application of the antibody can be achieved.

A limiting factor in the long-term use of cA2 is likely to be the induction of antiglobulin responses. Although cA2 has not proved especially immunogenic following single infusions (only one of 20 patients in the open-label trial developed an antiglobulin response), repeated use has resulted in a 50% incidence of antiglobulin responses, most of low titer (24). Further studies involving regular repeated therapy with cA2 will be required to determine whether antiglobulin responses to cA2 will limit therapy, either by interfering with antibody-TNF binding (and hence efficacy) or by the development of allergy.

A number of approaches could be adopted to circumvent the development of antiglobulin responses to cA2. The concurrent administration of traditional immunosuppressive drugs such as azathioprine or cyclosporine might be considered, but would be associated with the usual problems of nonspecific immunosuppression. An intriguing alternative approach involves the administration of cA2 together with a T-cell-directed MAb such as anti-CD4. Experiments performed in mice have demonstrated that such T-cell-directed therapy can result in tolerance to a coadministered antigen, possibly through the mechanism of suppressor-T-cell development (39). In the CIA model, coadministration of anti-CD4 and a hamster antimurine anti-TNF (TN3) was beneficial in a number of ways, including an enhancement of the clinical efficacy of suboptimal doses of TN3, a prolongation of the response duration, and most important, a reduction in the murine anti-TN3 antiglobulin response (40). Similar combination therapy in humans, perhaps utilizing a nondeleting anti-CD4 together with cA2, remains a high priority among future studies.

Other means of reducing the immunogenicity of cA2 might include the further "humanization" of the antibody such that only the antigen-binding sites remain of murine origin. Experience with a humanized MAb to the pan-leukocyte antigen CDw52 (CAMPATH-1H; 41), however, has shown a similar rate of antiglobulin development following repeated therapy to that seen with cA2, suggesting that humanization alone is unlikely to be effective. Finally, alternative means of blocking TNF function, including the use of soluble TNF receptors or of traditional pharmaceuticals that interfere with TNF/receptor binding, are under development and may emerge in the future.

VI. SUMMARY

The potent inflammatory and destructive properties of TNF, together with the results of TNF overexpression or inhibition in animal models of arthritis, have provided a strong rationale for the specific targeting of this cytokine in RA. We have exploited the properties of a chimeric human/mouse MAb to TNF-α (cA2) to investigate both the role of this cytokine in disease and the outlook for specific TNF-blocking therapy in clinical practice. The results of a small, open-label pilot study and of a larger multicenter, randomized, controlled clinical trial comparing cA2 with placebo have confirmed that this antibody is highly effective in the short-term suppression of disease, inducing rapid and marked changes in the clinical and biochemical indices of disease activity. Although clinical responses to a single cycle of treatment with cA2 are short-lived, averaging 3 months, we have demonstrated responses to repeated cycles of therapy, raising the possibility that regular treatment with cA2 may achieve long-term disease suppression.

The use of cA2 has been well tolerated in each of these studies, without evidence of opportunistic infection or other features of generalized immunosuppression. Interestingly, some patients treated with cA2 have developed lupus-associated autoantibodies, in keeping with the concept that TNF may be protective for the development of SLE.

Future studies with cA2 will aim to further define the position of this agent in different clinical settings in RA and will include intra-articular administration, use in the setting of extra-articular disease and secondary amyloidosis, and investigation of different strategies to avoid the development of sensitization with long-term use.

It is of some significance that inhibition of action of a single cytokine can induce such profound changes in disease activity in RA, where many inflammatory mediators are produced. The data strongly support the contention that TNF-α plays a key role in this disease.

REFERENCES

1. Gregersen PK, Silver J, Winchester RJ. The shared epitope hypothesis. Arthritis Rheum 1987; 30:1205–1213.
2. Saxne T, Palladino MA Jr, Heinegard D, Talal N, Wollheim FA. Detection of tumor necrosis factor α but not tumor necrosis factor β in rheumatoid arthritis synovial fluid and serum. Arthritis Rheum 1988; 31:1041–1045.
3. Buchan G, Barrett K, Turner M, Chantry D, Maini RN, Feldmann M. Interleukin-1 and tumour necrosis factor mRNA expression in rheumatoid: prolonged production of IL-1α. Clin Exp Immunol 1988; 73:449–453.
4. Husby G, Williams RC Jr. Synovial localization of tumor necrosis factor in patients with rheumatoid arthritis. J Autoimmun 1988; 1:363–371.
5. Chu CQ, Field M, Feldmann M, Maini RN. Localization of tumor necrosis factor α in synovial tissues and at the cartilage-pannus junction in patients with rheumatoid arthritis. Arthritis Rheum 1991; 34:1127–1132.
6. Deleuran BW, Chu C-Q, Field M, et al. Localization of tumor necrosis factor receptors in the synovial tissue and cartilage-pannus junction in patients with rheumatoid arthritis. Arthritis Rheum 1992; 35:1170–1178.
7. Cope AP, Aderka D, Doherty M, et al. Increased levels of soluble tumor necrosis factor receptors in the sera and synovial fluid of patients with rheumatoid diseases. Arthritis Rheum 1992; 35:1160–1169.
8. Vassalli P. The pathophysiology of tumor necrosis factors. Annu Rev Immunol 1992; 10:411–452.
9. Cooper WO, A. FR, Gates CA, Cremer MA, Townes AS. Acceleration of onset of collagen-induced arthritis by intra-articular injection of tumour necrosis factor or transforming growth factor-beta. Clin Exp Immunol 1992; 89:244–250.
10. Keffer J, Probert L, Cazlaris H, et al. Transgenic mice expressing human tumour necrosis factor: a predictive genetic model of arthritis. EMBO J 1991; 10:4025–4031.

11. Williams RO, Feldmann M, Maini RN. Anti-tumor necrosis factor ameliorates joint disease in murine collagen-induced arthritis. Proc Natl Acad Sci USA 1992; 89:9784–9788.
12. Thorbecke GJ, Shah R, Leu CH, Kuruvilla AP, Hardison AM, Pallandino MA. Involvement of endogenous tumor necrosis factor α and transforming growth factor β during induction of collagen type II arthritis in mice. Proc Natl Acad Sci USA 1992; 89:7375–7379.
13. Wooley PH, Dutcher J, Widmer MB, Gillis S. Influence of a recombinant human soluble tumor necrosis factor receptor Fc fusion protein on type II collagen-induced arthritis in mice. J Immunol 1993; 151:6602–6607.
14. Knight DM, Trinh H, Le J, et al. Construction and initial characterization of a mouse-human chimeric anti-TNF antibody. Mol Immunol 1993; 30: 1443–1453.
15. Elliott MJ, Maini RN, Feldmann M, et al. Treatment of rheumatoid arthritis with chimeric monoclonal antibodies to tumor necrosis factor α. Arthritis Rheum 1993; 36:1681–1690.
16. Paulus HE, Egger MJ, Ward JR, Williams HJ, Group CSSoRD. Analysis of improvement in individual rheumatoid arthritis patients treated with disease-modifying antirheumatic drugs, based on the findings in patients treated with placebo. Arthritis Rheum 1990; 33:477–480.
17. Gauldie J, Richards C, Harnish D, Lansdorp P, Baumann H. Interferon β_2/B-cell stimulatory factor type 2 shares identity with monocyte-derived hepatocyte-stimulating factor and regulates the major acute phase protein response in liver cells. Proc Natl Acad Sci USA 1987; 84:7251–7255.
18. Guerne P-A, Zuraw BL, Vaughan JH, Carson DA, Lotz M. Synovium as a source of interleukin 6 in vitro. J Clin Invest 1989; 83:585–592.
19. Elliott MJ, Maini RN, Feldmann M, et al. Randomised double-blind comparison of chimeric monoclonal antibody to tumour necrosis factor α (cA2) versus placebo in rheumatoid arthritis. Lancet 1994; 344:1105–1110.
20. Roodman GD, Johnson RA, Clibon U. Tumor necrosis factor alpha and the anemia of chronic disease: effects of chronic exposure to TNF on erythropoiesis in vivo. Adv Exp Med Biol 1989; 271:185–196.
21. Johnson RA, Waddelow TA, Caro J, Oliff A, Roodman GD. Chronic exposure to tumor necrosis factor in vivo preferentially inhibits erythropoiesis in nude mice. Blood 1989; 74:130–138.
22. Vreugdenhil G, Löwenberg B, Van Eijk HG, Swaak AJG. Tumor necrosis factor alpha is associated with disease activity and the degree of anemia in patients with rheumatoid arthritis. Eur J Clin Invest 1992; 22:488–493.
23. Elliott MJ, Maini RN. New directions for biological therapy in rheumatoid arthritis. Int Arch Allergy Immunol. 1994; 104:112–125.
24. Elliott MJ, Maini RN, Feldmann M, et al. Repeated therapy with monoclonal antibody to tumour necrosis factor α (cA2) in patients with rheumatoid arthritis. Lancet 1994; 344:1125–1128.
25. Havell EA. Evidence that tumor necrosis factor has an important role in antibacterial resistance. J Immunol 1989; 143:2894–2899.

26. Pfeffer K, Matsuyama T, Kündig TM, et al. Mice deficient for the 55 kd tumor necrosis factor receptor are resistant to endotoxic shock, yet succumb to *L. monocytogenes* infection. Cell 1993; 73:457–467.
27. Maini RN, Elliott MJ, Charles PJ, Feldmann M. Immunological intervention reveals reciprocal roles for TNFα and IL-10 in rheumatoid arthritis and SLE. Springer Semin Immunopathol 1994; 16:327–336.
28. Maini RN, Elliott MJ, Brennan FM, et al. Monoclonal anti-TNFα antibody as a probe of pathogenesis and therapy of rheumatoid disease. Immunol Rev 1995; 144:195–223.
29. Savage CO, Cooke SP. The role of endothelium in systemic vasculitis. J Autoimmun 1993; 6:237–249.
30. Robertson CR, McCallum RM. Changing concepts in pathophysiology of the vasculitides. Curr Opin Rheumatol 1994; 6:3–10.
31. Deguchi Y, Shibata N, Kishimoto S. Enhanced expression of the tumour necrosis factor/cachectin gene in peripheral blood mononuclear cells from patients with systemic vasculitis. Clin Exp Immunol 1990; 81:311–314.
32. Cook A, Gallagher G, Field M. Localisation of tumour necrosis factor α and its receptors in temporal arteritis. Clin Rheumatol 1994; 13:162.
33. Pai S, Helin H, Isomaki H. Frequency of amyloidosis in Estonian patients with rheumatoid arthritis. Scand J Rheumatol 1993; 22:248–249.
34. Suzuki A, Ohosone Y, Obana M, et al. Cause of death in 81 autopsied patients with rheumatoid arthritis. J Rheumatol 1994; 21:33–36.
35. Koota K, Isomäki HA, Mutru O. Death rate and causes of death in patients with rheumatoid arthritis. Scand J Rheumatol 1975; 4:205–208.
36. Baum J, Gutowska G. Death in juvenile rheumatoid arthritis. Arthritis Rheum 1977; 20(Suppl):253–255.
37. Husby G. Amyloidosis. Semin Arthritis Rheum 1992; 22:67–82.
38. Blackburn WD Jr. Validity of acute phase proteins as markers of disease activity. J Rheumatol 1994; 42(Suppl):9–13.
39. Waldmann H, Cobbold S. The use of monoclonal antibodies to achieve immunological tolerance. Immunol Today 1993; 14:247–251.
40. Williams RO, Mason LM, Feldmann M, Maini R. Synergy between anti-CD4 and anti-tumor necrosis factor in the amelioration of established collagen-induced arthritis. Proc Natl Acad Sci USA 1994; 91:2762–2766.
41. Issacs JD, Watts RA, Hazleman BL, et al. Humanised monoclonal antibody therapy for rheumatoid arthritis. Lancet 1992; 340:748–752.

11
Engineered Human Anti-Tumor Necrosis Factor-Alpha (TNFα) Antibody, CDP571, in Rheumatoid Arthritis

ERNEST H. S. CHOY and GABRIEL S. PANAYI
Guy's Hospital, London, England

I. INTRODUCTION

Tumor necrosis factor-alpha (TNF-α) is a proinflammatory cytokine produced mainly by monocytes/macrophages but also by B cells, T cells, and fibroblasts. TNF-α production can be stimulated by interferon-gamma (IFN-γ) and lipopolysaccharide (LPS). The latter is a component of gram-negative bacterial cell wall and hence TNF-α has been implicated in the pathogenesis of endotoxic shock syndrome (1). TNF-α is a 17-kDa protein composed of three subunits and it binds to TNF receptors (TNFR) on the surface of cell membranes. TNFR has two types: p55 and p75. Each contains four cysteine-rich repeats in its extracellular domain, a transmembrane domain, which is susceptible to proteinase lysis, and a cytoplasmic tail capable of signal transduction. The intracellular domains of p55 and p75 TNFR do not share any significant homology, which suggests they have different mechanisms of signal transduction pathways. Hence, blockage of binding of TNF-α to both p55 and p75 TNFR is essential to abolish its effects completely. Soluble forms of both TNFR are found and are thought to function as naturally occurring inhibitors of TNF-α.

In addition to the soluble form of TNF-α, there is also a 28-kDa membrane-bound form. Recently, it has been shown that the membrane-anchoring part is susceptible to lysis by a serine metalloproteinase and that inhibitors of the latter could suppress TNF-α release (2,3). Indeed, in an animal model of endotoxic shock, metalloproteinase inhibitors have been shown to be protective (4).

II. FUNCTION OF TNF-α AND ITS ROLE IN INFLAMMATION AND ARTHRITIS

TNFR are expressed by a large variety of cells including monocytes/macrophages, T cells, fibroblasts, osteoblasts, and endothelial cells. Consequently, TNF-α has diverse actions and is capable of stimulating a wide variety of cell types. Functionally, it shares many similarities with interleukin-1 (IL-1). TNF-α is a proinflammatory cytokine and it has been implicated as the pivotal cytokine in rheumatoid arthritis (RA) (5). We shall first discuss some of its actions and then its role in RA.

A. Basic Science

1. Mononuclear Cells

TNF-α is an autocrine stimulator as well as a potent inducer of other proinflammatory cytokines including IL-1, interleukin-6 (IL-6), interleukin-8 (IL-8), and granulocyte colony-stimulating factor (GM-CSF) (5). Hence, TNF-α release results in amplification of inflammation. IL-6 stimulates hepatocytes to release acute-phase reactants such as C-reactive protein (CRP), β_2 microglobulin, and serum amyloid A protein, which explains their close correlation with disease activity in RA (6). Furthermore, IL-6 stimulates megakaryocytes in the bone marrow to proliferate and differentiate into platelets resulting in the thrombocytosis commonly seen in acute inflammation (7). IL-8 is a potent neutrophil chemokine (8) and its level is raised in RA synovium (9). GM-CSF is a potent stimulator of monocytes/macrophages and leads to the production of both IL-1 and IL-1 receptor antagonists (10,11).

2. Endothelial Cells

Another mechanism through which TNF-α accentuates inflammation is stimulation of endothelial cells. Both IL-1 and TNF-α are potent inducers of adhesion molecules on the surface of endothelial cells (12). These include E-selectin (formerly ELAM-1), intercellular adhesion molecule-1 (ICAM-1), and vascular adhesion molecule-1 (VCAM-1). These interact with their respective ligands on the surface of leukocytes (Table 1) resulting in increased trafficking of the latter into inflammatory sites such as the rheumatoid joint.

Table 1 Adhesion Molecules Involved in Leukocyte Migration into the Joint in Rheumatoid Arthritis

Endothelial cell adhesion molecules	Leukocyte adhesion molecules
E-selectin	Sialyl Lewis X residues
Intercellular adhesion molecule (ICAM)-1 and 2	Leukocyte function–associated antigen-1 (LFA-1)
Vascular adhesion molecule (VCAM)-1	$\alpha_1\beta_4$ intergrin

The adhesion molecules are arranged as receptor/ligand pairs.

3. Neuroendocrine-Immune Axis

TNF-α, IL-1, and IL-6 are known to stimulate the hypothalamus to release corticotropin-releasing hormones (CRH) resulting in the release of adrenocorticotropic hormone from the pituitary (13). The latter stimulates the adrenal cortex to release cortisol. This relationship between the immune system and the neuroendocrine axis is important in the adaptive response to inflammation.

4. Mesenchymal Cells

TNF-α and IL-1 are potent stimulators of mesenchymal cells such as fibroblasts/ synoviocytes and chondrocytes, which release matrix metalloproteinases (MMP) (14,15). In RA, IL-1- and TNF-α-stimulated release of MMP-1 (collagenase) and MMP-3 (stromelysin), particularly at the cartilage-pannus junction, is thought to be responsible for joint damage. In addition, both TNF-α and IL-1 are known to inhibit the function of osteoblasts and may have an important role in osteoporosis (16,17). TNF-α stimulates the development of osteoclasts but this effect may be an indirect action through the production of interleukin-11 (18).

B. Studies in Arthritis

1. Rheumatoid Cachexia

During active RA, many patients feel unwell and may have systemic features such as fever and weight loss. The latter can be very drastic and is known as rheumatoid cachexia. TNF-α has been shown to play an important role in its pathogenesis (19).

2. TNF-α in Animal Models

Keffer et al. showed that TNF-α transgenic mice that expressed human TNF-α developed spontaneous inflammatory and destructive polyarthritis similar to RA (20). Treatment of these animals with anti-TNF-α antibody (Ab) inhibited development of arthritis completely. Anti-TNF-α MAbs have also been used in a number of animal models of RA, including collagen type II arthritis (21), and have ameliorated disease activity. Interestingly, in this model, anti-TNF and anti-CD4 Abs seem to synergistic (22).

3. TNF-α and RA

As a proinflammatory cytokine, TNF-α has an important role in rheumatoid synovitis. In vivo, serum levels of TNF-α are raised in patients with active disease and correlate closely with the acute-phase response (23). More importantly, there is an abundance of TNF-α in the rheumatoid synovial fluid and membrane (24). Brennan et al. have suggested that there is a cytokine hierarchy in RA with TNF-α as the main cytokine responsible for the induction of other proinflammatory cytokines. Hence TNF-α blockage ought to be more efficacious than inhibition of other cytokines (7).

III. CLINICAL STUDIES OF ANTI-TNF-α ANTIBODY

An open study of a chimeric anti-TNFα Ab, cA2, has suggested that doses of 10 or 20 mg/kg produce significant disease improvement (25). Subsequently, a multicenter, placebo-controlled trial involving 72 patients confirmed the results of the open study (26). When cA2 was given, it rapidly produced significant disease improvement, and clinical improvement lasted at least 4 weeks. In seven patients who received retreatment with cA2, the therapy was again effective (27). However, three of seven patients developed autoantibodies to double-stranded DNA. The mechanism for this is unknown.

We shall now describe our own study with engineered human anti-TNF-α antibody, CDP571, in greater detail.

IV. CDP571, AN ENGINEERED HUMAN ANTI-TNF-α ANTIBODY

CDP571 (Celltech Therapeutics Limited) is an engineered human anti-TNF-α Ab. It is a recombinant antibody containing the complementarity-determining regions of a murine anti-TNF-α Ab grafted into a human kappa light chain and IgG4 heavy chain. The serum half-life of CDP571 was about 10–14 days in a phase I study. In vitro, it binds to and neutralizes the biological effects of TNF-α.

A. Clinical Experience with CDP571

In collaboration with our colleagues Dr. Elizabeth Rankin and Prof. David Isenberg at the Bloomsbury Rheumatology Unit, London, we conducted a double-blind, placebo-controlled trial to investigate the therapeutic efficacy and safety of

CDP571. We recruited a total of 36 patients with active RA. All patients had failed at least one disease-modifying antirheumatic drug (DMARD) and had had RA for less than 10 years. DMARDS were stopped at least 4 weeks prior to treatment. Patients were randomized to receive either placebo or CDP571 at doses of 0.1, 1, or 10 mg/kg (Fig. 1) administered as a single intravenous infusion. Clinical assessments were performed at 0, 1, 2, 4, and 8 weeks. These were based on the European League Against Rheumatism (EULAR) core data set for assessment of RA disease activity (28): the number of tender joints (TJ) and swollen joints (SJ) (maximum 28 joints), visual analog scale of pain (VASP), and patient global assessment (PGA). In addition, we also measured the duration of early-morning stiffness (minutes), erythrocyte sedimentation rate (ESR), and C-reactive protein (CRP). After the initial double-blind period, patients were treated openly with either 1 or 10 mg/kg of anti-TNF-α Ab.

B. Clinical Results

Twelve patients received placebo infusions and eight patients in each group received the three doses of CDP571. Treatment was well tolerated. Two placebo-treated patients withdrew from the study because of worsening of disease. Response to CDP571 was rapid in that most assessments showed a reduction in disease activity by week 1. The 10 mg/kg group showed the greatest improvement. Patients who received 0.1 mg/kg of CDP571 failed to show any clinical improvement. Those who received 1 mg/kg showed some clinical improvement, but the clinical response was clearly inferior to the response of those who had received 10 mg/kg. In the latter group, by week 2, there were statistically

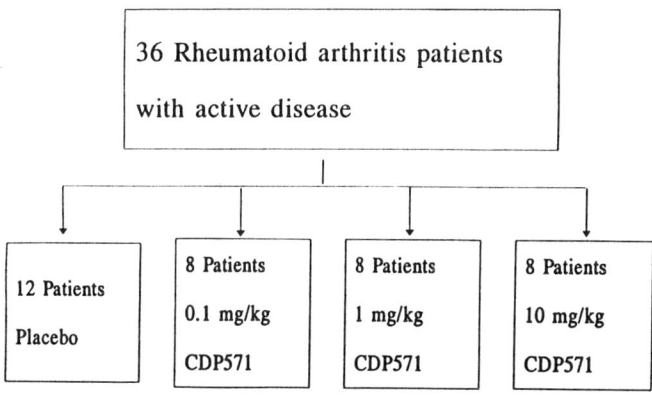

Figure 1 Treatment regimen.

significant reductions in the VASP and in the number of TJ versus the placebo group. There was a trend for clinical improvement in both PGA and SJ although this did not reach statistical significance. The acute-phase response, as measured by ESR and CRP, showed rapid reduction, the ESR showing a statistically significant reduction by week 2 and the CRP by week 1, which lasted at least 4 weeks. A significant reduction in serum IL-6 was also seen. Using the EULAR-defined disease activity score (28) to summarize patients' clinical response, patients who received 10 mg/kg of CDP571 showed a statistically significant clinical improvement at week 2 (Fig. 2) versus the placebo group; those who received 1 mg/kg showed a trend toward improvement, although this did not reach statistical significance, while those on 0.1 mg/kg showed no change.

Serum levels of CDP571 showed a clear dose response and the half-life of CDP571 was 1–2 weeks. After a single treatment, a low-level anti-idiotypic response toward CDP571 was detected in the majority of patients. Interestingly, none of the 24 patients developed anti-dsDNA Ab, unlike the experience with the Kennedy Institute group where three of eight patients who received cA2 did so (27).

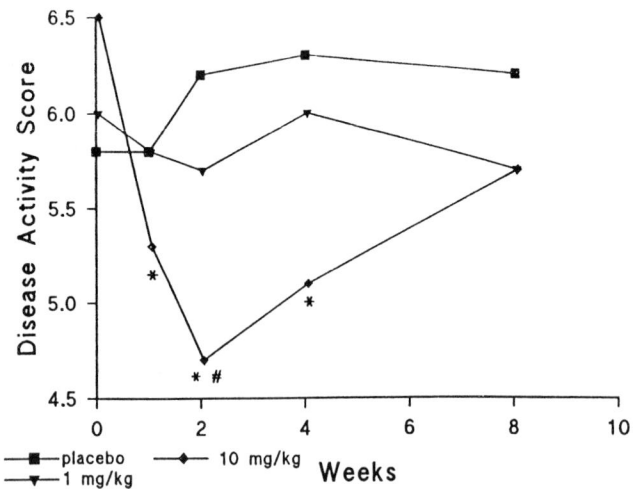

Figure 2 Changes in disease activity score in placebo, 1 mg/kg, and 10 mg/kg CDP571-treated patients. ■, placebo-treated patients; ▼, patients treated with 1 mg/kg of CDP571; ♦, patients treated with 10 mg/kg of CDP571. *, p < 0.05; #, p < 0.01.

V. SUMMARY

Treatment with the engineered human TNF-α Ab CDP571 produced rapid and significant clinical disease improvement lasting approximately 4 weeks. This confirmed the results of the multicenter, double-blind, placebo-controlled trial by Elliott et al., which showed that a chimeric anti-TNF-α Ab, cA2, was clinically efficacious at a dose of 10 mg/kg (26). The results of these studies support the hypothesis that TNF-α is a proinflammatory cytokine in RA and that its inhibition can produce clinical improvement. However, a number of important questions must be addressed before the exact role of anti-TNF-α Ab in the treatment of RA can be determined. First, since disease relapse occurs approximately 4–6 weeks after treatment, the efficacy of repeated treatments needs to be established. Studies using chimeric anti-TNF-α Ab suggested that the duration of clinical efficacy may shorten with repeated treatments (27). Second, since TNF-α is an important cytokine in the body's defense against bacteria and parasitic infections, the long-term effect of TNF-α blockage needs to be assessed. Third, since TNF-α is involved in the immune response, the development of autoantibodies to dsDNA on repeated treatments raises important concerns regarding the safety of long-term repeated administration at least of cA1 antibody (27); none of the patients treated with CDP571 have developed anti-dsDNA antibodies so far.

VI. CONCLUSION

TNF-α is an important anti-inflammatory target in RA. Engineered human anti-TNF-α Ab, CDP571, produced short-term disease improvement. However, more studies are necessary to establish its potential role in the long-term management of RA.

REFERENCES

1. Carswell EA, Old LJ, Kassel RL, Green S, Fiore N, Williamson B. An endotoxin-induced serum factor that causes necrosis of tumors. Proc Natl Acad Sci USA 1975; 72:3666–3670.
2. McGeehan GM, Becherer JD, Bast RC Jr, Boyer CM, Champion B, Connolly KM, Conway JG, Furdon P, Karp S, et al. Regulation of tumour necrosis factor-alpha processing by a metalloproteinase inhibitor. Nature 1994; 370:558–561.
3. Gearing AJ, Beckett P, Christodoulou M, Churchill M, Clements J, Davidson AH, Drummond AH, Galloway WA, Gilbert R, Gordon JL, et al. Processing of tumour necrosis factor-alpha precursor by metalloproteinases. Nature 1994; 370:555–557.
4. Mohler KM, Sleath PR, Fitzner JN, Cerretti DP, Alderson M, Kerwar SS, Torrance DS, Otten-Evans C, Greenstreet T, Weerawarna K, et al. Protection against a lethal dose of endotoxin by an inhibitor of tumour necrosis factor processing. Nature 1994; 370:218–220.

5. Brennan FM, Maini RN, Feldmann M. TNF alpha—a pivotal role in rheumatoid arthritis: Br J Rheumatol 1992; 31:293–298.
6. Cohick CB, Furst DE, Quagliata S, Corcoran KA, Steere KJ, Yager JG, Lindsley HB. Analysis of elevated serum interleukin-6 levels in rheumatoid arthritis: correlation with erythrocyte sedimentation rate or C-reactive protein. J Lab Clin Med 1994; 123:721–727.
7. van Snick J. Interleukin 6: an overview. Annu Rev Immunol 1990; 8:253–278.
8. Huber AR, Kunkel SL, Todd RF, Weiss SJ. Regulation of transendothelial neutrophil migration by endogenous interleukin-8. Science 1991; 254:99–102.
9. Deleuran B, Lemche P, Kristensen M, Chu CQ, Field M, Jensen J, Matsushima K, Stengaard-Pedersen K. Localisation of interleukin 8 in the synovial membrane, cartilage-pannus junction and chondrocytes in rheumatoid arthritis. Scand J Rheumatol 1994; 23:2–7.
10. Janson RW, Joslin FG, Arend WP. The effects of differentiating agents on IL-1 beta production in cultured human monocytes. J Immunol 1990; 145:2161–2166.
11. Kasinrerk W, Baumruker T, Majdic O, Knapp W, Stonkinger H. CD1 molecule expression on human monocytes induced by granulocyte-macrophage colony-stimulating factor. J Immunol 1993; 150:579–584.
12. Picker LJ. Mechanisms of lymphocyte homing. Curr Opin Immunol 1992; 4: 277–286.
13. Jones TH, Kennedy RL. Cytokines and hypothalamic-pituitary function. Cytokine 1993; 5:531–538.
14. Shinmei M, Masuda K, Kikuchi T, Shimomura Y. The role of cytokines in chondrocyte mediated cartilage degradation. J Rheumatol 1989; 18(Suppl):32–34.
15. Shingu M, Nagai Y, Isayama T, Naono T, Nobunaga M. The effects of cytokines on metalloproteinase inhibitors (TIMP) and collagenase production by human chondrocytes and TIMP production by synovial cells and endothelial cells. Clin Exp Immunol 1993; 94:145–149.
16. Pacifici R, Brown C, Puscheck E, Friedrich E, Slatopolsky E, Maggio D, McCracken R, Avioli LV. Effect of surgical menopause and estrogen replacement on cytokine release from human blood mononuclear cells. Proc Natl Acad Sci USA 191; 88: 5134–5138.
17. Zarrabeitia MT, Riancho JA, Amado JA, Napal J, Gonzalez-Macias J. Cytokine production by peripheral blood cells in postmenopausal osteoporosis. Bone Mineral 1991; 14:161–167.
18. Girasole G, Passeri G, Jilka RL, Manolagas SC. Interleukin-11: a new cytokine critical for osteoclast development. J Clin Invest 1994; 93:1516–1524.
19. Roubenoff R, Roubenoff RA, Cannon JG, Kehayias JJ, Zhuang H, Dawson-Hughes B, Dinarello CA, Rosenberg IH. Rheumatoid cachexia: cytokine-driven hypermetabolism accompanying reduced body cell mass in chronic inflammation. J Clin Invest 1994; 93:2379–2386.
20. Keffer J, Probert L, Cazlaris H, Georgopoulos S, Kaslaris E, Kioussis D, Kollias G. Transgenic mice expressing human tumour necrosis factor: a predictive genetic model of arthritis. EMBO J 1991; 10:4025–4031.

21. Williams RO, Feldmann M, Maini RN. Anti-tumor necrosis factor ameliorates joint disease in murine collagen-induced arthritis. Proc Natl Acad Sci USA 1992; 89:9784–9788.
22. Williams RO, Mason LJ, Feldmann M, Maini RN. Synergy between anti-CD4 and anti-tumor necrosis factor in the amelioration of established collagen-induced arthritis. Proc Natl Acad Sci USA 1994; 91:2762–2766.
23. Espersen GT, Vestergaard M, Ernst E, Grunnet N. Tumour necrosis factor alpha and interleukin-2 in plasma from rheumatoid arthritis patients in relation to disease activity. Clin Rheumatol 1991; 10:374–376.
24. Chu CQ, Field M, Allard S, Abney E, Feldmann M, Maini RM. Detection of cytokines at the cartilage/pannus junction in patients with rheumatoid arthritis: implications for the role of cytokines in cartilage destruction and repair. Br J Rheumatol 1992; 31: 653–661.
25. Elliott MJ, Maini RN, Feldmann M, Long-Fox A, Charles P, Katsikis P, Walker J, Bijl H, Ghrayeb J, et al. Treatment of rheumatoid arthritis with chimeric monoclonal antibodies to tumor necrosis factor alpha. Arthritis Rheum 1993; 36:1681–1690.
26. Elliott MJ, Maini RN, Feldmann M, Kalden JR, Antoni C, Smolen JS, Leeb B, Breedveld FC, Macfarlane JD, Bijl H, et al. Randomised double-blind comparison of chimeric monoclonal antibody to tumour necrosis factor α (cA2) versus placebo in rheumatoid arthritis. Lancet 1994; 344:1105–1110.
27. Elliott MJ, Maini RN, Feldmann M, Long-Fox A, Charles P, Bijl H, Woody JN. Repeated therapy with monoclonal antibody to tumour necrosis factor α (cA2) in patients with rheumatoid arthritis. Lancet 1994; 344:1125–1127.
28. Scott DL, Panayi GS, Van Riel PLCM, Smolen J, van-de PB. Disease activity in rheumatoid arthritis: preliminary report of the Consensus Study Group of the European Workshop for Rheumatology Research. Clin Exp Rheumatol 1992; 10:521–525.

12
Clinical Experience with Recombinant Human Interleukin-1 Receptor Type I (Rhu IL-1RI) in Patients with Rheumatoid Arthritis

RICHARD M. POPE, BARBARA DREVLOW, JENNIFER CAPEZIO, and ROSA LOVIS
Northwestern University Medical School, Chicago, Illinois

ALAN LANDAY
Rush-Presbyterian–St. Luke's Medical Center, Chicago, Illinois

I. INTRODUCTION

Interleukin-1 (IL-1) is a mediator of the inflammation that may be important in rheumatoid arthritis (RA). It is produced locally by synovial lining cells and by synovial tissue macrophages. Locally produced IL-1 may contribute to the pathogenesis of RA by stimulation of the secretion of stromelysin and collagenase by fibroblasts. Furthermore, IL-1 contributes to the induction of other inflammatory cytokines by macrophages and fibroblasts, including IL-8, TNF-α, and IL-6 (reviewed in Ref. 1). Natural inhibitors of IL-1, including IL-1 receptor (IL-1R) and IL-1R antagonist (IL1-Ra), are produced locally in inflammatory sites in an attempt to down-regulate the activity of IL-1 (2–4). Nevertheless, an imbalance between IL-1 and its inhibitors may still occur and contribute to the chronic inflammation.

Receptors for IL-1 are found in a wide variety of cells, such as synovial fibroblasts, macrophages, chondrocytes, osteoblasts, and T cells (5,6). Both IL-1α and IL-1β bind to, and mediate their biological activities, through two specific membrane receptors (types I and II) (3,6). In an attempt to inhibit the inflammation mediated by IL-1, the cDNA encoding the type I receptor was isolated and truncated to produce a soluble protein representing the extracellular ligand-binding domain (7). Quantitative experiments showed that binding of soluble recombinant human type I IL-1 receptor (rhu IL-1RI) to IL-1α and IL-1β was comparable to that of the membrane-bound receptor, suggesting that the soluble form of rhu IL-1RI might be employed as a competitive inhibitor of binding of IL-1α and IL-1β to the membrane-associated IL-1R.

Models of acute and chronic inflammation, including two models of arthritis, have been effectively suppressed in experimental animals by treatment with IL-1RI (8,9). Beneficial effects of soluble IL-1RI were observed in the adjuvant-induced arthritis model, even when the receptor was administered after the onset of the synovitis (9). In addition, antibodies to type II collagen were reduced in the treated animals. In MRL mice that spontaneously develop autoimmune disease including nephritis and synovitis, survival, proteinuria, and synovitis were all improved in the animals that received murine IL-1R prior to disease onset (8). Rheumatoid factor concentrations were also reduced in those animals that received soluble IL-1R (8).

II. TREATMENT PROTOCOLS

We recently completed two protocols to examine the safety and efficacy of soluble rhu IL-1RI in the treatment of active RA (10,11). In an attempt to directly modulate the local environment of the joint, a single dose of rhu IL-1RI was injected intra-articularly into the inflamed knees of patients with chronic rheumatoid synovitis (10). The second study employed rhu IL-1RI, injected subcutaneously daily for 4 weeks (11).

Both protocols were double-blinded, randomized, phase I studies. All patients fulfilled the 1987 American Rheumatism Association criteria for the diagnosis of RA (12). They were classified by Steinbrocker's criteria as functional criteria I, II, or III (13). Patients in the first protocol were allowed to remain on a stable dose of methotrexate. No DMARDS were allowed in the second protocol. Patients on corticosteroids were included if they were on ≤10 mg/day of prednisone or its equivalent. Patients were maintained on a stable dose of nonsteroidal anti-inflammatory drugs.

To be included in the first protocol, active RA, as defined by swelling of at least one knee joint for which intra-articular therapy with IL-1R would be possible, was required (10). Sixteen patients were included in this study, four at each dose. The four escalating doses of rhu IL-1RI were 25, 100, 250, and 500 μg/knee. At each dose level three patients were injected with active drug and one with placebo. At the time of intra-articular injection, synovial fluid was aspirated if present. Each dose level was completed before escalation to the next. Prior to escalation, development of circulating antibodies to rhu IL-1RI was excluded.

III. RESULTS

The results of the clinical assessments of the target joint prior to rhu IL-1RI injection and at days 2 and 7 are presented in Table 1. A trend toward improvement in each parameter was noted following aspiration and injection of placebo ($n = 4$), particularly between days 0 and 2. However, the only significant change in the placebo-treated group was in walk time between days 0 and 7.

To evaluate the patients injected with the rhu IL-1RI, patients receiving the 25- and 100-μg doses were considered the low-dose treatment group, while those receiving the 250- and 500-μg doses were considered the high-dose treatment group. In the low-dose group, no change was observed for any parameter except walk time, which was significantly improved on day 2. In the high-dose group, no significant change was noted between days 0 and 2. However, by day 7, improvement in tenderness and walk time was observed. Although this improvement was significant, it was not obvious clinically to the examining physicians, such as might be observed following an intra-articular corticosteroid injection. No trends were observed at any dose level in target joint flexion, extension, measured

Table 1 Target Joint Assessment at the Time of Intra-articular Injection and After 2 and 7 Days

	Dose	Day 0	Day 2	Day 7
Pain	Placebo ($n = 4$)	1.5 ± 0.6	1.0 ± 0.0	1.3 ± 0.5
	Low ($n = 6$)	1.0 ± 0.0	0.8 ± 0.4	1.1 ± 1.0
	High ($n = 6$)	1.2 ± 1.2	1.0 ± 1.3	0.8 ± 1.2
Tenderness	Placebo	1.3 ± 1.0	0.5 ± 0.6	0.8 ± 1.0
	Low	0.8 ± 0.4	0.7 ± 0.5	1.0 ± 0.6
	High	1.0 ± 1.1	1.3 ± 1.0	0.5 ± 0.8[a]
Swelling	Placebo	2.0 ± 1.2	1.5 ± 0.6	1.8 ± 1.0
	Low	1.8 ± 1.0	1.8 ± 1.0	2.0 ± 0.9
	High	2.0 ± 0.9	2.2 ± 0.8	1.8 ± 1.2
Walk time	Placebo	12.3 ± 4.3	9.5 ± 1.2	10.8 ± 4.1[b]
	Low	14.7 ± 3.6	11.7 ± 2.6[c]	12.2 ± 2.7
	High	10.8 ± 2.7	10.7 ± 4.6	9.2 ± 2.7[b]

[a] Improved from day 2, $p = .042$.
[b] Improved from day 0, $p < .05$.
[c] Improved from day 0, $p = .003$.

circumference, or pain at rest. No effects were observed on overall disease activity as assessed by joint swelling or by the erythrocyte sedimentation rate (10).

Two patients noted worsening of the target joint following intra-articular injection of rhu IL-1RI, one at 100 µg/joint and the other at 250 µg/joint. The worsening of pain and swelling in the patient at 100 µg/joint was not apparent in the clinical scores because this individual started out at the most severe score for these parameters. However, this individual required repeat aspirations at days 2 and 7, to allow the patient to walk. Cultures of this joint fluid were repeatedly negative. Injection with intra-articular methylprednisolone on day 7 resulted in marked clinical improvement. It is not clear whether aspiration alone, of a fairly large effusion, resulted in worsening in this patient. Alternatively, it is possible that the injection of rhu IL-1RI may have contributed to the exacerbation. This might occur if the injected dose of rhu IL-1RI preferentially bound to and neutralized IL-1 receptor antagonist (IL-1Ra). This will be discussed later.

In the second protocol, rhu IL-1RI or placebo was self-administered subcutaneously, daily, for 4 weeks (11). Rhu IL-1RI was employed at four dose levels: 125, 250, 500, and 1000 $\mu g/m^2$/day. Additional patients were recruited for those patients who did not complete 4 weeks of the trial and for the highest dose level. In addition to the criteria noted in the first protocol, none of the patients were on DMARDS and at least five actively swollen joints were required for entry into the study. As in the first protocol, prior to escalation to the next dose level, the sera of treated patients were examined to exclude the development of circulating antibodies to rhu IL-1RI. No dose-limiting clinical toxicity was noted prior to escalation to the highest dose employed. Of the patients included in the second protocol, all but one (in the placebo group) had received one or more DMARDS prior to entering the study. Patients were examined weekly during the study and at 2 and 4 weeks after the medication was stopped.

The data were analyzed in two ways: the number of patients experiencing clinically relevant improvement by predetermined criteria and the mean change from baseline 4 weeks after initiation of therapy. The primary criterion of clinically relevant improvement was ≥40% improvement of the total swollen joint score (total score was the sum of each swollen joint multiplied by the severity of its swelling on a scale of +1 to +3) (11). None of the control placebo-treated and only one of the 19 patients treated at any dose of rhu IL-1RI showed ≥40% improvement of joint swelling. This patient had been treated with the highest dose (1000 $\mu g/m^2$/day) of IL-1RI.

Additional parameters used to define clinically relevant change in this protocol included 40% improvement in the number of tender joints or joints painful on motion, a two-integer improvement of the global assessments of the patient and physician (on a 1–10 scale, 1 being the best), 20% improvement of the modified Health Assessment Questionnaire (HAQ), 30% decrease of the 50-ft walk time (in seconds), a 40% reduction in the duration of morning stiffness (in hours), and a 30% reduction of the Westergren erythrocyte sedimentation rate (ESR). Prior to initiation of the study, improvement of three of these secondary criteria was also considered clinically relevant improvement. None of the placebo-treated patients and no patients treated with less than 1000 µg/m^2/day of rhu IL-1RI experienced clinically relevant improvement of any of these secondary parameters. Three of the nine patients treated with 1000 µg/m^2/day rhu IL-1RI experienced clinically relevant improvement of one parameter, while one patient improved in four parameters. The later patient was the same patient who improved ≥40% in joint swelling.

When the patients were examined by dosage group, no significant improvement in the mean scores was observed (Tables 2–4) for any clinical parameter or the ESR (11). In fact, there was a trend for worsening of many of the clinical measures. The ESR tended toward worsening in all treatment groups (Table 2). Morning stiffness and 50-ft walk time also tended toward worsening in each group. Despite the minimal reduction of the joint swelling in the 250 µg/m^2/day group, there was significant ($p < .05$) worsening of the number of tender or painful joints compared to other groups (Table 2). A number of other parameters also fared more poorly in the 250 µg/m^2/day rhu IL-1RI group. There was significant deterioration of patient global assessment in the 250 µg/m^2/day rhu IL-1RI group compared to others (Table 3). In addition, the difficulty, satisfaction, and physical

Table 2 Effects of Treatment with Rhu IL-1RI and Placebo on Pain, Swelling, and ESR

	Pain		Swelling		ESR	
	Day 1	Day 29	Day 1	Day 29	Day 1	Day 29
125 µg/m^2/day	31.1[a]	31.7	29.3	33.0	63.7	83.0
($n = 3$)	(9.1)	(12.7)	(12.5)	(8.7)	(38.1)	(55.2)
250 µg/m^2/day	6.7	17.3[b]	24.0	20.0	69.0	77.7
($n = 3$)	(5.0)	(4.6)	(8.9)	(7.2)	(61.6)	(60.6)
500 µg/m^2/day	23.7	30.3	32.0	38.0	48.3	74.3
($n = 3$)	(15.9)	(8.6)	(13.5)	(9.0)	(12.6)	(8.6)
1000 µg/m^2/day	57.1	67.8	33.8	31.8	29.6	42.3
($n = 9$)	(35.2)	(46.1)	(16.2)	(15.0)	(16.8)	(27.4)
Placebo	5.3	8.3	21.5	21.0	25.0	42.8
($n = 4$)	(3.0)	(6.1)	(6.6)	(7.2)	(25.4)	(27.5)

[a]Mean values for pain, swelling, and ESR on day 1 and day 29. The values in parentheses represent the standard deviations.
[b]Change from day 1 to day 29 significantly ($p < .05$) worse than at 125 or 1000 µg/m^2/day.

Table 3 Effects of Treatment with Rhu IL-1RI or Placebo on Stiffness, Walk Time, and Global Assessment

	A.M. stiffness		Walk time		Physician		Patient	
	Day 1	Day 29	Day 1	Day 29	Day 1	Day 29	Day 1	Day 29
125 µg/m^2/day	0.8^a	9.2	9.7	11.1	6.0	6.7	5.0	6.7
($n = 3$)	(0.8)	(12.9)	(2.31)	(2.2)	(1.0)	(1.5)	(1.7)	(2.5)
250 µg/m^2/day	0.9	9.8	11.5	13.4	5.3	7.7	5.3	9.3^b
($n = 3$)	(0.9)	(12.4)	(2.3)	(5.8)	(2.5)	(1.5)	(3.5)	(1.2)
500 µg/m^2/day	12.7	20.0	13.0	13.5	7.7	9.0	7.7	10.0
($n = 3$)	(11.0)	(7.0)	(2.1)	(0.7)	(0.6)	(1.0)	(1.2)	(0)
1000 µg/m^2/day	8.2	10.8	18.8	21.8	6.6	6.8	6.7	6.9
($n = 9$)	(9.7)	(11.0)	(11.9)	(13.5)	(2.7)	(3.0)	(2.8)	(3.3)
Placebo	0.1	6.4	9.3	11.8	6.0	5.8	5.5	5.8
($n = 4$)	(0.1)	(11.7)	(1.7)	(4.2)	(1.2)	(1.5)	(1.3)	(1.5)

^{a}Mean values for A.M. stiffness (hours), walk time (seconds), and global assessments (0–10, 0 is the best). The values in parentheses are the standard deviations.
^{b}Change in patient global assessment worse ($p < .05$) at 250 compared to 500 or 1000 µg/m^2/day or placebo.

Table 4 Effects of Treatment with Rhu IL-1RI or Placebo of Modified HAQ Between Days 0 and 29

	Satisfaction		Difficulty		Function	
	Day 1	Day 29	Day 1	Day 29	Day 1	Day 29
125 µg/m^2/day	13.0^a	17.3	16.7	21.3	2.7	3.0
($n = 3$)	(3.0)	(5.0)	(4.0)	(4.9)	(1.2)	(1.0)
250 µg/m^2/day	12.7	20.0^b	12.0	23.0^c	2.3	3.3^d
($n = 3$)	(3.8)	(4.4)	(4.6)	(2.0)	(0.6)	(0.6)
500 µg/m^2/day	15.7	18.3	22.0	21.0	3.3	2.7
($n = 3$)	(4.0)	(7.0)	(2.0)	(6.6)	(0.6)	(2.3)
1000 µg/m^2/day	19.2	20.3	22.0	24.9	3.2	3.2
($n = 9$)	(6.6)	(6.9)	(9.1)	(8.1)	(0.8)	(0.8)
Placebo	13.0	14.5	13.8	17.0	2.3	2.3
($n = 4$)	(5.6)	(4.4)	(7.1)	(6.9)	(0.5)	(0.5)

^{a}Mean values for HAQ subsections for difficulty, satisfaction, and ability to function (0 is the best for each). The standard deviations are in parentheses.
^{b}Change from day 1 significantly ($p < .05$) worse than noted for the 1000 µg/m^2/day group.
^{c}Change from day 1 significantly ($p < .05$) worse than noted in the 500 and 1000 µg/m^2/day groups.
^{d}Change from day 1 significantly ($p < .05$) worse than noted in the 500 µg/m^2/day group.

function sections of the HAQ also significantly worsened in those treated with 250 µg/m^2/day rhu IL-1RI compared to other groups (Table 4).

Although the administration of rhu IL-1RI subcutaneously was relatively safe, two patients stopped treatment prematurely because of toxicity. Both patients were receiving 1000 µg/m^2/day of rhu IL-1RI. One developed hives and the other a diffuse erythematous rash. Two other patients receiving rhu IL-1RI developed injection site reactions that did not require premature termination of the medication. One placebo-treated patient also developed an urticarial rash, which did not result in premature cessation of therapy.

Since no appreciable clinical benefit was noted in this study, was there any evidence that the rhu IL-1RI was biologically active in these patients? At the highest doses employed, there was a trend toward reduction of in vitro proliferative responses to the mitogen PHA and mitogenic anti-CD3 (11). No changes of proliferative responses by peripheral blood mononuclear cells to recall antigens or mitogens were noted at the 125, 250, or 500 µg/m^2/day doses of rhu IL-1RI (11). Additionally, no change in monocyte or lymphocyte numbers or subsets was noted at any dose (11). In contrast, monocyte cell surface IL-1α was decreased at all dosage levels of rhu IL-1RI but not in the placebo-treated patients (11). This suggests that rhu IL-1RI was able to bind the cell surface IL-1α of circulating monocytes and either block the recognition of IL-1α or cause its down-modulation.

There are a number of potential explanations for the lack of benefit noted in these studies. One possibility is that IL-1 might not be responsible for driving the chronic synovitis seen in the majority of patients studied. It is possible that the IL-1 produced locally is already adequately neutralized by locally synthesized IL-1RI and II, and IL-1Ra (14). It is also possible that rhu IL-1RI may be effective but that it was not adequately concentrated in the synovial tissue, despite local injection and systemic administration for 1 month.

Rhu IL-1RI binds IL-1α and IL-1β less avidly than to IL-1Ra, and the interaction of IL-1RI with IL-1Ra is essentially irreversible (14,15). Therefore, it is possible that rhu IL-1RI, particularly at lower concentrations, might exacerbate disease activity by binding to IL-1Ra, making more IL-1 biologically available and active. An exacerbation of joint swelling was observed in two of 12 patients treated intra-articularly with rhu IL-1RI, but in none of the four control placebo-injected patients (10). In the systemic protocol, no significant exacerbation of joint swelling was observed. However, a number of the secondary parameters worsened significantly in those treated with 250 µg/m^2/day (11).

IV. CONCLUSION

These preliminary observations of the treatment with RA with rhu IL-1RI were disappointing. It is possible that type II IL-1R might be more effective at modulating disease activity. Type II IL-1R does not function in signal transduction (16). Its major function is to serve as a decoy molecule to adsorb IL-1 (16). Whether or not inhibition of IL-1 or the use of receptor-based therapy as means of modulating RA will be effective will require further study.

ACKNOWLEDGMENTS

This study was supported in part by Immunex Corporation, the Multipurpose Arthritis and Musculoskeletal Diseases Center Grant AR30692, and the Veterans Administration Medical Research Service.

REFERENCES

1. Dinarello CA. The biology of interleukin-1. Interleukins: molecular biology and immunology. Chem Immunol 1992; 51:1–32.
2. Foxwell BMJ, Barrett K, Feldman M. Cytokine receptors: structure and signal transduction. Clin Exp Immunol 1992; 90:161–169.
3. Slack J, McMahon CJ, Waugh S, Schooley K, Spriggs MK, Sims JE, Dower SK. Independent binding of interleukin-1α and interleukin-1β to type I and type II interleukin-1 receptors. J Biol Chem 1993; 268:2513–2524.
4. Symons JA, Eastgate JA, Duff GW. Purification and characterization of a novel soluble receptor for interleukin-1. J Exp Med 1991; 174(5):1251–1254.
5. Koch AE, Kunkel SL, Chensue SW, Haines KG, Streiter RM. Expression of interleukin-1 and interleukin-1 receptor antagonist by human rheumatoid synovial tissue macrophages. Clin Immunol Immunopathol 1992; 65:23–29.
6. Dower SK, Sims JE, Cerretti DP, Bird TA. The interleukin-1 system: receptors, ligands and signal. Interleukins: molecular biology and immunology. Chem Immunol 1992; 51:33–64.
7. Dower SK, Wignall JM, Schooley K, McMahan CJ, Jackson JL, Prickett KS, Lupton S, Cosman D, Sims JE. Retention of ligand binding activity by the extracellular domain of the IL-1 receptor. J Immunol 1989; 142:4314–4320.
8. Schorlemmer HU, Kanzy EJ, Langner KD, Kurrle R. Immunoregulation of SLE-like disease by the IL-1 receptor: disease modifying activity of BDF1 hybrid mice and MRL autoimmune mice. Agents Actions 1993; 39:c117–c120.
9. Schorlemmer HU, Kanzy EJ, Langner KD, Kurrle R. Immunomodulatory activity of recombinant IL-1 receptor (IL-1-R) on models of experimental rheumatoid arthritis. Agents Actions 1993; 39:c113–c116.
10. Pope R, Drevlow B, Capezio J, Lovis R, Jacobs C, Blosch C, Beck C, Haag MA, Landay A. Intra-articular administration of recombinant human interleukin-1 receptor-type I in patients with active rheumatoid arthritis. In: Biological Agents in Autoimmune Disease IV (in press).
11. Drevlow B, Lovis R, Haag MA, Sinacore JM, Jacobs C, Blosche C, Landay A, Moreland LW, Pope RM. Recombinant human interleukin 1 receptor, type I (rHu IL-1RI) in the treatment of patients with active rheumatoid arthritis. Arthritis Rheum (in press).

12. Arnett FC, Edworthy SM, Block DA, McShane DJ, Fries JF, Cooper NS, Healey LA, Kaplan SR, Liang MH, Luthra HS, Medsgar TA Jr., Mitchell DM, Nuistaar DH, Pinals RS, Schaller JG, Sharp JT, Wilder RL, Hunder GG. The American Rheumatism Association 1987 revised criteria for the classification of rheumatoid arthritis. Arthritis Rheum 1988; 31:315–324.
13. Steinbrocker O, Traeger CH, Batterman RC. Therapeutic criteria in rheumatoid arthritis. JAMA 1949; 140:659–662.
14. Arend WP, Malyak M, Smith MF Jr, Whisenend TD, Slack JL, Sims JE, Giri JG, Dower SK. Binding of the IL-1α, IΛ–1β, and IL-1 receptor antagonist by soluble IL-1 receptors and levels of soluble IL-1 receptors by synovial fluids. J Immunol 1994; 4766–4774.
15. Svenson M, Hansen MB, Heegard P, Abell K, Bendtzen K. Specific binding of interleukin-1 (IL-1)β and IL-1 receptor antagonist (IL-1ra) to human serum: high affinity binding of IL-Ira to soluble IL-1 receptor type 1. Cytokine 1993; 5(5):427–435.
16. Re F, Muzio M, De Rossi M, Polentarutti N, Giri JG, Mantovani A, Colotta F. The type II "receptor" as a decoy target for interleukin 1 in polymorphonuclear leukocytes: characterization of induction by dexamethasone and ligand binding properties of the released decoy receptor. J Exp Med 1994; 179:739–743.

13

Treatment of Rheumatoid Arthritis with Soluble Tumor Necrosis Factor Receptor

GARY R. MARGOLIES
Lexington, Kentucky

WILLIAM J. KOOPMAN and LARRY W. MORELAND
University of Alabama at Birmingham, Birmingham, Alabama

I. INTRODUCTION

Tumor necrosis factor alpha (TNF-α) plays a central role in the sequence of cellular and molecular events underlying the inflammatory response (1). TNF-α is a homotrimer consisting of three identical 17 kDa polypeptide subunits. It is expressed primarily by stimulated mononuclear phagocytes. Increased levels of TNF-α are observed in many infectious, neoplastic, and autoimmune diseases. In addition to its participation in the physiological response to infection and neoplasm, local TNF-α expression may contribute to the pathogenesis of chronic inflammatory diseases, such as rheumatoid arthritis (RA).

Evidence supporting a role for TNF-α in the pathogenesis of RA includes: the presence of TNF-α at the cartilage-pannus junction in RA patients (1) and increased levels of TNF-α in RA synovial fluid (2,3). Furthermore, increased TNF-α production has been found to be increased in the synovial cells of patients with active RA, but not in synovial cells from patients with inactive RA. Several proinflammatory actions of TNF-α may contribute to its role in the pathogenesis of RA. In addition to stimulating the release of other proinflammatory cytokines, including IL-6, IL-8, IL-1β, and leukemia inhibitory factor (LIF), TNF-α also induces the release of proteases from neutrophils, fibroblasts, and chondrocytes (4–6). These enzymes, including collagenase and other neutral metalloproteinases, are likely to be responsible for joint destruction in RA. TNF-α also induces the expression of endothelial adhesion molecules (e.g., ICAM-1 and E-selectin, etc.), leading to rapid transmigration of leukocytes into extravascular sites (7).

While IL-1 share many activities with TNF-α, the latter appears to represent a more attractive therapeutic target. This view is supported by observations that inhibition of TNF-α suppresses spontaneous production of IL-1, IL-6, and granulocyte-macrophage colony-stimulating factor by RA synovial cells, whereas inhibition of IL-1 does not diminish expression of TNF-α (8,9). Thus, TNF-α may be a "pivotal" cytokine in regulating expression of other inflammatory mediators in RA. This view has led to therapeutic interest in developing strategies to modulate TNF-α activity in patients with RA.

II. STRUCTURE AND BINDING OF TNF-α TO TNF RECEPTORS

Biological activities of TNF-α require binding to specific membrane-bound TNF receptors. There are two known membrane receptors for TNF-α, designated p60 (p55) and P80 (p75) (10,11). Receptors for TNF-α are expressed by numerous different cell types, including polymorphonuclear leukocytes, vascular endothelial cells, and fibroblasts (1). TNF binding to its receptors mediates a wide variety of actions, including its proinflammatory effects. Soluble TNF receptors (sTNFR) have been isolated (12,13) and demonstrated to arise from the shed extracellular portion of the membrane-bound type I (p55 or p60) and type II (p75 or p80) molecules. Both sTNFR p55 and sTNFR p75 have been detected in the synovial lining layers as well as in deeper layers (14). The sTNFR-expressing cells are in the vicinity of TNF-α-containing cells (14). sTNFR-expressing cells have also been detected in osteoarthritis synovial tissue (15) and chondrocytes (15). Binding of sTNF p75 on RA synovial fibroblasts is increased by IL-1, IL-4, and interferon-γ (IFN-γ) (16). These same cytokines also cause an increase of TNF-α receptor (TNFR) shedding in inflamed joints (16).

sTNFR levels are increased in sera and synovial fluid (17–19) and can be detected at the cartilage pannus junction (20) of RA patients and patients with active systemic lupus erythematosus (SLE) (21,22). In SLE patients, serum levels of sTNFR correlated better with disease activity than any other laboratory parameter (21,22). There is a good correlation of the plasma levels of sTNFR p55 with levels of sTNFR p75 in patients with SLE, progressive systemic sclerosis (PSS), and mixed connective tissue disease (MCTD) (21). Both types of sTNFRs have been detected in sera of RA and osteoarthritis (OA) patients even when there were no detectable levels of TNF-α (23,24). Other conditions (health and disease) in which increased levels of either sTNFR p75, sTNFR p55, or both have been reported include human immunodeficiency virus (HIV)-infected individuals (25), acute Kawasaki's disease (26), ascites (27), chronic renal failure (28), hemodialysis patients (29), hair cell leukemia (30), chronic lymphocytic leukemia (30), solid tumors (31), sepsis syndrome (32), experimental endotoxemia (33), transplantation (34), cerebrospinal fluid (CSF) of multiple sclerosis (MS) patients (35), serum of pregnant women (36), urine of newborns (37), amniotic fluid (37), and meningiococcemia (38).

The role of sTNFRs in modulating the inflammatory and immune reactions by inhibiting the effects of TNF-α are currently being investigated (39). These two different TNF receptors mediate distinct cellular responses (40). The p55 TNFR mediates cell cytotoxicity, whereas the p75 TNFR stimulates thymocyte proliferation (40). The soluble forms of these receptors may play a physiological role in protecting against the harmful effects of TNF. For example, when neutrophils adhere to the vessel wall, they release both types of sTNFRs, which correlate with a decrease in neutrophil responses to TNF (41). In vitro, sTNFRs inhibit the TNF-α-mediated respiratory burst of neutrophils (42). In persons undergoing IL-2 immunotherapy, interleukin-1 receptor antagonist (IL-1ra) and sTNFR p75 markedly down-regulate IL-2-induced IL-8 synthesis (43). IL-4, previously shown to down-regulate the production of proinflammatory cytokines such as TNF-α, up-regulates both types of sTNFRs on synovial cells in culture (44). Both sTNFR p75 and sTNFR p55 inhibit the cytolytic activity of human TNF-α in vitro (45). At low concentrations, sTNFRs may indeed stabilize TNF and augment its activity (46). Thus, the sTNFRs may in some situations inhibit the effects of TNF and in other situations serve as carriers for TNF and may augment the effects of TNF by prolonging its function.

III. POTENTIAL PROBLEMS WITH ANTAGONISM OF TNF

Potential problems exist with the proposed clinical use of TNF-α antagonists in RA. First, the physiological role(s) of TNF-α is not completely understood. While TNF-α is at least partly responsible for some of the deleterious effects associated with septic shock, including increased vascular permeability, myocardial depression, and intravascular coagulation, it also appears to play a beneficial role in facilitating appropriate host responses to infection. Antibodies to TNF-α as well as recombinant human soluble tumor necrosis factor receptor Fc fusion protein (rhu sTNFR:Fc) decrease the acute effects of LPS in normal subjects (47). Preliminary studies in patients with septic shock have indicated no survival benefit for anti-TNF monoclonal antibody (MAb) in treated patients compared to controls (48,49).

In addition to its role in infection, TNF-α is also known to play a role in resistance to neoplasms (50). Whether TNF-α participates in the surveillance functions of the immune system is not yet known. Future long-term studies of

TNF-α antagonists will need to focus on outcomes associated with both infections and malignancy.

A second potential obstacle in the development of TNF-α antagonists is the induction of host immunity against the agent itself. This is anticipated to be a problem with anti-TNF MAb. Theoretically, these antibody responses may result in adverse clinical effects such as serum sickness–type reactions. In addition, host immune responses may decrease the activity and half-lives of these therapeutic molecules if they are repeatedly administered.

IV. TNF-α IN ANIMAL MODELS OF RA

Studies in animal models support the idea that increased TNF-α secretion plays a role in the pathogenesis of RA. Keffer et al. demonstrated that mice transgenic for the human TNF-α gene developed chronic inflammatory arthritis (51). Treatment of these arthritic mice with a MAb against human TNF abrogated the arthritis. Antibodies (IgG1 MAb) to TNF have been shown to be effective in collagen-induced arthritis (52). When administered before the onset of disease, anti-TNF-α antibodies reduced the histological severity of the arthritis as well as decreasing joint swelling, while histological severity of the disease was reduced when they were given after the onset of arthritis (52). More recently, combined treatment with an anti-TNF antibody and an anti-CD4 antibody in collagen-induced arthritis revealed significantly greater reductions in pain swelling and joint erosion than that achieved by anti-TNF alone (53).

A recombinant human soluble TNF receptor Fc fusion protein (rhu sTNFR:Fc) significantly reduced both the incidence and severity of collagen-induced arthritis in preventive and therapeutic protocols (54). This fusion protein was tested by Wooely et al. for antirheumatic activity in mice with type II collagen-induced arthritis (54). Mice were administered rhu TNFR:Fc before challenge with collagen; the resulting outcome was appearance of arthritis in 25% of the treated mice versus 85% of controls. Similarly, rhu sTNFR:Fc administered to animals that had already developed collagen-induced arthritis resulted in 40% improvement in the arthritis score versus controls at 10 weeks. Piguet et al. demonstrated that both a MAb to TNF-α and the recombinant sTNFR p55 fusion protein prevented development of arthritis in DBA/1 mice immunized with type II collagen (55).

V. TNF-α ANTAGONISTS AS THERAPEUTIC AGENTS IN RA

If local production of TNF-α contributes to inflammation and joint destruction in RA, does antagonism of TNF-α mitigate or reverse the process? Elliott et al. recently reported on 20 RA patients treated with a chimeric mouse/human monoclonal anti-TNF-α antibody (56). The subjects, each treated with 20 mg/kg of the MAb parenterally, exhibited significant improvement in several clinical and laboratory measures of disease activity, including the Ritchie articular index, swollen joint count, and C-reactive protein (CRP). The trial was open-labeled and not placebo-controlled. A follow-up randomized, phase II, double-blind, placebo-controlled study involving 73 patients confirmed these preliminary results (57). These trials provide further evidence for the critical role of TNF-α in the pathogenesis of RA. This same antibody is being evaluated in Crohn's disease (58). A humanized anti-TNF MAb (CDP571) has also been evaluated in a phase I dose-escalating trial in patients with RA (59). Dose-dependent clinical benefit was noted within 1 week and lasted for up to 8 weeks.

sTNFRs are thought to be natural inhibitors of the activity of TNF. There appears to be an imbalance of the sTNFRs and TNF-α at inflammatory sites. Therefore, an approach for treating RA is to administer recombinant sTNFR fusion proteins and increase the amount of sTNFR that is present to inhibit the activities of TNF. Two recombinant sTNFR fusion protein products, sTNFR type I fusion protein (p60) and sTNFR type II fusion protein (p80), have been evaluated as potential therapy for RA.

Immunex Corporation constructed a recombinant human TNFR fusion protein (rhu sTNFR:Fc p80) for therapeutic neutralization of TNF-α. DNA encoding the soluble portion of the human p80 TNFR was linked to DNA encoding the Fc portion of the human IgG molecule and expressed in a mammalian cell line. The resulting immunoglobulin-like dimer (rhu sTNFR:Fc) possesses several attractive features as a TNF-α antagonist agent. First, the dimeric receptor fusion protein has substantially higher affinity for TNF-α than the monomeric soluble receptor (60). Second, the immunoglobulin-like Fc structure results in a longer half-life of the molecule in vivo. Third, the immunoglobulin-like structure of rhu TNFR:Fc may afford more rapid clearance or neutralization of the resulting complex once the molecule is bound to TNF-α.

VI. USE OF rhu sTNFR:Fc FUSION PROTEIN IN RA

Experience with rhu sTNFR:Fc fusion protein in humans has been limited. Safety studies in normal human volunteers demonstrated no adverse events following intravenous administration (61). A phase I study using rhu sTNFR:Fc in patients with severe RA was recently completed at our center (62). Sixteen patients with severe, refractory RA were treated for 4 weeks and observed for an additional month. Patients had to have failed at least one disease-modifying antirheumatoid drug (DMARD), have active disease (≥5 swollen joints and ≥9 painful joints), and be functional class I, II, or III. Concomitant treatment with an NSAID and stable (≥1 month) dose of prednisone (≤10 mg/day) was allowed. Patients were enrolled in groups of four; three in each group received active drug and one received placebo in a double-blind fashion. Rhu sTNFR:Fc was given as an intravenous load followed by 4 weeks of twice-weekly subcutaneous administration. The groups were as follows:

	IV loading dose	SO maintenance dose (give 2x/week × 4 weeks)
Group I	4 mg/m^2	2 mg/m^2
Group II	8 mg/m^2	4 mg/m^2
Group III	16 mg/m^2	8 mg/m^2
Group IV	32 mg/m^2	16 mg/m^2

Patient characteristics are described in Table 1. The mean age of the patients was 52.8 years (range 21–73), and the mean disease duration 8.5 years (range 1–49). Concomitant treatment with NSAIDS and/or low-dose prednisone (≤10 mg/day) was permitted. All patients had failed at least one DMARD.

Adverse events included mild injection site rashes in four patients that did not necessitate discontinuation of the drug. There were no serious adverse effects and all patients completed 4 weeks of treatment.

Table 2 lists the major clinical parameters measured. There was no clear-cut dose response among the treatment groups. Therefore, analysis included all treated patients ($n = 12$) grouped together and compared with placebo-treated ($n = 4$) patients. The joint score (painful or swollen) was the total score of each joint multiplied by the severity of pain or swelling on a scale of 1–3. At day 31, there was a 44% mean improvement in total pain and total joint scores in patients receiving active drug, compared to 22% improvement in the patients receiving placebo. Average morning stiffness improved by 55% in treated patients. Compared to baseline, there was a significant ($p < .05$) decrease in Westergren ESR (32%). CRP levels also decreased (27%) significantly in the treated patients compared to 13% in the placebo-treated patients; this was most pronounced in the highest dose group (57, 85, and 100% decrease in CRP at 31 days in the three patients in group IV).

Inadequate data exist to determine whether rhu sTNFR:Fc is immunogenic. In our study, none of the 12 patients who received rhu sTNFR:Fc developed measurable antibody to the agent. These preliminary data indicate that rhu sTNFR:Fc is well tolerated. Efficacy of rhu sTNFR:Fc in RA is being further evaluated in a multicenter phase II randomized, controlled clinical trial.

Table 1 Patient Characteristics at Baseline

Patient number	Treatment	Dose[a] group	Gender	Age	Disease duration (years)	ESR (mm/hr) (day 0)	CRP (mm/hr) (day 0)	Swollen joint count (day 0)	Tender joint count (day 0)
2	Active drug	I	F	73	8	45	5.3	26	40
3	Active drug	I	M	50	3	25	2.6	50	64
4	Active drug	I	F	46	5	11	1.6	7	25
5	Active drug	II	F	21	3	35	3.5	8	9
6	Active drug	II	M	32	5	49	3.6	38	62
7	Active drug	II	M	60	5	25	<0.5	17	31
9	Active drug	III	F	48	3	75	5.9	13	26
10	Active drug	III	F	50	4	6	1.0	11	56
12	Active drug	III	F	62	7	39	4.9	26	71
13	Active drug	IV	F	67	2	67	4.5	19	41
15	Active drug	IV	F	64	1	83	7.4	23	37
16	Active drug	IV	M	54	16	47	1.8	9	25
1	Placebo	I	M	65	49	15	2.8	33	64
8	Placebo	II	F	43	2	8	<.05	53	67
11	Placebo	III	M	47	13	35	2.5	39	55
14	Placebo	IV	F	62	15	40	3.5	20	22

[a]See text for dose level.

Table 2 Average Percent Improvement at Day 31 in Patients Receiving rhu sTNFR:Fc or Placebo

Measurement	Active treatment ($n = 12$)	Placebo ($n = 4$)
Total painful joint score[a]	44	23
Total swollen joint score[a]	40	25
Total joint score[a]	44	22
Morning stiffness	55	−10
Erythrocyte sedimentation rate (Westergren method)	32	12
C-reactive protein	27	13

[a]See text for criteria on determining scores.

VII. SUMMARY

Considerable evidence suggests that TNF-α contributes to the pathogenesis of RA. Initial experience with a rhu sTNFR:Fc fusion protein in RA indicates the molecule is well tolerated. Improvement observed in patients receiving rhu sTNFR:Fc justifies further evaluation of this agent as a potential therapeutic agent in RA.

ACKNOWLEDGMENTS

The clinical studies using rhu sTNFR:Fc were supported by a financial grant from Immunex, Inc. The expert technical assistance by Betsy Perry in preparation of the manuscript is acknowledged. The outstanding clinical care by the study coordinator, Diane Horton, is also much appreciated. The collaborative help and support from Dr. Consuelo Blosch (Immunex), Dr. Lou Heck, and Dr. Gary Margolies, and all Clinical Faculty and Fellows at The University of Alabama at Birmingham is also noted.

REFERENCES

1. Brennan FM, Feldmann M. Cytokines in autoimmunity. Curr Opin Immunol 1992; 4:754–759.
2. Chu CQ, Field M, Feldmann M, Maini RN. Localization of tumor necrosis factor α in synovial tissues and at the cartilage-pannus junction in patients with rheumatoid arthritis. Arthritis Rheum 1991; 34:1125–1132.
3. Saxne T, Palladino MA Jr, Heinegard D, Talal N, Wollheim FA. Detection of tumor necrosis factor alpha but not tumor necrosis factor beta in rheumatoid arthritis synovial fluid and serum. Arthritis Rheum 1988; 31:1041–1045.

4. Shingu M, Nagai Y, Isayama T, Naono T, Nobunaga M, Nagai Y. The effects of cytokines on metalloproteinase inhibitors (TIMP) and collagenase production by human chondrocytes and TIMP production by synovial cells and endothelial cells. Clin Exp Immunol 1993; 94:145–149.
5. MacNaul KL, Chartrain N, Lark M, Tocci MJ, Hutchinson NI. Differential effects of IL-1 and TNF alpha on the expression of stromelysin, collagenase and then natural inhibitor, TIMP, in rheumatoid synovial fibroblasts. Matrix 1992; 1(Suppl): 198–199.
6. Ahmadzadeh N, Shingu M, Nobunaga M. The effect of recombinant tumor necrosis factor-alpha on superoxide and metalloproteinase production by synovial cells and chondrocytes. Clin Exp Rheumatol 1990; 8:387–391.
7. Moser RB, Schleiffenbaum B, Groscurth P, Fehr J. Interleukin 1 and tumor necrosis factor stimulate human vascular endothelial cells to promote transendothelial neutrophil passage. J Clin Invest 1989; 83:444–455.
8. Brennan FM, Chantry D, Jackson A, Maini FN, Feldmann M. Inhibitory effects of TNF alpha antibodies on synovial cell interleukin-1 production in rheumatoid arthritis. Lancet 1989; 2:244–247.
9. Haworth C, Brennan FM, Chantry D, Turner M, Maini RN, Feldmann M. Expression of granulocyte-macrophage colony-stimulating factor in rheumatoid arthritis: regulation by tumor necrosis-alpha. Eur J Immunol 1991; 21:2575–2579.
10. Banner DW, D'Arcy A, Janes W, Gentz R, Schoenfeld HJ, Broger CL, Lesslauer W. Crystal structure of the soluble human 55 kd TNF receptor-human TNF beta complex: implications for TNF receptor activation. Cell 1993; 73:431–445.
11. Seckinger P, Zhang J, Hauptmann B, Dayer J: Characterization of a tumor necrosis factor α (TNF-α) inhibitor: evidence of immunological cross-reactivity with the TNF receptor. Proc Natl Acad Sci USA 1990; 87:5188–5192.
12. Engelmann H, Aderka D, Rubinstein M, Rotman D, Wallach D. A tumor necrosis factor binding protein purified to homogeneity from human urine protects cells from tumor necrosis factor toxicity. J Biol Chem 1989; 264:11974–11980.
13. Olsson I, Lantz M, Nilsson E, Peetre C, Thysell H, Grubb A, Adolf G. Isolation and characterization of a tumor necrosis factor binding protein from urine. Eur J Haematol 1989; 42:270–275.
14. Deleuran BW, Chu CQ, Field M, Brennan FM, Mitchell T, Feldmann MM, Miani RN. Localization of tumor necrosis factor receptors in the synovial tissue and cartilage pannus junction in patients with rheumatoid arthritis: implications for local actions of tumor necrosis factor alpha. Arthritis Rheum 1992; 35:1170–1178.
15. Westacott CI, Atkins RM, Dieppe PA, Elson CJ. Tumor necrosis factor-α receptor expression on chondrocytes isolated from human articular cartilage. J Rheumatol 1994; 21:1710–1715.
16. Taylor DJ. Cytokine combinations increase p75 tumor necrosis factor receptor binding and stimulate receptor shedding in rheumatoid synovial fibroblasts. Arthritis Rheum 1994; 37:232–235.
17. Roux-Lombard P, Punzi L, Hasler F, Bas S, Todesco S, Gallati H, Guerne PA, Dayer JM. Soluble tumor necrosis receptors in human inflammatory synovial fluids. Arthritis Rheum 1993; 36:485–489.

18. Barrera P, Boerbooms AM, Janssen EM, Sauerwein RW, Gallati H, Mulder J, de Boo T. Circulating soluble tumor necrosis factor receptors, interleukin-2 receptors, tumor necrosis factor alpha, and interleukin-6 levels in rheumatoid arthritis: longitudinal evaluation during methotrexate and azathioprine therapy. Arthritis Rheum 1993; 36:1070–1079.
19. Cope AP, Aderka D, Doherty M, Engelmann H, Gibbons D, Jones ACB, Maini RN, Wallach D, Feldmann M. Increased levels of soluble tumor necrosis factor receptors in the sera and synovial fluid of patients with rheumatic diseases. Arthritis Rheum 1992; 35:1160–1169.
20. Deleuran BW, Chu CQ, Field M, Brennan FM, Mitchell T, Feldmann MM. Localization of tumor necrosis factor receptors in the synovial tissue and cartilage-pannus junction in patients with rheumatoid arthritis: implications for local actions of tumor necrosis factor alpha. Arthritis Rheum 1992; 35:1170–1178.
21. Heilig B, Fiehn C, Brockhaus M, Gallati H, Pezzutto A, Hunstein W. Evaluation of soluble tumor necrosis factor (TNF) receptors and TNF receptors antibodies in patients with systemic lupus erythematodes, progressive systemic sclerosis, and mixed connective tissue disease. J Clin Immunol 1993; 13:321–328.
22. Aderka D, Wysenbeek A, Engelmann H, Cope AP, Brennan FM, Molad Y, Hornik V, Levo Y, Maini RN, Feldmann M, Wallach D. Correlation between serum levels of soluble tumor necrosis factor receptor and disease activity in systemic lupus erythematosus. Arthritis Rheum 1993; 36:1111–1120.
23. Chikanza IC, Roux-Lomabard P, Dayer JM, Panayi GS. Tumour necrosis factor soluble receptors behave as acute phase reactants following surgery in patients with rheumatoid arthritis, chronic osteomyelitis and osteoarthritis. Clin Exp Immunol 1993; 92:19–22.
24. Roux-Lombard P, Punzi L, Hasler F, Bas S, Todesco S, Gallati H, Guerne PA, Dayer JM. Soluble tumor necrosis receptors in human inflammatory synovial fluids. Arthritis Rheum 1993; 36:485–489.
25. Zangerle R, Gallati H, Sarcletti M, Weiss G, Denz H, Wachter HF, Fuchs D. Increased serum concentrations of soluble tumor necrosis factor receptors in HIV-infected individuals are associated with immune activation. J AIDS 1994; 7:79–85.
26. Furukawa S, Masubara T, Umezawa Y, Okumura K, Yabuta K. Serum levels of p60 soluble tumor necrosis factor receptor during acute Kawasaki disease. Pediatrics 1994; 124:721–725.
27. Andus T, Gross V, Holstege A, Ott M, Weber M, David M, Gallati HG, Scholmerich J. High concentrations of soluble tumor necrosis factor receptors in ascites. Hepatology 1992; 16:749–755.
28. Brockhaus M, Bar-Khayim Y, Gurwicz S, Frendsdorff A, Haran N. Plasma tumor necrosis factor soluble receptors in chronic renal failure. Kidney Int 1992; 42:663–667.
29. Ward RA, Gordan L. Soluble tumor necrosis factor receptors are increased in hemodialysis patients. ASAIO J 1993; 39:M782-6.
30. Digel W, Porzsolt F, Schmid M, Herrmann F, Lesslauer W, Brockhaus M. High levels of circulating soluble receptors for tumor necrosis factor in hairy cell leukemia and type B chronic lymphocytic leukemia. J Clin Invest 1992; 89:1690–1693.

31. Elsasser-Beile U, Gallati H, Weber W, Wild ED, Schulte Monting J, vonKleist S. Increased plasma concentrations for type I and II tumor necrosis factor receptors and IL-2 receptors in cancer patients. Tumor Biol 1994; 15:17–24.
32. Froon AH, Bemelmans MH, Greve JW, van der Linden CG, Buurman WA. Increased plasma concentrations of soluble tumor necrosis factor receptors in sepsis syndrome: correlation with plasma creatinine values. Crit Care Med 1994; 22:803–809.
33. Spinas GA, Keller U, Brockhaus M. Release of soluble receptors for tumor necrosis factor (TNF) in relation to circulating TNF during experimental endotoxinemia. J Clin Invest 1992; 90:533–536.
34. Kraus T, Mehrabi A, Arnold J, Wermann M, Klar E, Otto G, Herfarth C, Heilig B. Evaluation of soluble tumor necrosis factor receptors with orthotopic liver transplantation. Transplant Proc 1992; 24:2539–2541.
35. Tsukada N, Matsuda M, Miyagi K, Yanagisawa N. Increased levels of intercellular adhesion molecule-1 (ICAM-1) and tumor necrosis factor receptor in the cerebrospinal fluid of patients with multiple sclerosis. Neurology 1993; 43:2679–2682.
36. Austgulen R, Liabakk NB, Lien E, Espevik T. Increased levels of soluble tumor necrosis factor-alpha receptors in serum from pregnant women and in serum and urine samples from newborns. Pediatr Res 1993; 33:82–87.
37. Baumann P, Romero R, Berry S, Gomez R, McFarlin B, Araneda H, Cotton DB, Fidel P. Evidence of participation of the soluble tumor necrosis factor receptor I in the host response to intrauterine infection in preterm labor. Am J Reprod Immunol 1993; 30:184–193.
38. Girardin E, Roux-Lombard P, Grau GE, Suter P, Gallati H, Dayer JM. Imbalance between tumour necrosis factor-alpha and soluble TNF receptor concentrations in severe meningococcaemia: the J5 Study Group. Immunology 1992; 76:20–23.
39. Pennica D, Lam VT, Mize NK, Weber RF, Lewis M, Fendly BM, Lipari MT, Goeddel DV. Biochemical properties of the 75-kDa tumor necrosis factor receptor: characterization of ligand binding, internalization, and receptor phosphorylation. J Biol Chem 1992; 267:21172–21178.
40. Tartaglia LA, Weber RF, Figari IS, Reynolds C, Palladino MA Jr. The two different receptors for tumor necrosis factor mediate distinct cellular responses. Proc Natl Acad Sci USA 1991; 88:9292–9296.
41. Lantz M, Bjornberg F, Olsson I, Richter J. Adherence of neutrophils induces release of soluble tumor necrosis factor receptor forms. J Immunol 1994; 152:1362–1369.
42. Ferrante A, Hauptmann B, Seckinger P, Dayer J-M. Inhibition of tumour necrosis factor alpha (TNF-α)-induced neutrophil respiratory burst by a TNF inhibitor. Immunology. 1991; 72:440–442.
43. Tilg H, Shapiro L, Atkins MB, Dinarello CA, Mier JW. Induction of circulating and erythrocyte-bound IL-8 by IL-2 immunotherapy and suppression of its in vitro production by IL-1 receptor antagonist and soluble tumor necrosis factor receptor (p75) chimera. J Immunol 1993; 151:3299–3307.
44. Cope AP, Gibbons DL, Aderke D, Foxwell BM, Wallach D, Maini RN, Feldmann M, Brennan FM. Differential regulation of tumour necrosis factor receptors (TNF-R) by IL-4; upregulation of p55 and p75 TNF-R on synovial joint mononuclear cells. Cytokine 1993; 5:205–212.

45. Gatanga T, Hwang CD, Kohr W, Cappuccini F, Lucci JA, Jeffes EWL, Lentz R, Tomich J, Yamamoto RS, Granger GA. Purification and characterization of an inhibitor (soluble tumor necrosis factor receptor) for tumor necrosis factor and lymphotoxin obtained from the serum ultrafiltrates of human cancer patients. Proc Natl Acad Sci USA 1990; 87:8781–8784.
46. Aderka D, Engelmann H, Maor Y, Brakebusch C, Wallach D. Stabilization of the bioactivity of tumor necrosis factor by its soluble receptors. J Exp Med 1992; 175: 323–329.
47. Ashkenazi A, Marsters SA, Capon DJ, Chamow SM, Figari IS, Pennica D, Goeddel DV, Palladino MA, Smith DH. Protection against endotoxic shock by a tumor necrosis factor receptor immunoadhesion. Proc Natl Acad Sci USA 1991; 88:10535–10539.
48. Vincent JL, Bakker J, Marecaux G, Schandene L, Kahn RJ, Dupont E. Administration of anti-TNF antibody improves left ventricular function in septic shock patients: results of a pilot study. Chest 1992; 101:810–815.
49. Fisher CJ Jr, Opal SM, Dhainaut JF, Stephens S, Zimmerman JL, Nighingale P, Harris SJ, Schein RM, Panacek EA, Vincent JL. Influence of an anti-tumor necrosis factor monoclonal antibody on cytokine levels in patients with sepsis. the CB0006 Sepsis Syndrome Study Group. Crit Care Med 1993; 21:318–327.
50. Schneider J, Hofman FM, Apuzzo ML, Hinton DR. Cytokines and immunoregulatory molecules in malignant glial neoplasms. J Neurosurg 1992; 77:265–273.
51. Keffer J, Probert L, Cazlaris H, Georgopoulous S, Kaslaris E, Kioussis D, Kollias G. Transgenic mice expressing human tumour necrosis factor: a predictive genetic model of arthritis. EMBO J 1991; 10:4025–4031.
52. Williams RO, Feldmann M, Maini RN. Anti-tumor necrosis factor ameliorates joint disease in murine collagen-induced arthritis. Proc Natl Acad Sci USA 1992; 89: 9784–9788.
53. Williams RO, Mason LJ, Feldmann M, Maini RN. Synergy between anti-CD4 and anti-tumor necrosis factor in the amelioration of established collagen-induced arthritis. Proc Natl Acad Sci USA 1994; 91:2767–2766.
54. Wooley PH, Dutcher J, Widmer MB, Gillis S. Influence of a recombinant human soluble tumor necrosis factor receptor FC fusion protein on type II collagen-induced arthritis in mice. J Immunol 1993; 11:6602–6607.
55. Piguet PF, Grau GE, Vesin C, Loetscher H, Gentz R, Lesslauer W. Evolution of collagen arthritis in mice is arrested by treatment with anti-tumour necrosis factor (TNF) antibody or a recombinant soluble TNF receptor. Immunology 1992; 77: 510–514.
56. Elliott MJ, Maini RN, Feldmann M, Long-Fox A, Charles P, Katsikis P, Brennan FM, Walker J, Bijl H, Ghrayeb J, Woody JN. Treatment of rheumatoid arthritis with chimeric monoclonal antibodies to TNF-α. Arthritis Rheum 1993; 36:1681–1690.
57. Elliott MJ, Maini RN, Feldmann M, Kalden JR, Antoni C, Smolen JS, Leeb B, Breedveld FC, Macfarlane JD, Bijl H, Woody JN. Randomized double-blind comparison of chimeric monoclonal antibody to tumour necrosis factor alpha (cA2) versus placebo in rheumatoid arthritis. Lancet 1994; 344:1105–1110.

58. Derkx B, Taminiau J, Radema S, Stronkhorst A, Wortel C, Tytgat GV, Deventer S. Tumour-necrosis-factor antibody treatment in Crohn's disease. Lancet 1992; 342: 173–174 (letter).
59. Rankin ECC, Choy EHS, Kassimos D, Sopwith M, Kingsley GH, Sopwith AM, Isenberg DA, Panayi GS. Therapeutic effects of an engineered human anti-tumour necrosis factor alpha antibody (CDP571) in rheumatoid arthritis. Br J Rheumatol 1995; 34:334–342.
60. Mohler KM, Torrance DS, Smith CA, Goodwin RG, Stremier KE, Fung VP, Madani H, Widmer MB. Soluble tumor necrosis factor (TNF receptors are effective therapeutic agents in lethal endotoxemia and function simultaneously as both TNF Carriers and TNF antagonists. J Immunol 1993; 151:1548–1561.
61. Nam MH, Reda D, Boujouko S, Agosti J, Suffredini AF. Recombinant human chimeric tumor necrosis factor (TNF) receptor (TNFR:Fc): safety and pharmacokinetics in humans. Clin Res 1993; 41 (abstract).
62. Moreland LW, Margolies GR, Heck LW, Saway PA, Jacobs C, Beck C, Blosch C, Koopman WJ. Soluble tumor necrosis factor receptor (sTNFR): results of a phase I dose-escalation study in patients with rheumatoid arthritis. Arthritis Rheum 1994; 37(Suppl):R27.

IV
Adhesion Molecule Targeted Therapies

A variety of agents designed to interfere with the interaction of cell adhesion molecules (CAMs) and their cell surface ligands are currently under development, to prevent the migration of leukocytes to sites of inflammation. They include monoclonal antibodies, soluble receptors, oligosaccharide analogs of selectins to block binding and antisense molecules. Most are in clinical trials in indications such as transplant rejection, reperfusion injury, and acute respiratory distress syndrome. A murine monoclonal antibody to ICAM-1 has been studied in rheumatoid arthritis (Chapter 14), and humanized MAbs to leukointegrins are under evaluation in multiple sclerosis. These agents may acutely alter the inflammatory response; whether this results in sustained improvement in disease activity without an increased incidence of infection remains to be seen.

14
Treatment of Rheumatoid Arthritis with a Monoclonal Antibody to Intercellular Adhesion Molecule-1

ARTHUR F. KAVANAUGH
University of Texas Southwestern Medical Center at Dallas, and Dallas Veterans Affairs Medical Center, Dallas, Texas

PETER E. LIPSKY
University of Texas Southwestern Medical Center at Dallas, Dallas, Texas

I. INTRODUCTION

Rheumatoid arthritis (RA) is a chronic, systemic inflammatory disease that causes substantial morbidity and accelerated mortality in affected patients (1,2). Beginning with the identification of rheumatoid factor in the middle of this century, it has become increasingly evident that the hallmark pathophysiological abnormality in RA is impaired regulation of the immune system. Although alterations in various cell types and soluble mediators have been identified in patients with RA, it appears that T cells are particularly important in this "autoimmune disease." Data from a myriad of studies suggest that CD4+ T lymphocytes subserve a pivotal role in the initiation and perpetuation of the immunologically driven inflammation characteristic of RA (1,3,4). Therefore, a conceptually attractive approach to the treatment of RA targets T cells. There are sundry techniques by which the number or function of T cells may be modulated. This includes medications and biological agents that limit T-cell cytokine function and monoclonal antibodies (MAb) that diminish T-cell number or function (5,6). Experiences with several such agents are reviewed in other chapters.

Parts of this chapter are reproduced from *Arthritis and Rheumatism* (36) with the permission of the publisher.

A novel therapeutic approach for systemic inflammatory diseases such as RA is directed toward adhesion receptors. These cell surface molecules mediate heterotypic intercellular interactions. Because such interactions govern the recruitment of inflammatory cells into extravascular sites as well as the activation of these cells, adhesion receptors may be useful therapeutic targets (7). In RA, for example, adhesion receptor–directed therapy might hinder the ability of arthritogenic T cells to access the synovium and perpetuate the inflammatory response. Alternatively, inhibition of adhesion receptors might modulate the function of immunocompetent cells already present in the inflamed synovium, thereby attenuating the progression of disease. The potential value of adhesion receptor directed therapy has been established in numerous animal models of human inflammatory disease, including models that resemble RA (8). Inhibition of adhesion receptor function might be achieved by several methods. To date, the most widely employed approach has been the use of MAb directed against specific adhesion receptors.

Because T cells are presumably critical in orchestrating rheumatoid inflammation, adhesion receptor–directed therapy in RA should target the most prominent T-cell adhesion receptors. As is true of all circulating cells, T cells possess an array of adhesion receptors that they may potentially utilize to effect intercellular interactions. Of the repertoire of adhesion receptors, one receptor-counterreceptor pair plays a central role in adhesive interactions that mediate the transendothelial migration of T lymphocytes: leukocyte-function associated antigen-1 (LFA-1; CD11a/CD18) and one of its counterreceptors, intercellular adhesion molecule-1 (ICAM-1; CD54) (9). Of note, this receptor-counterreceptor interaction is also critical in the cell-to-cell contacts leading to T-cell activation (10). These adhesion receptors are therefore attractive therapeutic targets for T-cell-mediated diseases such as RA. We report on the experience with an anti-ICAM-1 MAb used to treat patients with RA.

II. METHODS AND RESULTS

A. Patients

The study population consisted of patients with an established diagnosis of RA, as defined by the 1987 American College of Rheumatology (ACR) criteria (11). Because of its preliminary nature, the initial phase of the study was designed to assess the safety and efficacy of the anti-ICAM-1 MAb in patients with relatively refractory disease. Thus, to be eligible for enrollment in this part of the study, patients were required to have had: (1) disease duration of ≥4 years at the time of study entry, and (2) failure of previous therapeutic trials with at least two disease-modifying antirheumatic drugs (DMARDS). Later, in an effort to assess the response to therapy in a group of patients with less long-standing and less refractory disease, the eligibility criteria were liberalized. In the later phase of the study, patients who had used ≤1 DMARD were treated. Patient demographics are shown in Table 1.

Although the primary endpoint of the study was the evaluation of safety, a secondary endpoint was the evaluation of efficacy. Therefore, to be eligible for treatment, patients were required to have active RA. For purposes of the study, active disease was defined by: (1) the presence of six or more swollen joints, plus the presence of two of the following three criteria; (a) ≥9 tender joints, (b) morning stiffness ≥45 min duration, and (c) erythrocyte sedimentation rate (ESR; Westergren method) ≥28 mm/hr. Initial evaluation parameters for the patients are shown in Table 1. Both groups of patients had active disease and were comparable in terms of joint counts, laboratory parameters, and subjective global assessments.

Table 1 Patient Demographics and Initial Evaluation Parameters

Parameter	Refractory patients[a] ($n = 32$)	New patients[b] ($n = 10$)
Age	48.3 ± 11.3	41.3 ± 9.8
Sex	26 ♀, 6 ♂	9 ♀, 1 ♂
Duration of RA (years)	15.2 ± 9.2	1.4 ± 2.1[d]
DMARDS previously used[c]	4.3 ± 1.5	0.3 ± 0.5
Tender joint score	26 (11–83)	30 (19–96)
Swollen joint score	28 (9–116)	15 (10–37)
A.M. stiffness (minutes)	180 (20–960)	180 (60–750)
ESR (mm/hr)	47 (10–120)	66 (26–104)
CRP (µg/dl)	2.9 (<0.8–14.4)	1.7 (<0.8–6.4)
RF (IU/ml)	290 (<30–1840)	350 (35–2000)
Patient global assessment	3 (1–4)	2 (2–3)
Physician global assessment	2 (1–4)	2 (1–3)
HAQ score	2.2 (1.0–3.5)	2.0 (1.2–3.2)

Data are reported as mean ± SD, or as median and (range).
[a] Refractory patients were those ($n = 32$) recruited in the initial phase of the study, as described in Methods. Patients had to have ≥4 years of disease duration and had to have failed ≥2 DMARDS.
[b] New patients were those ($n = 10$) recruited in the latter phase of the study, who had used ≤1 DMARD.
[c] DMARD = disease-modifying antirheumatic drug. For the refractory patients, this included: methotrexate (30 patients), intramuscular gold (28 patients), hydroxychloroquine (22 patients), sulfasalazine (17 patients), D-penicillamine (15 patients), auranofin (13 patients), azathioprine (12 patients), and cyclosporine (one patient). For the new patients, two have received methotrexate, one had received sulfasalazine, and seven had not been treated with any DMARD.
[d] Seven of the 10 patients had a disease duration of <12 months at entry.
Source: Modified from Ref. 36 with the permission of the publisher.

Exclusion criteria included functional class IV according to Steinbrocker's criteria (12), serious pulmonary, cardiac, or renal disease, intercurrent infection, or allergy to murine products. All DMARDS were discontinued at least 1 month before treatment, but patients were allowed to remain on stable doses of non-steroidal anti-inflammatory drugs (NSAID) and low-dose corticosteroids (maximum dose: ≤10 mg prednisone/day). An investigated new drug (IND) was applied for and obtained by Dr. Peter Lipsky, local Institutional Review Board approval was obtained, and all patients gave informed consent before entry into the study.

B. MAb Preparation and Administration

The anti-ICAM-1 MAb employed (BIRR1; Enlimomab) was a murine IgG2a MAb (previously designated R6.5). It is directed against extracellular domain 2 of the ICAM-1 molecule and is able to inhibit the binding of ICAM-1 by both CD11a/CD18 (LFA-1) and CD11b/CD18 (13).

The initial protocol was designed for a 5-day (study days 1–5) inhospital administration, following a dose-escalation schedule. Anti-ICAM-1 was administered intravenously over a period of approximately 20 min. Because the only previous experience in humans had been obtained in studies of renal allograft recipients, the amounts of antibody administered were based on this experience (14). The dose regimens that were assessed included total doses of 140 mg (a 60-mg loading dose followed by four daily 20-mg doses), 280 mg (1 120-mg loading dose followed by four daily 40-mg doses), and 560 mg (a 240-mg loading dose followed by four daily 80-mg doses). An initial goal of therapy was to achieve a sustained serum concentration of ≤10 µg/ml. This target was chosen because in various in vitro assays of ICAM-1-dependent adhesion, such a concentration of anti-ICAM-1 MAb achieves nearly complete inhibition.

In some patients, the protocol was modified to a shorter treatment period. These patients received a 2-day regimen with a total of 240 mg of BIRR1 administered (120 mg/day).

C. Clinical Assessment

For evaluation of the clinical response, a modification of the composite criteria established by Paulus et al. was employed (15). These criteria were established to help distinguish clinical responses related to a therapeutic intervention in RA from placebo responses. In this evaluation process, six parameters of disease activity are measured: (1) tender joint score, rated on a scale of 0 → 3 for each of 69 diarthrodial joints (maximum score 207); (2) swollen joint score, rated on a scale of 0 → 3 for each of 66 diarthrodial joints (maximum score 198); (3) duration of morning stiffness in minutes; (4) patient global assessment of disease activity, rated on a scale of 0 → 4; (5) physician global assessment of disease activity, rated on a scale of 0 → 4; and (6) the ESR (mm/hr) by the Westergren method. Using these criteria, a complete response to treatment is defined by a 75–100% improvement in at least four of the six parameters tested, a marked response as a 50 → 74% improvement in at least four of six parameters, and a moderate response as 20 → 49% improvement in at least four of six parameters. All other changes are considered to be no response. In addition to these parameters of disease activity, a modified health assessment questionnaire (HAQ) assessment of functional status (16), serum C-reactive protein (CRP) concentration, and serum rheumatoid factor (RF) concentrations were also measured. Evaluation was performed at study entry, at days 8, 15, 29, and monthly until the patient exited the protocol. Patients were exited from the protocol if they withdrew consent, or if there was any alteration in their treatment regimen.

D. Clinical Outcomes

The number of patients who experienced a response to therapy is shown in Table 2. For purposes of analysis, refractory RA patients (Table 2) have been grouped into those who received 5 days of therapy ($n = 23$) and those who received 1 or 2 days of therapy ($n = 9$). Thirteen of the 23 patients receiving the 5-day protocol had a marked or moderate response to treatment at day 8 of follow-up (56%). A marked or moderate response was sustained for 13/23 (56%) patients at days 15, 13/23 (56%) at day 29, 9/23 (39%) at day 60, and 3/23 (13%) through day 90. Patients not achieving a clinical response to therapy at day 29 were allowed to be started on other therapies at that time. For purposes of analysis, these patients are counted as nonresponders to anti-ICAM-1 throughout the time points indicated in Table 2. In addition, three patients who initially achieved a clinical response were subsequently considered nonresponders although they exited the protocol for reasons not strictly related to failure of therapy. The reasons included inability to continue follow-up ($n = 1$), undergoing an elective hip arthroplasty ($n = 1$), and being required to observe strict bed rest for unrelated reasons on the advice of another physician ($n = 1$). The patients responding through day 60 of follow-up included one of the two patients receiving the lowest dose, seven of the 20 receiving the intermediate dose, and the only patient who completed the high-dose regimen. Treatment with the shorter regimens was less efficacious. Although 3/9 (33%) patients demonstrated a moderate response at day 8, the response was sustained beyond that time for only one patient (11%).

Ten new RA patients were treated on the 5-day protocol (Table 2). The response of patients in this group appeared somewhat better than in the group of patients with more long-standing disease. Thus, 7/10 patients achieved a clinical response through day 29 of follow-up and 5/10 patients sustained their response through day 60. Three of the 10 patients treated had long-term clinical benefit. One patient

Table 2 Response to Therapy

	Refractory Patients				
	Day of treatment				
Response	8	15	29	60	90
5-day regimen ($n = 23$)					
Marked	6	5	5	4	2
Moderate	7	8	8	5	1
1- or 2-day regimen ($n = 9$)					
Marked	0	0	0	0	0
Moderate	3	1	1	1	1

	Response to Therapy: New Patients								
	Day of treatment								
	8	15	29	60	90	120	150	240	310
Evaluable patients	10	10	10	10	10	10	10	9	8
Responders									
Complete response	0	0	0	0	0	1	1	1	1
Marked response	3	4	4	3	3	1	2	1	0
Moderate response	3	2	3	2	0	1	0	0	0

Source: Modified from Ref. 36 with the permission of the publisher.

met ACR criteria for a complete remission (17) for almost a year after treatment. The changes in several disease parameters for this patient are shown in Figure 1. Two other patients are still being actively followed and have maintained a marked clinical response for 5 and 8 months, respectively.

E. Adverse Effects

During therapy, 33/42 (79%) patients experienced some type of adverse event (Table 3). The most common untoward reactions included headache, fever (T max ≤39.7°C), and nausea or vomiting. Adverse effects generally occurred on the first or second day of therapy, abated despite continued therapy, responded to symptomatic treatment, were mild to moderate in severity, and resolved without sequelae. Importantly, there were no changes in hemodynamic parameters during

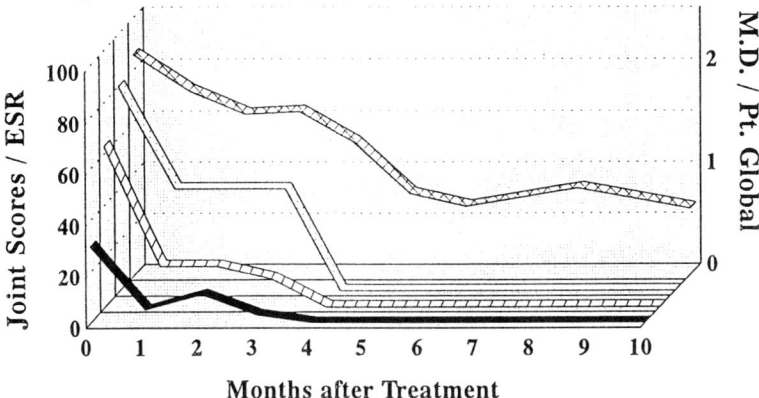

Figure 1 Changes in various parameters of clinical response in a patient with early RA as a result of treatment with anti-ICAM-1 MAb. ■, swollen joints; ◨, tender joints; ⊟, physician/patient global assessment; ⊠, ESR.

or after treatment. Furthermore, no infectious complications related to anti-ICAM-1 therapy were noted throughout the follow-up period, which ranged from 1 month to 1 year posttreatment. The occurrence of adverse reactions appeared to correlate with initial loading dose. For the first 32 patients treated, adverse events were noted in 0/2 patients receiving the 60-mg loading dose, 21/27 receiving the 120-mg dose, and 3/3 receiving the 240-mg dose.

Table 3 Adverse Effects Related to Anti-ICAM-1 MAb Therapy

Adverse effect	Number ($n = 42$)
Headache	26
Fever (<39.7°C)	17
Nausea/vomiting	15
Pruritus	8
Dizziness	5
Urticaria	5
Malaise	2
Photophobia	2
No adverse event	9

Source: Reprinted from Ref. 36 with the permission of the publisher.

Because ICAM-1 is abundantly expressed on normal endothelium, it was hypothesized that treatment with an anti-ICAM-1 MAb might result in activation of the endothelium. Conceivably, such activation could alter the normally antithrombotic properties of the endothelium. Of note, none of the 42 RA patients treated to date has experienced signs or symptoms suggestive of thrombosis or embolic phenomena. Nevertheless, to address the question of thrombotic diathesis related to therapy, a panel of coagulation studies was obtained before treatment, during treatment, and after treatment from 10 RA patients receiving anti-ICAM-1 MAb (Table 4). As can be seen, there were no significant alterations in the values obtained for PT/PTT, fibrinogen, D-dimers, proteins C or S, or ATIII. There was a trend toward an increase in the antigenic and functional activity of factor VIII/von Willebrand's factor on day 3 as compared to day 0 and day 15. In addition, there was an increase in factor VIII activity during treatment. Because ICAM-1 is abundantly expressed on endothelium, and because EC are a major source of vWF, it is hypothesized that treatment with anti-ICAM-1 MAb may cause a temporary perturbation of the endothelium, which may result in a transiently increased release of vWF. As the availability of circulating factor VIII depends to a significant extent on vWF transport, the increase in factor VIII activity noted during treatment would also be consistent with increased vWF release. Although it appeared that treatment with anti-ICAM-1 effected some

Table 4 Coagulation Parameters During Anti-ICAM-1 MAb Therapy[a]

	Day 0 (pretreatment)	Day 3 (during treatment)	Day 15 (posttreatment)
PT	11.5 (10.9–11.9)	11.2 (10.3–11.9)	11.3 (11.0–11.9)
PTT	27.6 (25.4–35)	25.2 (21.7–32.8)	26.5 (22.3–30.7)
Fibrinogen	379.6 ± 21	418.2 ± 29	381.2 ± 25
D-dimer	>05 <1.0 (<.05–>2 <4)	>0.5 <1.0 (<0.5–>1 <2)	>0.5 <1.0 (<0.5–>2 <4)
Protein C activity	110% (84–170%)	107% (89–165%)	131% (94–180%)
Protein S activity	106% (77–198%)	99% (70–175%)	118% (73–169%)
Antithrombin III	105% (93–142%)	103% (85–127%)	118% (110–162%)
vWF Ag	103% (54–410%)[b]	176% (80–262%)	80% (60–226%)[c]
vWF activity	106% (92–174%)[b]	118% (90–197%)	122% (90–197%)
Factor VIII activity	172% (59–351%)[d]	263% (97–414%)	185% (104–446%)[d]

Data represent median and range, except for fibrinogen (mean ± standard deviation).
Data were analyzed utilizing the Wilcoxon signed rank test.
[a] $n = 10$ patients studied.
[b] $p \leq .14$, day 0 compared to day 3.
[c] $p \leq .027$, day 15 compared to day 3.
[d] $p < .005$, day 0 compared to day 3.

stimulation of the endothelium, it did not appear to modify the intrinsic antithrombogenic characteristic of the endothelium substantially.

F. Laboratory Parameters

1. Pharmacokinetics

Pharmacokinetic analysis revealed that all patients had detectable serum anti-ICAM-1 during treatment. Furthermore, anti-ICAM-1 MAb could be detected in the synovial fluid during therapy. In addition, anti-ICAM-1 MAb could be detected by histochemistry on the vascular endothelium and also on perivascular leukocytes from skin biopsy specimens. Finally, the anti-ICAM-1 MAb administered to the patients could be detected bound to ICAM-1 on the surface of circulating leukocytes (18). Thus, administration of anti-ICAM-1 resulted in detectable serum levels that had biological activity, were present in synovial fluid, and coated ICAM-1 present on endothelium and circulating leukocytes.

Levels of anti-ICAM-1 MAb attained in the serum of treated patients were dose-dependent. There was a significant correlation between serum anti-ICAM-1 concentrations and an in vitro test of ICAM-1-dependent adhesive interactions, namely the capacity of treated patients' sera to inhibit the homotypic aggregation of JY cell ($\rho = -0.81; p < .01$).

Pharmacokinetic data from the first 13 patients treated are shown in Figure 2. Those patients receiving the intermediate dose regimen (120 mg/40 mg/40 mg/40 mg/40 mg) had a mean serum trough anti-ICAM-1 concentration of 17± 6 µg/ml maintained during the treatment period. This exceeded the initial goal of 10 µg/ml. As noted above, this was based on previous observations that concentrations of anti-ICAM-1 MAb ≥10 µg/ml cause nearly complete inhibition of LFA-1/ICAM-1-dependent adhesive interactions in vitro (19). However, there was no correlation between clinical improvement and any of the measured pharmacokinetic parameters. Pharmacokinetic analysis of patients who received only 1 or 2 days of anti-ICAM-1 MAb revealed that similar peak values were obtained on the days of therapy, but that serum concentrations of MAb fell to undetectable levels within a few days. Thus, given their apparently lesser clinical response, it appears likely that realization of the desired clinical endpoints requires more prolonged blockade of ICAM-1.

All patients who were tested developed IgG human anti-mouse antibodies (HAMA). HAMA were detectable by day 15 of follow-up.

2. DTH Testing

Serial DTH testing was performed on 14 patients. DTH tests were done during the course of therapy and at intervals 1 month after and, in some cases, 1 month before therapy. Patients demonstrating cutaneous anergy before treatment were not tested

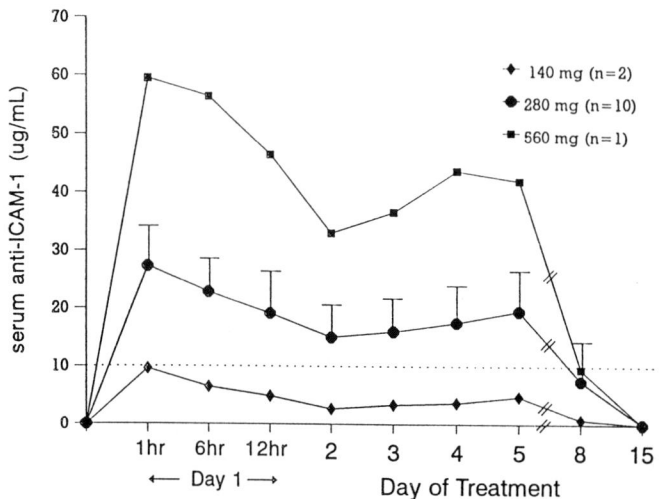

Figure 2 Pharmacokinetics of anti-ICAM-1 MAb administration. Serum concentrations on day 1 are shown at the indicated time points after infusion. Values shown for days 2–5 represent trough values.

at subsequent time points. Six patients were anergic both during and after therapy. However, three patients who were anergic during therapy displayed reactivity when tested 1 month after therapy. In addition, three patients were reactive before therapy, anergic during therapy, and reactive after therapy. This suggests that transient cutaneous anergy may be a result of anti-ICAM-1 MAb therapy. This finding was not universal, however, as two patients remained reactive during therapy. In addition, there were three patients who were reactive before treatment, but remained anergic both during and subsequent to therapy, suggesting that the cutaneous anergy may persist beyond the duration of treatment in some patients.

3. Cell Numbers and Circulating Subpopulations

The absolute cell counts for circulating lymphocytes, neutrophils, and monocytes during the different days of treatment are shown in Figure 3. There was a significant increase in the number of lymphocytes on days 2–5 of therapy, as compared to values at enrollment (day 0) and immediately prior to treatment (day 1). There was no consistent, significant change in the absolute counts of neutrophils or monocytes.

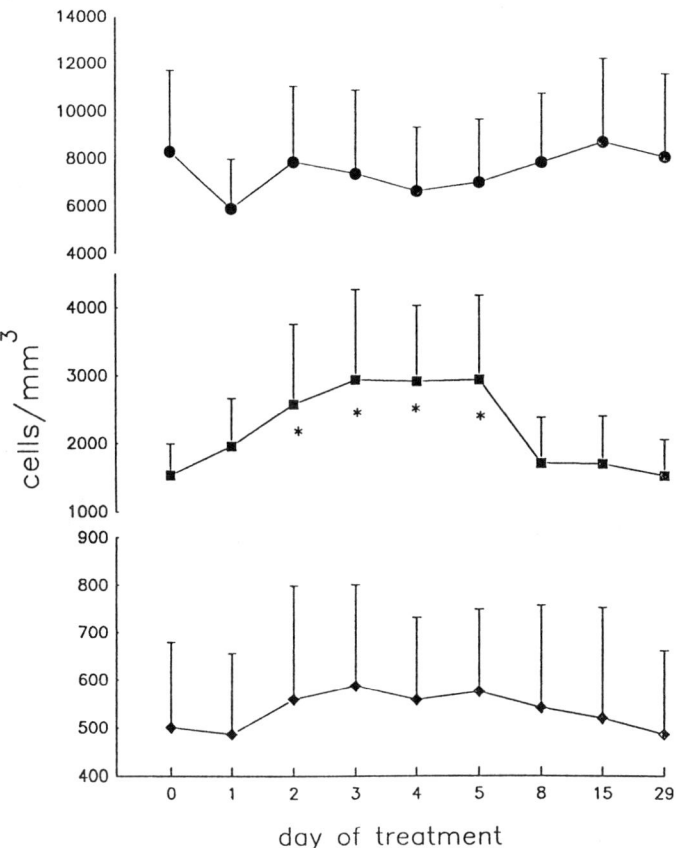

Figure 3 Alterations in leukocyte numbers as a result of therapy with anti-ICAM-1. Results are the mean for the 32 patients with refractory RA. Error bars indicate standard deviation. *Significant at $p < .0005$, compared to values on day 0 and day 1 (pretreatment. ●, neutrophils; ■, lymphocytes; ◆, monocytes. (Reprinted from Ref. 36 with the permission of the publisher.)

Analysis of the phenotype of circulating cells by fluorescein-activated cell sorter analysis indicates that during therapy, there was a significant increase in the numbers of circulating CD3+ T cells (Fig. 4). There was no significant change in circulating B-cell numbers. Further analysis revealed that the increase in T-cell numbers primarily reflected an increase in CD4+ T cells. Although most of the

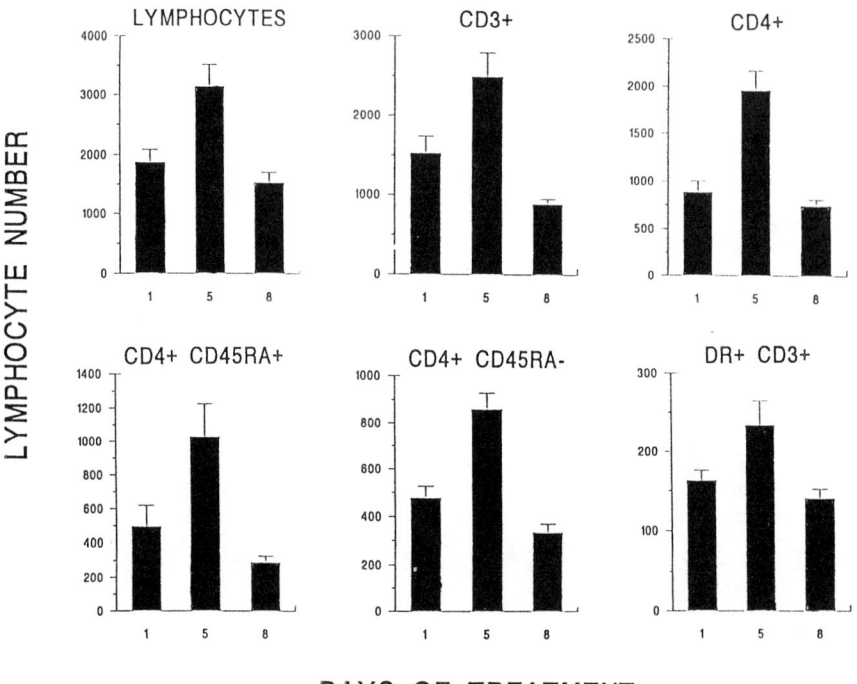

Figure 4 Lymphocyte phenotype during anti-ICAM-1 treatment. Results are for seven patients with refractory disease treated on the 5-day regimen. Numbers represent absolute number of cells/mm^3. Significant increases in total number of lymphocytes, CD3+ cells, and CD4+ cells were seen ($p < .001$). Both CD45RA+ and CD45RA− subsets of CD4+ cells were increased, as were CD3+/DR+ T cells (all $p < .05$). (Reprinted from Ref. 36 with the permission of the publisher.)

increase in circulating T cells could be accounted for by CD4+ cells, there was a small but statistically significant increase in CD8+ T cells during therapy (absolute number of CD8+ T cells/mm^3; 724 ± 122 on day 1, 971 ± 181 on day 5, 623 ± 160 on day 8; the change from day 1 to day 5 was statistically significant, $p < .05$). In the CD4+ population, there was an increase in both memory (CD45RA−) and naïve (CD45RA+) T cells. In addition, in the T-cell population there was an increase in the numbers of activated circulating T cells, as evidenced by the increased expression of HLA-DR (Fig. 2; $p < .05$) and the α chain of the IL-2R (CD25; $p < .05$).

III. DISCUSSION AND CONCLUSION

RA is a systemic inflammatory disease in which CD4+ T cells orchestrate chronic synovial inflammation. Several lines of investigation support this thesis, including analysis of the phenotypic characteristics of the cellular infiltrate in the rheumatoid synovium (20), the association of RA with certain alleles of the class II major histocompatibility complex (21), and extrapolation of data from animal studies (8). Perhaps the most compelling evidence supporting the role of T cells in RA is the established clinical efficacy of various therapeutic interventions that have in common a profound impact on T-cell numbers or function. Such therapies include lymphapheresis (22), thoracic duct drainage (23), total lymphoid irradiation (24), medications such as cyclosporine (25), and, most recently, anti-T-cell MAb (6,26). Although these interventions have all shown some efficacy, long-term remission was not usually achieved. This fact, along with various other concerns such as toxicity, has kept them from widespread utility.

A novel approach to the treatment of inflammatory diseases such as RA employs adhesion receptors as therapeutic targets. These cell surface molecules mediate the heterotypic intercellular interactions that govern not only cellular adhesion and traffic, but also cellular activation. Because they perform such a crucial role in the initiation and propagation of inflammatory reactions, therapy directed at adhesion receptors might be anticipated to modulate inflammatory responses. Indeed, in several animal models of human inflammatory disease, including antigen-induced and adjuvant arthritis (27,28), ischemia-reperfusion injury (8), inflammatory asthma (29), and renal allograft rejection (13), anti-adhesion therapy has been shown to abrogate inflammation effectively. Transient inhibition of inflammation would be expected to be beneficial in acute human inflammatory diseases as it has been in animal models. However, RA is a chronic disease, characterized by persistent inflammation, and presumably driven by the continuous activation of autoreactive T cells. Importantly, it can reasonably be postulated that therapy directed against adhesion receptors could effect a long-term modulation of the activity of inflammatory disease. It has been established that adhesion receptors serve an important role as accessory molecules in the propagation of immune responses (7,10). During the generation of immune responses, inhibition of interactions with accessory molecules has been suggested to be a mechanism through which immunological tolerance can be induced (30). Interference with adhesion receptor function might, therefore, induce tolerance to the arthritogenic antigen(s) and thereby achieve long-term mitigation of disease activity. In this regard, treatment with anti-LFA-1 and anti-ICAM-1 MAb resulted in long-term tolerance to a cardiac allograft in an animal model (31). Such treatment might therefore be an important therapeutic modality in RA. In our study, for some of the patients treated with the anti-ICAM-1 MAb a state of T-cell hyporesponsiveness appeared to be achieved. Thus, T-cell proliferative responses to mitogens were impaired in some patients after therapy, whereas proliferative responses to recall antigens were preserved. Of note, there was a correlation between this T-cell hyporesponsiveness and clinical outcome. Therefore, in this study a form of peripheral T-cell anergy may have been induced, which might account for some of the clinical improvement noted (18).

In animal studies, the ability to induce specific tolerance can be modulated by a number of variables. Among the most important factors is the time interval between the initial generation of an immune response and the subsequent attempts at tolerance induction (30). The ability to engender tolerance varies indirectly with this interval, such that it may no longer by possible to induce tolerance at extended time periods after introduction of the antigenic stimulus. In various animal models, for example experimental allergic encephalomyelitis (EAE), tolerance induction and biological benefit are most readily achieved when disease induction and specific therapy are in close temporal approximation. This has significant implications for the treatment of a chronic systemic inflammatory disease such as RA. In patients with refractory RA of many years' duration, it may be difficult for any immunomodulatory agent to achieve a long-standing remission of disease activity when used as monotherapy. Therefore, the clinical results seen in the cohort of patients with refractory RA who received anti-ICAM-1 MAb are encouraging. In addition, the extended clinical benefit noted from the treatment of patients with early disease is promising. It must be remembered, however, that even in those patients we label as having "early" RA, the immunologically driven systemic inflammatory process is certainly well established.

Conceivably, any adhesion receptor could be a target for anti-inflammatory therapy. The leukocyte integrins, which play a particularly important role in leukocyte extravasation, have been a prominent target of adhesion receptor–directed therapy. One integrin, LFA-1, plays a crucial role in the transendothelial migration of T cells (9) and is an important accessory molecule for T-cell activation (10). Because T cells appear to be the relevant etiopathogenic cell type in RA, LFA-1 and one of its counterreceptors, ICAM-1, are attractive targets of anti–adhesion receptor therapy. Furthermore, it has been demonstrated that ICAM-1 is expressed at high concentrations in the rheumatoid synovium (32), and that serum concentrations of circulating ICAM-1, presumably shed from the surface of cells, are increased in patients with RA compared with controls (33). Therefore, therapy directed against this adhesion receptor would be anticipated to be more specific for active inflammatory processes, and less nonspecifically immunosuppressive. Indeed, animal studies have suggested that interference with the function of the β chain of LFA-1 (CD18) may be associated with an increased susceptibility to severe infectious complications (34). However, studies in rabbits have not demonstrated a similar reduced resistance to infection with either gram-negative or gram-positive organisms after administration of anti-ICAM-1 MAb (Mileski, W. J., and Lipsky, P. E., unpublished observations). In addition, although mice genetically deficient in ICAM-1 can be shown to have defects in neutrophil migration as well as contact hypersensitivity, they develop normally and do not appear unduly susceptible to life-threatening infections (35). Finally, anti-ICAM-1 MAb has been administered safely to human renal allograft recipients (13).

Therapeutic use of an anti-ICAM-1 MAb was well tolerated in this study. Side effects related to therapy were similar to those reported with the use of other MAb in various diseases. In all instances, untoward effects were transient. Importantly, no infectious sequelae related to anti-ICAM-1 therapy were noted. In addition, although there was some evidence for activation of the endothelium as a result of treatment, there was no evidence that anti-ICAM-1 therapy induced a thrombotic diathesis.

Therapeutic use of anti-ICAM-1 resulted in clinical improvement in a subset of the treated patients. Although the data were from an uncontrolled trial, the criteria used for determination of clinical effect were a modification of the Paulus criteria (14). These criteria were established to differentiate true therapeutic responses in patients with RA from those related to a placebo effect. Utilizing these criteria, patients showing improvement in only one or a few indices of disease activity are not considered responders. It is therefore anticipated that therapeutic efficacy might reasonably be estimated, even in uncontrolled trials.

The mechanisms of action of anti-ICAM-1 therapy have not been unequivocally defined. However, the data suggest that administration of the MAb may cause an alteration in lymphocyte recruitment and recirculation. ICAM-1 is expressed by a variety of cell types, including endothelial cells, antigen-presenting cells, and activated T cells (7). The rationale for the use of anti-ICAM-1 was to inhibit the function of ICAM-1 on endothelial cells of the inflamed synovium such that T cells would be inhibited from accessing this site. This may have been one of the mechanisms of action, as during therapy there was a peripheral lymphocytosis. Phenotypic analysis suggested that the increase in circulating cells resulting from therapy consisted primarily of CD3+CD4+ T cells, with both memory and naïve populations being increased. There was no consistent significant increase in the numbers of circulating B cells, monocytes, or neutrophils. Of note, some of the T cells that may have been retained in the circulation as a result of anti-ICAM-1 therapy bore an activated phenotype, as demonstrated by the expression of HLA-DR and IL-2R. In addition, several of the patients demonstrated transient cutaneous anergy during therapy. These results suggest that treatment with anti-ICAM-1 inhibited the recruitment of T cells, including those with an enhanced migratory capacity. Although no attempt was made to characterize the synovial cellular infiltrate serially during therapy, the data suggest that anti-ICAM-1 MAb therapy may have caused a redistribution of T cells out of the rheumatoid synovium. Interestingly, similar changes have been noted during thoracic duct drainage, with peripheral lymphocytosis and egress of lymphocytes from the inflamed synovium (23). The results suggest that temporary interference with lymphocyte recruitment might have effected a long-lasting attenuation of synovial inflammation. However, it should be noted that whereas transient lymphocytosis was seen nearly universally, clinical efficacy was achieved in a smaller subset of patients. Therefore, mechanisms other than modulation of lymphocyte traffic are probably also involved. An attractive hypothesis is that anti-ICAM-1 MAb may have exerted an immunomodulatory effect by interfering with the interaction of T cells and antigen-presenting cells at the site of inflammation. In support of this, there was some evidence that T-cell hyporesponsiveness subsequent to treatment correlated with clinical efficacy (18). The various mechanisms by which anti-ICAM-1 exerted its immunomodulatory effects are currently being examined.

REFERENCES

1. Harris ED. Rheumatoid arthritis: pathophysiology and implications for treatment. N Engl J Med 1990; 322:1277–1289.
2. Pincus T, Callahan LF, Sale WG, Brooks AL, Payne LE, Vaughan WK. Severe functional declines, work disability, and increased mortality in seventy-five rheumatoid arthritis patients studied over nine years. Arthritis Rheum 1984; 27:864–872.
3. Janossy G, Panayi GS, Duke O, Bofill M, Poulter LW, Goldstein G. Rheumatoid arthritis: a disease of T-lymphocyte/macrophage immunoregulation. Lancet 1981; 2:839–842.
4. Strober S, Holoshitz J. Mechanisms of immune injury in rheumatoid arthritis: evidence for the involvement of T cells and heat-shock protein. Immunol Rev 1990; 118:233–255.
5. Brooks PM. Clinical management of rheumatoid arthritis. Lancet 1993; 341: 286–290.
6. Isaacs JD, Watts RA, Hazleman BL, et al. Humanised monoclonal antibody therapy for rheumatoid arthritis. Lancet 1993; 340:748–752.
7. Springer TA. Adhesion molecules of the immune system. Nature 1990; 346:425–433.
8. Carlos TM, Harlan JM. Membrane proteins involved in phagocyte adherence to endothelium. Immunol Rev 1990; 114:5–28.
9. Kavanaugh A, Lightfoot E, Lipsky P, Oppenheimer-Marks N. The role of CD11/CD18 in adhesion and transendothelial migration of T cells: analysis utilizing CD18 deficient T cell clones. J Immunol 1991; 146:4149–4156.
10. Wacholtz MC, Patel SS, Lipsky PE. Leukocyte function-associated antigen 1 is an activation molecule for human T cells. J Exp Med 1989; 170:431–448.
11. Arnett FC, Edworthy SM, Bloch DA, et al. The American Rheumatism Association 1987 revised criteria for the classification of rheumatoid arthritis. Arthritis Rheum 1988; 31:315–324.

12. Steinbrocker O, Traegre CH, Batterman RC. Therapeutic criteria in rheumatoid arthritis. JAMA 1949; 140:659–662.
13. Cosimi AB, Conti D, Delmonico FL, et al. In vivo effects of monoclonal antibody to ICAM-1 (CD54) in nonhuman primates with renal allografts. J Immunol 1990; 144:4604–4611.
14. Haug CE, Colvin RB, Delmonico FL, et al. Phase I trial of immunosuppression with anti-ICAM-1 (CD54) MAb in renal allograft recipients. Transplantation 1993; 55:766–773.
15. Paulus HE, Egger MJ, Ward JR, Williams HJ, and the Cooperative Systematic Studies of Rheumatic Diseases Group. Analysis of improvement in individual rheumatoid arthritis patients treated with disease modifying antirheumatic drugs, based on the findings in patients treated with placebo. Arthritis Rheum 1990; 33: 477–489.
16. Pincus T, Callahan LF, Brooks RH, Fuchs HA, Olsen NJ, Kaye JJ. Self-report questionnaire scores in rheumatoid arthritis compared with traditional physical, radiographic, and laboratory measures. Ann Intern Med 1989; 110:259–266.
17. Pinals RS, Baum J, Bland J, et al. Preliminary criteria for clinical remission in rheumatoid arthritis. Arthritis Rheum 1981; 24:1308–1315.
18. Davis L, Kavanaugh A, Nichols L, Lipsky P. The induction of persistent T cell hyporesponsiveness in vivo by monoclonal antibody to ICAM-1 in patients with rheumatoid arthritis. J Immunol (submitted).
19. Rothlein R, Dustin ML, Marlin SD, Springer TA. A human intercellular adhesion molecule (ICAM-1) distinct from LFA-1. J Immunol 1986; 137:1270–1279.
20. Cush JJ. Lipsky PE. Phenotypic analysis of synovial tissue and peripheral blood lymphocytes isolated from patients with rheumatoid arthritis. Arthritis Rheum 1988; 31:1230–1239.
21. Stastny P. Association of the B-cell alloantigen DRw4 with rheumatoid arthritis. N Engl J Med 1978; 298:869–873.
22. Karsh J, Klippel JH, Plotz PH, Decker JL, Wright DG, Flye MW. Lymphapheresis in rheumatoid arthritis. Arthritis Rheum 1981; 24:867–873.
23. Ueo T, Tanaka S, Tominaga Y, Ogawa H, Sakurami T. The effect of thoracic duct drainage on lymphocyte dynamics and clinical symptoms in patients with rheumatoid arthritis. Arthritis Rheum 1979; 22:1405–1412.
24. Kotzin BL, Strober S, Engleman EG, et al. Treatment of intractable rheumatoid arthritis with total lymphoid irradiation. N Engl J Med 1981; 305:969–976.
25. Yocum E, Klippel JH, Wilder RL. Cyclosporine A in severe, treatment refractory rheumatoid arthritis. Ann Intern Med 1988; 109:863–869.
26. Reiter C, Kakavand B, Rieber EP, Schattenkirchner M, Riethmüller G, Krüger K. Treatment of rheumatoid arthritis with monoclonal CD4 antibody M-T151. Arthritis Rheum 1991; 34:525–535.
27. Jasin HE, Lightfoot E, Davis LS, Rothlein R, Faanes B, Lipsky PE. Amelioration of antigen-induced arthritis in rabbits treated with monoclonal antibodies to leukocyte adhesion molecules. Arthritis Rheum 1992; 35:541–549.
28. Iigo Y, Takashi T, Tamatani T, et al. ICAM-1-dependent pathway is critically involved in the pathogenesis of adjuvant arthritis in rats. J Immunol 1991; 147:4167–4171.

29. Wegner CD, Gundel RH, Reilly P, Haynes N, Letts LG, Rothlein R. Intercellular adhesion molecule-1 (ICAM-1) in the pathogenesis of asthma. Science 1990; 247: 456–460.
30. Waldmann H, Cobbold S. The use of monoclonal antibodies to achieve immunological tolerance. Immunol Today 1993; 14:247–251.
31. Isobe M, Yagita H, Okumura K, Ihara A. Specific acceptance of cardiac allograft after treatment with antibodies to ICAM-1 and LFA-1. Science 1992; 255: 1125–1127.
32. Lindsley HB, Smith DD, Davis LS, Koch AE, Lipsky PE. Regulation of the expression of adhesion molecules by human synoviocytes. Semin Arthritis Rheum 1992; 21:330–334.
33. Cush JJ, Rothlein R, Lindsley HB, Mainolfi EA, Lipsky PE. Increased levels of circulating intercellular adhesion molecule 1 in the sera of patients with rheumatoid arthritis. Arthritis Rheum 1993; 36:1098–1102.
34. Sharar SR, Winn RK, Murray CE, Harlan JM, Rice CL. A CD18 monoclonal antibody increases the incidence and severity of subcutaneous abscess formation after high-dose *Staphylococcus aureus* injection in rabbits. Surgery 1991; 110:213–220.
35. Sligh JE, Ballantyne CM, Rich SS, Hawkins H, Smith CW, Bradley A, Beaudet AL. Inflammatory and immune responses are impaired in mice deficient in intercellular adhesion molecule 1. Proc Natl Acad Sci USA 1993; 90:8529–8533.
36. Kavanaugh A, Davis L, Nichols L, Norris S. Rothlein R, Scharschmidt L, Lipsky P. Treatment of refractory rheumatoid arthritis with a monoclonal antibody to intercellular adhesion molecule-1. Arthritis Rheum 1994; 37:992–999.

V
Potential Antigen-Specific Therapies

With the possible exception of myasthenia gravis, multiple sclerosis (MS), and type I diabetes mellitus (IDDM), the antigen (Ag) which initiates and/or perpetuates the inflammatory process in most human autoimmune diseases has not been identified. Nonetheless, potential therapies that can interfere with the trimolecular complex of major histocompatibility complex II (MHC)-Ag-T cell receptor (Tcr) are under development.

Vaccination with putatively antigen-specific attentuated T cells has been evaluated in three pilot studies with equivocal results (1,2,3). A refinement of this approach is to immunize with peptides based on Tcr β variable region expression in disease-causing (or overexpressed) T cell clones. Clinical trials are under way in multiple sclerosis (MS) (Chapter 15), rheumatoid arthritis (RA) (Chapter 16), and psoriasis evaluating vaccination with a cocktail of Tcr Vβ peptides. Alternatively, administering peptides which compete for MHC II binding may also prevent complexing of the Ag-Tcr to MHC. Studies in RA employing vaccination with a peptide based on HLA DR4/1 are currently under way (4).

Oral administration of antigen can induce specific immunologic unresponsiveness to that antigen. Although this phenomenon is well described, its underlying mechanism remains not completely understood. Several trials have evaluated oral tolerance induction in MS, RA, and uveitis (Chapters 17 and 18).

Each immunoglobulin is specific to its antigen; this specificity resides within the hypervariable region. Many of the immunomodulatory effects attributed to intravenous immunoglobulin (IVIg) treatment may occur because of interactions between autoantibody variable regions (idiotypes) and variable regions present on the administered immunoglobulin G (IgG) (anti-idiotypes). Autoantibodies to a broad variety of self antigens are present in the sera of healthy individuals, as well as those with autoimmune disease. IVIg preparations are pooled from 10,000 to 15,000 donors and presumably contain a vast assortment of idiotypes and

V: Potential Antigen-Specific Therapies

anti-idiotypes. Although unproven, id-anti-id interactions best explain many of the acute and long-term effects observed after IVIg administration in autoimmune disease (Chapter 19).

These therapies offer significant promise and they are well tolerated. Presumably, as we learn more about the inciting and perpetuating processes underlying autoimmunity we will be able to develop effective disease-specific therapies that do not alter underlying immune responsiveness.

REFERENCES

1. Kingsley G, Panayi G. Intervention with immunomodulatory agents: T cell vaccination. Bailliers's Clin Rheum 1992; 2:435–454.
2. van Laar JM, Miltenburg AMM, Verdonk MJA, et al. Effects of inoculation with attenuated autologous T cells in patients with rheumatoid arthritis. J Autoimmun 1993; 6:159–167.
3. Smith BJ, Fort JG. Treatment of rheumatoid arthritis by immunization with mononuclear white blood cells: results of a preliminary trial. J Rheumatol 1996; 23:220–225.
4. Pratt W, Heck L, Moreland L, et al. Safety and immunogenicity of a single intramuscular injection of a synthetic HLA-DR4/1 peptide vaccine with alum adjuvant in rheumatoid arthritis patients. Arth Rheum 1995; 38:S281.

15
T-Cell-Receptor Peptide Therapy for Multiple Sclerosis

ARTHUR A. VANDENBARK, DENNIS N. BOURDETTE, and
RUTH H. WHITHAM
*Portland Veterans Affairs Medical Center and Oregon Health Sciences University,
Portland, Oregon*

HALINA OFFNER
Oregon Health Sciences University, Portland, Oregon

I. INTRODUCTION

This chapter describes the clinical application of T-cell-receptor (TCR) peptides for inducing systemic autoregulation in humans with multiple sclerosis (MS). The human trials represent a direct extension of work carried out concomitantly in rodents with experimental autoimmune encephalomyelitis (EAE), a paralytic disease with a number of similarities with MS. As there has been considerable practical knowledge gained from the EAE model, we include a discussion of relevant preclinical data.

II. PRECLINICAL ANIMAL STUDIES

A. Rationale

Rats and mice developing EAE tend to overutilize certain V genes in the TCR of encephalitogenic T cells specific for myelin basic protein (BP). The common expression of germline TCR V-gene sequences by the pathogenic T cells raised the possibility that synthetic TCR peptides might be able to induce regulatory T cells, assuming there was no natural tolerance to the TCR determinants. Compared with whole-cell vaccination, the use of synthetic peptides as autoregulatory agents had the advantages of increased precision and ease of production and administration. However, in our experience, it is crucial to be able to identify which V genes are involved in the autoimmune disease, and which peptides from the V-gene sequence are actually antigenic on the available MHC background.

1. Identification of Disease-Relevant V Genes

In the Lewis rat model, Vβ8.2 is well established as the predominant V gene involved in EAE (1). We evaluated Vβ8.2 expression in lymph nodes, blood, central nervous system (CNS), and cerebrospinal fluid (CSF) during the course of EAE in Lewis rats to determine the optimal time and place to detect Vβ8.2 bias (2). We found that the highest expression of Vβ8.2 occurred on activated T cells within the affected tissue (the CNS) just prior to onset of EAE. Vβ8.2 expression was also observed on nonactivated T cells from the CSF prior to EAE onset, and a later study confirmed that both CNS and CSF Vβ8+ populations contained a high percentage of T cells with typical encephalitogenic CDR3 sequences (3).

As the clinical signs of EAE diminished during the recovery phase, Vβ8.2 expression on freshly isolated CNS T cells decreased almost to background (2). At this time, pathogenic T cells could only be distinguished by expanding activated CNS or peripheral lymphocytes with IL-2 and then focusing the response by restimulating the cells with the inducing encephalitogen, BP. These studies established that relevant V-gene biases could be detected without enhancement or knowledge of the inducing autoantigen during the acute stages of EAE. As the acute disease resolved, however, identification of disease-relevant V genes required expansion of encephalitogenic T-cell clones with the inducing autoantigen. This was accomplished most easily with cells obtained from the CNS and to a lesser degree from the CSF, which became rapidly depleted of inflammatory T cells during recovery. From blood, the most accessible compartment in humans, identification of disease-relevant V genes required isolation of T-cell lines or clones specific for BP. As presented below, BP-specific T cells from the blood of MS patients has thus far provided the most useful information in choosing target V-gene sequences for TCR peptide therapy.

2. Choice of TCR Sequence

A second important variable is to identify potentially antigenic regions within the disease-relevant V-gene sequence. We used two algorithms to predict antigenic sequences (4,5) and also reasoned that complementarity-determining regions that putatively contact antigen/MHC might provide external exposure to any antibodies that might be induced. On this basis, we chose the CDR2 peptide, residues 39–59 of the Vβ8.2 sequence, for our initial studies. As described previously (6), preimmunization with this Vβ8.2 peptide or its truncated version, residues 44–54, induced strong delayed-type hypersensitivity (DTH) reactions and complete protection against EAE in Lewis rats. In contrast, the Vβ8.2 CDR1 peptide, a Vβ8.2 CDR3 peptide patterned after an encephalitogenic clone, and CDR1 and CDR2 peptides from other Vβ sequences were not protective against EAE (7). A complete set of overlapping Vβ8.2 peptides has now been prepared to identify other potentially regulatory regions. At present, however, the CDR2 region from Vβ8.2 and other disease-relevant V genes (including Vβ6 in rats and Vβ8.2 in mice) have consistently been found to be both antigenic and highly protective against EAE, although in some instances (i.e., the 1–17 peptide of Vβ17a), a peptide outside of CDR2 may be protective. These data indicate that the peptides must be immunogenic to be protective, and that inclusion of a CDR, particularly CDR2, may improve the regulatory properties of the peptide. CDR3 peptides reported to be protective by others have not been active in our hands and are difficult to rationalize, due to the heterogeneity within this region among clones of identical fine specificity.

B. Therapeutic Effects on EAE

To be useful in human trials, TCR peptides needed to have therapeutic effects when administered after onset of clinical signs. Indeed, we found that i.d., s.c., i.p., and i.m. but not i.v. injection of relatively low doses of peptide without adjuvant prevented worsening of and speeded recovery from clinical EAE (8). Histological lesions persisted in the CNS of clinically well rats, however (7). Additionally, we found that the optimal dose was about 100–300 µg/rat, with a single dose being sufficient for a full clinical effect. Higher doses of 500–1000 µg had reduced effects. In mice given two TCR peptides (Vβ4 and 17) corresponding to disease-associated V genes, administration of a single or multiple doses reduced relapse rate and severity (9). The reduced clinical scores in rats were reflected by a significantly reduced frequency of BP-specific T cells in the blood and CNS of treated animals and "epitope switching" to nonencephalitogenic specificities of BP-specific T cells in the draining lymph nodes (10).

C. Mechanism

The ability of TCR peptides to affect the clinical course of EAE within 48 hr suggested a boosting or recall phenomenon. Further studies in recovered rats that were not treated with the TCR peptide demonstrated the natural induction of TCR peptide–specific DTH and proliferation responses that presumably arose as a consequence of the focused induction of encephalitogenic Vβ8.2+ T cells during EAE.

As is discussed in more detail elsewhere (11), immunization with the Vβ8 peptides induced T cells and antibodies, both of which could transfer protection against EAE. The T cells were CD4+CD8lo and were MHC I restricted. In vitro, these T cells inhibited the ability of BP-specific T cells to proliferate, to produce message for IL-3 and, to a lesser extent, IL-2 and IFN-γ, and to transfer EAE (12). The inhibitory effect could occur by cell-cell contact, implying the display of naturally processed, MHC-restricted TCR epitopes on the surface of Vβ8.2+ T cells, and was potentiated by addition of the synthetic peptide to the reaction mixture. Of particular interest, inhibition was mediated by soluble factors secreted by activated TCR peptide–specific T cells. These factors have not yet been identified, but could include TGF-β or IL-4, which have been shown by Swanborg's group to be produced by CD4+ suppressor T cells specific for the Vβ8-2-39-59 peptide (13,14). The major implication of this result is that antigen-specific activation may release soluble factors that mediate antigen-nonspecific "bystander suppression" of all activated T cells in the local vicinity, including T cells bearing V genes for which the regulatory T cells are not specific. This suggests that the TCR peptide approach could be useful in situations where heterogeneous V genes are involved, if there is a sufficient frequency of regulatory T cells directed at one of the disease-associated V genes.

Antibodies, mainly IgG, specific for the Vβ8 CDR2 peptides were found to weakly stain Vβ8.2+ T cells but not T cells bearing other Vβ genes (15). These antibodies appeared after 30–40 days of immunization with the TCR peptide and could suppress the EAE disease course when administered after injection of the encephalitogenic emulsion. The weak staining pattern suggests recognition of the TCR epitope on the T-cell surface, but cannot discriminate between a processed, MHC-associated peptide and a partially exposed linear determinant on the intact TCR β chain.

III. CLINICALLY IMPORTANT VARIABLES

To apply the principles of TCR peptide therapy learned in the EAE model to humans with MS, there are a number of prerequisites. Foremost, there needs to be a substantial T-cell response to a potential autoencephalitogen that involves biased V-gene expression. Our prior studies demonstrated increased frequencies of T cells specific for BP in blood and CSF of MS patients that occurred episodically (16,17). At the peak of response, these frequencies approximated those found in animals with EAE and those to recall antigens sufficient to induce DTH responses (approximately 10 cells/million). However, during periods of relatively low response, the BP-specific T-cell frequencies returned to baseline values (approximately 1 cell/million). Importantly, there appeared to be more clinical disease activity in patients during periods of elevated BP responses than during baseline responses, suggesting involvement of BP as one of the target antigens in MS.

A. Screening for Vβ Bias: Antigen-Specific T Cells from Blood and CSF

To identify target V-gene sequences for human studies, we characterized the Vβ-gene expression of a number of BP-specific T-cell clones from the blood of MS and control donors with elevated BP frequencies. Overall, a striking preferential usage of Vβ5.2, and to a lesser degree Vβ6.1, was found in BP-specific T-cell clones from seven MS patients compared with controls (18). Of 48 BP-specific clones analyzed from the MS donors, 27 used Vβ5.2 and 10 used Vβ6.1. In contrast, only two of 41 BP-specific clones from normal donors were Vβ5.2+ ($p < .0001$), and four of 41 were Vβ6.1; moreover, 0 of 15 non-BP-specific clones from MS donors were Vβ5.2+ ($p < .0001$), and two of 15 were Vβ6.1. Three of these initial MS study donors (patients NL, MR, and WS), included in the TCR peptide trial described below, were DR2+ and had strongly biased expression of Vβ5.2 and Vβ6.1 (Table 1). The common appearance of BP-specific Vβ5.2+ and Vβ6.1+ clones among DR2+ donors was remarkable in light of an independent study by Oksenberg et al. (19), who found rearranged Vβ5.2 and Vβ6.1 genes with CDR3 motifs characteristic of BP-specific T cells in DR2 (Dw2)+ cadaver brain tissue from MS patients but not controls. Together, these studies provide evidence that BP-specific T cells in the periphery may be a reflection of cells involved in CNS damage, and that BP may be a component of the pathogenic response in MS. However, it seems probable that factors such as patient selection, location, or T-cell isolation techniques may influence V-gene bias, since other studies have identified different V genes (20,21) or no V-gene bias (22) in MS responses to BP.

Table 1 Summary of Vβ Use in Clones from MS Patients and Controls

				Clones, no. positive		
Patient	HLA type	Tested	Specificity	Vβ5.2	Vβ6.1	Other Vβs[a]
NL	DR1,2; DQw1	14	BP	13	0	Vβ5.1
MR	DR2; DQw1	10	BP	4	2	Vβ9(2×),-3,-4,-15,-19
		3	PPD	0	1	Cβ8.1,-13.2
WS	DR2; DQw1	6	BP	3	2	Vβ12
		7	Nonspecific	0	1	Vβ13.1(2×),-10,-13.2,-15,-19

[a]In this analysis, only the predominant Vβ band for each clone is included. However, for some clones, two bands of nearly equal intensity are both recorded and, therefore, the total number of Vβs may be greater than the number of clones analyzed.
PPD, purified protein derivative; 2×, twice.

The studies described above in Lewis rats demonstrated that Vβ8.2+ T cells derived from CSF at onset of EAE reflected the pathogenic BP-specific T cells found in the inflamed CNS tissue. On this basis, the CSF might be preferable to blood as a source of potentially pathogenic T cells from which to identify disease-relevant V genes. To address this possibility, we carried out serial comparisons of BP-specific T cells from paired CSF and blood samples from two patients with MS (23). In one patient with relapsing MS, the CSF, sampled during a relapse, contained elevated cell numbers that included Vβ5+ T cells specific for a variety of BP epitopes. The paired blood sample also contained Vβ5+ T cells specific for BP, but fewer epitopes were recognized. Four months later, during remission, the CSF cellularity was significantly decreased, with no evidence of BP-reactive T cells, whereas Vβ5+ BP-specific T cells persisted in the blood. In the second patient with chronic progressive MS, the CSF cellularity remained slightly elevated in the initial and two successive samplings taken 1 and 4 months later. BP-reactive T cells were present in all three samplings, and as expected, the V-gene bias present in the first CSF sample was reflected in the paired blood sample. Surprisingly, the initial V-gene bias switched to a different pattern in the second and third CSF samples, and the paired blood samples then expressed both the initial and subsequent biased V-gene pattern. These data suggest that BP-specific T cells from the blood provide a partial but long-lasting reflection of BP-specific T cells from the CSF. Put a different way, the CSF most likely provides a more representative view of inflammatory T cells in the CNS than does blood, but only during periods of disease activity when there is an elevated CSF cellularity. In contrast, most of the relevant T-cell specificities can be obtained from blood samplings, even months after a clinical relapse. In that TCR peptide–specific T cells may be able to regulate T cells expressing a variety of different TCRs by "bystander" suppression, immunoregulation induced by TCR peptides from the major disease-associated V genes identified in blood samples will likely be sufficient to substantially inhibit systemic BP reactivity.

B. Selection of TCR Epitopes

Based on the overexpression of Vβ5.2 and Vβ6.1 by BP-specific T cells from the blood of MS patients but not control donors, and the concomitant demonstration of Vβ5.2 and Vβ6 message from plaque tissue obtained from the CNS of DR2 (Dw2) donors, we decided to inject peptides from the Vβ5.2 and Vβ6.1 sequences into MS patients. As in the rat Vβ8.2 sequence, TCR residues 39–59, containing the CDR2 of human Vβ5.2 and Vβ6.1 sequences, were predicted to be antigenic, and thus, these two peptides were synthesized and tested in a phase I clinical trial (24). In retrospect, TCR CDR2 peptides from Vβ5.2 and Vβ6.1 were the optimal choice for inducing regulatory T cells, because other regions of the germline

Vβ5.2 sequences were subsequently found to be much less active (in preparation). Moreover, CDR2 peptides from three additional Vβ genes have been used successfully to boost TCR-specific T-cell responses in MS patients overexpressing the same V genes in response to BP. The consistent immunogenicity of residues 39–59 suggests that the CDR2 may generally represent the most biologically important region of the V-gene sequence.

C. Injection Regime

The variables for administering TCR peptides to humans (dose, route, carrier, timing) were based largely on information from the preclinical animal studies described above. First, we assumed that specific responses could only be induced by boosting an already-present, naturally primed set of TCR-specific T cells directed against the V genes used in response to BP. *Thus, we chose to start with a relatively low dose (100 µg) of TCR peptides, given intradermally in buffered saline once per week for 4 weeks and once monthly thereafter.* If responses failed to develop at the initial dose, the dose was escalated progressively to 200 and 300 µg and, in some patients, to 600, 1500, and 3000 µg. The intradermal route was chosen to allow for the development of DTH responses at the injection site, which could provide complementary information to the T-cell frequency analysis from blood for measuring T-cell recognition of TCR peptides. Serum was also collected for analysis of anti-TCR antibodies, as measured by ELISA. Eleven patients, most with fairly advanced progressive MS, were included in the study, including the three patients, NL, MR, and WS, in whom we had demonstrated the overexpression of Vβ5.2 and Vβ6.1 by BP-specific T cells.

IV. ASSESSMENT OF EFFECTS

A. Changes in the Frequencies of TCR Peptide–Specific T Cells

All 11 patients had measurable circulating T cells specific for Vβ5.2-39-59 and Vβ6.1-39-59 prior to injection, ranging from 0.2 to 2.8×10^{-6} (24). After boosting with the peptides, most patients had significantly increased T-cell frequencies. Seven patients responded to Vβ5.2-39-59, and six of these patients also responded to Vβ6.1-39-59 (Fig. 1). Four patients failed to respond to either peptide. Interestingly, one of these patients later responded to a CDR2 peptide from Vβ9, which was preferentially used in this patient's T-cell response to BP. Among responders, maximal T-cell frequencies for either peptide ranged between 2.2 and 25×10^{-6}. Significant boosting was detected 1–8 weeks after the first peptide injection and peaked between 1 and 15 weeks. Significant responses persisted from 4 to more than 60 weeks, in some cases even after cessation of the peptide injections.

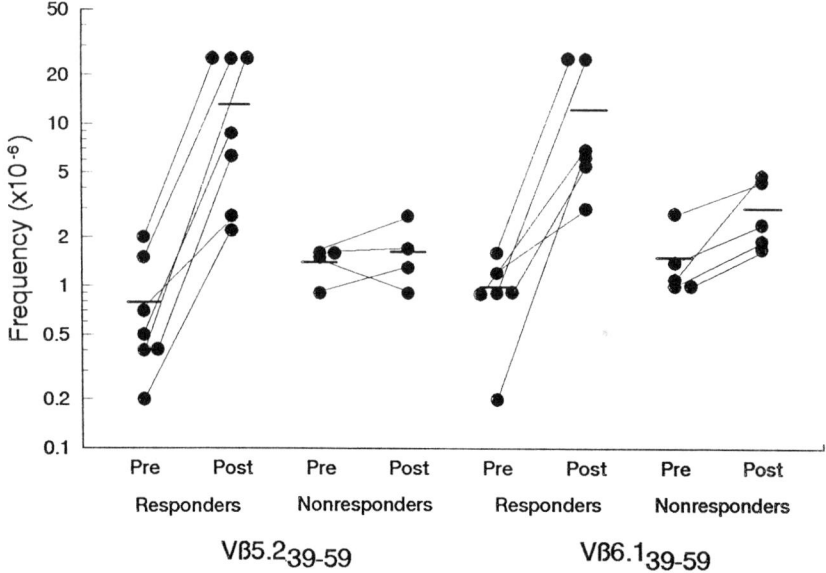

Figure 1 Preimmunization and maximum postimmunization T-cell frequencies for Vβ5.2$_{39-59}$ and Vβ6.1$_{39-59}$ for each patient. The arithmetic mean for each group is represented by a solid bar. (From Ref. 24.)

Longitudinal changes in T-cell frequencies to Vβ5.2-39-59, Vβ6.1-39-59, BP, and herpes simplex virus (HSV) are illustrated during immunization for patients NL (Fig. 2) and MR (Fig. 3). Of note, increases in dose above 600 μg resulted in decreased T-cell frequencies (see Fig. 3). Biopsy-proven DTH responses were observed at both injection sites for Vβ5.2-39-59 and Vβ6.1-39-59 in one of the responders. DTH responses tended to occur in patients when T-cell frequencies were >8 × 10^{-6}. An increased titer of TCR peptide–specific antibody, predominantly IgG, was detected in only one responder to each peptide.

B. Changes in BP Frequencies

All three patients whose BP-specific T cells preferentially used Vβ5.2 or Vβ6.1 were peptide responders (24). The majority of BP-specific T-cell clones from two of these patients (MR and WS) used either Vβ5.2 or Vβ6.1, whereas for the other (NL), 13 of 14 clones used Vβ5.2, one used Vβ5.1, and none used Vβ6.1. In these patients, changes in the frequencies of BP-specific T cells occurred coincident with peptide immunization. In NL, BP-specific T-cell frequencies did not change

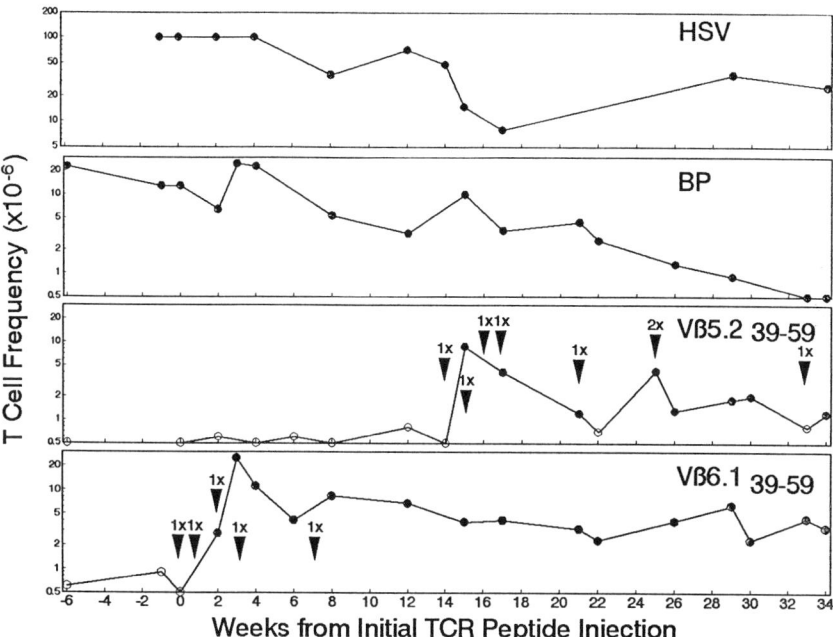

Figure 2 Serial T-cell frequencies from patient NL for HSV, MBP, Vβ5.2$_{39-59}$, and Vβ6.1$_{39-59}$. Arrowheads indicate time when the patient received peptide. Doses were as follows: 1×, 100 µg; 2×, 200 µg. A significant rise in T-cell frequencies to Vβ6.1$_{39-59}$ occurred beginning 2 weeks after the first injection of peptide. Significantly elevated T-cell frequencies for Vβ6.1$_{39-59}$ persisted for over 27 weeks after the last injection with Vβ6.1$_{39-59}$. A significant rise in T-cell frequencies to Vβ5.2$_{39-59}$ also occurred after the first injection of this peptide. MBP-specific T-cell frequencies were initially high and remained elevated during immunization with Vβ6.1$_{39-59}$. MBP-specific T-cell frequencies fell to low levels after immunization with Vβ5.2$_{39-59}$. The frequency of T cells specific for HSV fluctuated over the course of the study. (From Ref. 24.)

substantially while she was receiving the Vβ6.1 peptide, which her BP-specific T-cell clones did not use, and then fell to levels significantly below her preimmunization frequency once she was immunized with the Vβ5.2 peptide (Fig. 2). A different pattern was seen in the two patients whose BP-specific T cells utilized both Vβ5.2 and Vβ6.1. In MR, the frequency of BP-specific T cells rose above preimmunization levels approximately 16 weeks after initiation of treatment with Vβ5.2-39-59 and then returned to preimmunization levels after he was immunized with Vβ6.1-39-59 along with Vβ5.2-39-59 (Fig. 3). Assuming regulation of the Vβ5.2+ T cells upon initial immunization, it seems possible that

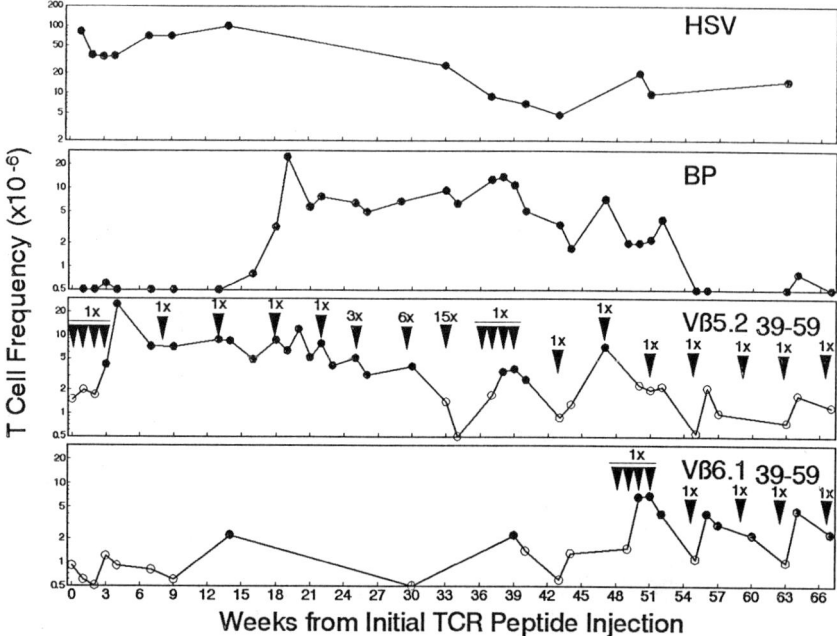

Figure 3 Serial T-cell frequencies from patient MR for HSV, MBP, Vβ5.2$_{39-59}$, and Vβ6.1$_{39-59}$. Arrowheads indicate time when the patient received peptide. Doses were as follows: 1×, 100 µg; 3×, 300 µg; 6× 600 µg; 15× 1500 µg. A significant rise in T-cell frequencies to Vβ6.1$_{39-59}$ occurred beginning 3 weeks after the first injection. After dose escalation to 300–1500 µg, the frequency of T cells specific for Vβ5.2$_{39-59}$ decreased. A significant rise in T cell frequencies to Vβ6.1$_{39-59}$ also occurred after immunization with this peptide. MBP-specific T-cell frequencies were initially low but rose significantly above basline approximately 16 weeks after initiation of treatment with Vβ5.2$_{39-59}$. MBP-specific T-cell frequencies returned to baseline levels after immunization with Vβ6.1$_{39-59}$. Frequencies of T cells specific for HSV fluctuated over the course of the study. (From Ref. 24.)

the BP response that arose during immunization might have been dominated by Vβ6.1$^+$ T cells that were subsequently regulated by coinjection of Vβ6.1-39-59 peptide. A similar pattern occurred in WS; BP-specific T-cell frequencies rose above preimmunization levels after initiation of the Vβ6.1-39-59 peptide and then fell as he was immunized with Vβ5.2-39-59 (data not shown). The frequencies of T cells specific for HSV, a recall Ag, fluctuated over the course of the study, but did not parallel the changes in BP- or TCR peptide–specific T-cell frequencies (Figs. 2 and 3).

C. Toxicity of TCR Peptides

TCR peptide therapy did not appear to alter general immunity (24). None of the patients developed cell-mediated immune skin test anergy on therapy, nor did any manifest changes in response of their blood cells to test mitogens or recall Ag. No abnormalities were seen in any patient on complete blood count, 24-channel chemistry panel, or urinalysis. Seven patients experienced no side effects. One patient had a vasovagal reaction after the second injection, and this did not recur with subsequent injections. Two patients who developed vigorous DTH reactions experienced itching at the injection site, but did not require treatment. One patient, who did not have detectable T-cell responses to either TCR peptide, developed a biopsy-proven leukocytoclastic vasculitis involving the skin of his lower extremities after receiving 3000 µg of Vβ5.2-39-59. Further peptide injections were withheld; he received a short course of prednisone, and the vasculitis resolved completely. No antibodies or immune complexes were detected that could account for the vasculitis, and since he was a peptide nonresponder, we could not definitely link the vasculitis with TCR peptide administration.

D. Clinical Changes

Twelve months after the initiation of TCR peptide immunization, a determination was made as to whether each patient was worse, stable, or improved clinically, as indicated in Table 2. Overall, of the seven TCR peptide responders, two were improved, two were stable, and three were worse. In contrast, all four of the nonresponders were worse. Of particular importance, the three patients who had biased expression of Vβ5.2 and Vβ6.1 in response to BP were either improved (NL and MR) or stable (WS). It is also noteworthy that the two patients who improved during TCR peptide therapy had the least disease disability at the initiation of the trial. These observations suggested that the TCR peptide responders may have had a more favorable clinical course than the nonresponders, and that TCR peptide therapy may be most beneficial in patients with less disability. However, no statistical conclusion may be drawn owing to the small sample size.

V. CONCLUSIONS

There is now a large body of research data on the use of TCR peptides in the treatment of EAE, an animal model for MS. Lessons from the EAE model have been and will continue to be invaluable in guiding the development of this potentially powerful treatment approach as a therapeutic tool for MS. EAE has taught us that treatment of MS as well as other human autoimmune diseases with specific TCR peptides will require (1) identification and characterization of

Table 2 Clinical Outcome in MS Patients Treated with TCR Peptides Vβ5.2–39–59 and Vβ6.1–39–59

Patients	EDSS		25' Walk (sec)		9HPT (R/L) (sec)		B&BT (R/L) (#/min)		Clinical outcome
	0 mos	12 mos	0 mos	12 mos	0 mos	12 mos	0 mos	12 mos	
Responders									
MR[a]	3.5	3.5	6	5	36/25	37/27	44/56[b]	52/68[b]	Improved
NL[a]	4.5[b]	3.0[b]	6	8	26/33	30/36	49/39	56/44	Improved
WS[a]	6.5	6.5	101	70	85/39	98/36	40/60	40/56	Stable
OB	7.5	7.5	—	—	41/55	37/56	37/37	40/40	Stable
JM	6.0[b]	6.5[b]	28	36	28/33	27/31	64/56	60/60	Worse
JS	6.5	6.5	9[b]	16[b]	55/93[b]	61/—[b]	31/20	28/26	Worse
KJ	6.5	6.5	20[b]	34[b]	31/26	33/25	56/68[b]	52/52[b]	Worse
Nonresponders									
HK	6.0	6.0	12[b]	21[b]	27/21	27/22	56/76	52/68	Worse
RS	6.0[b]	6.5[b]	8[b]	63[b]	18/23[b]	21/29[b]	81/66	68/68	Worse
CC	6.0[b]	7.0[b]	37[b]	—[b]	31/42	34/43	40/40[b]	52/52[b]	Worse
JC	6.0	6.0	7	6	26/30[b]	23/37[b]	64/60	68/60	Worse

[a]Patient had known V-gene bias for Vβ5.2 or Vβ6.1, as previously described.
[b]Parameter that had changed and that determined the clinical outcomes.

Pt, patient; EDSS, Expanded Disability Status score; 25' Walk, timed 25-ft walk; 9HPT, nine-hole peg test; B&BT, box and block test; R, right upper extremity; L, left upper extremity; sec, seconds; #/min, number of blocks per minute; mos, months.

Patients were immunized with Vβ5.2–39–59 and Vβ6.1–39–59 and classified as responders or nonresponders, as previously described. At baseline (0 months) when peptide therapy was started and 12 months after starting peptide therapy, patients underwent a clinical evaluation that included determination of their EDSS, measurement of the time to walk 25 ft, and measurement of the 9HPT and B&BT for each extremity. These clinical parameters were considered to have improved or worsened significantly at 12 months based on the following: EDSS change of 1.0 or greater if the baseline EDSS was <6.0 or a change of 0.5 or greater if the baseline EDSS was ≥6.0; timed 25-ft walk change of 50% or greater from the baseline measurement; 9HPT or B&BT change of 20% or greater in either hand. Clinical outcome at 12 months was determined as follows: patients classified as improved had one or more clinical parameters that had improved and no parameters that had worsened; patients classified as stable had no clinical parameters that had improved or worsened; patients classified as clinically worse had one or more clinical parameters that had worsened. Dashes indicate that the patient was unable to perform the test.

disease-causing T cells, (2) preferential TCR V-gene usage of the disease-causing T cells, and (3) successful immunization of patients with synthetic peptides from the disease-associated TCR V genes.

While T cells with other specificities may be important, there is considerable evidence that BP-specific T cells may be relevant to the pathogenesis of MS. For now, we believe it is reasonable to continue to focus efforts on regulating BP-specific T cells in MS. BP-specific T cells, isolated from the blood of most patients, usually demonstrate a restricted use of one to three TCR Vβ genes, which is a necessary precondition for TCR peptide therapy to be beneficial. However, there is considerable interpatient variability and it appears unlikely that one or two TCR peptides will be effective in treating the majority of patients. TCR peptide therapy thus will probably need to be individualized because of the interpatient variability in V-gene bias.

Our initial work in humans suggests that TCR peptide immunization is safe and well tolerated and does not induce broad-spectrum immunosuppression. In addition, in patients in whom we had demonstrated a Vβ5.2 or Vβ6.1 bias, immunization with CDR2 peptides from these V genes appeared to alter the circulating frequency of BP-specific T cells, similar to what occurs in Lewis rats. At present, we do not know whether TCR peptide immunization can alter the clinical course of MS. However, our initial clinical data did not indicate that TCR peptide immunization worsened MS, and patients who were successfully immunized, particularly the three patients who had known V-gene biases, appeared to do better clinically than the nonresponders. A large, placebo-controlled trial of TCR peptide immunization eventually will be needed before any conclusions about clinical efficacy can be made.

Peptides from the CDR2 region appear to be the most consistently immunogenic among TCR Vβ peptides. Our initial trial in humans indicated that we could immunize about 60% of patients with peptides from the CDR2 regions of Vβ5.2 and Vβ6.1. This was accomplished using low doses (100–300 μg) and without the use of adjuvants. It is unclear whether using adjuvants will increase the frequency of responders or intensity of responses to a given TCR peptide. However, thus far, we have been consistently able to boost immunity in patients when we challenge them with a CDR2 peptide from a V gene that is biased among their BP-specific T cells. Therefore, TCR peptide treatment based on knowledge of V-gene bias may be more effective than immunizing with a single given peptide.

In summary, TCR peptide therapy remains a promising, new treatment approach for MS and other human autoimmune diseases. The major obstacle to the development of TCR peptide therapy for MS and other human diseases may be the need to devise an easy method for determining V-gene bias in individual patients. Once the V-gene bias is known for a given patient, it appears likely that the patient can be safely immunized with a CDR2 peptide from the appropriate

V-gene family, activating regulatory T cells that can down-regulate disease-causing cells. Devising easy methods of identifying V-gene biases is a solvable problem. Once solved, TCR peptide therapy should rapidly become a safe and effective treatment approach for use in humans.

ACKNOWLEDGMENTS

This work was supported by the Department of Veterans Affairs and by NIH Grants NS23221 and NS23444. The authors wish to acknowledge the scientific contributions of our Portland collaborators, Abigail C. Buenafe, Bozena Celnik, Yuan K. Chou, Richard E. Jones, Margarita R. Vainiene, and Andrew D. Weinberg, and the secretarial assistance of Ms. Eva Jarvie.

REFERENCES

1. Heber-Katz E, Acha-Orbea H. The V-region disease hypothesis: evidence from autoimmune encephalomyelitis. Immunol Today 1989; 10:164–169.
2. Offner H, Buenafe AC, Vainiene M, Celnik B, Weinberg AD, Gold DP, Hashim G, Vandenbark AA. Where, when, and how to detect biased expression of disease-relevant Vβ genes in rats with experimental autoimmune encephalomyelitis. J Immunol 1993; 151:506–517.
3. Buenafe AC, Vainiene M, Celnik B, Vandenbark AA, Offner H. Analysis of Vβ8-CDR3 sequences derived from the CNS of Lewis rats with experimental autoimmune encephalomyelitis. J Immunol 1994; 153:396–394.
4. Margalit H, Spouge JL, Cornette JL, Ease KB, DeLisi C, Bersofksy JA. Prediction of immunodominant helper T cell antigenic sites from the primary sequence. J Immunol 1987; 138:2213–2229.
5. Rothbard JB, Taylor WR. A sequence pattern common to T cell epitopes. EMBO J 1988; 7:93–100.
6. Vandenbark AA, Hashim G, Offner H. Immunization with a synthetic T-cell receptor V-region peptide protects against experimental autoimmune encephalomyelitis. Nature 1989; 341:541–544.
7. Offner H, Hashim G, Chou YK, Bourdette D, Vandenbark AA. Prevention, suppression and treatment of EAE with a synthetic T cell receptor V region peptide. In: Alt FW, Vogel MG, eds. Molecular Mechanisms of Immunological Self-Recognition. Orlando, FL: Academic Press 1993; 199–230.
8. Offner H, Hashim GA, Vandenbark AA. T cell receptor peptide therapy triggers autoregulation of experimental encephalomyelitis. Science 1991; 251:430–432.
9. Whitham RH, Kotzin BL, Buenafe AC, Weinberg AD, Jones RE, Hashim GA, Hoy CM, Vandenbark AA, Offner H. Treatment of relapsing experimental autoimmune encephalomyelitis with T cell receptor peptides. J Neurosci Res 1993; 35:115–128.
10. Vandenbark AA, Vainiene M, Celnik B, Hashim G, Offner H. TCR peptide therapy decreases the frequency of encephalitogenic T cells in the periphery and the central nervous system. J Neuroimmunol 1992; 39:251–261.

11. Vandenbark AA, Hashim G, Offner H. TCR peptide therapy in autoimmune diseases. Int Rev Immunol 1993; 9:251–276.
12. Offner H, Vainiene M, Celnik B, Weinberg AD, Buenafe A, Vandenbark AA. Co-culture of TCR peptide-specific T cells with basic protein-specific T cells inhibits proliferation, IL-3 mRNA, and transfer of experimental autoimmune encephalomyelitis. J Immunol 1994; 153:4988–4996.
13. Karpus W, Swanborg R. CD4+ suppressor cells inhibit the function of effector cells of EAE through a mechanism involving transforming growth factor-beta. J Immunol 1991; 146:1163–1168.
14. Karpus WJ, Gould KE, Swanborg RH. CD4+ suppressor cells of autoimmune encephalomyelitis respond to T cell receptor-associated determinants on effector cells by interleukin-4 secretion. Eur J Immunol 1992; 22:1757–1763.
15. Hashim GA, Vandenbark AA, Galang AB, Diamanduros T, Carvalho E. Srinivasan J, Jones R, Vainiene M, Morrison WJ. Offner H. Antibodies specific for a Vβ8 T cell receptor peptide suppress experimental autoimmune encephalomyelitis. J Immunol 1990; 144:4621–4627.
16. Chou YK, Bourdette DN, Offner H, Whitham R, Wang R, Hashim GA, Vandenbark AA. Frequency of T cells specific for myelin basic protein and myelin proteolipid protein in blood and cerebrospinal fluid in multiple sclerosis. J Neuroimmunol 1992; 38:105–114.
17. Vandenbark AA, Bourdette DN, Whitham R, Hashim GA, Chou YK, Offner H. Episodic changes in T cell frequencies to myeling basic protein in patients with multiple sclerosis. Neurology 1993; 43:2416–2417.
18. Kotzin BL, Karuturi S, Chou YK, Lafferty J, Forrester JM, Better M, Nedwin GE, Offner H, Vandenbark AA. Preferential T cell receptor Vβ gene usage in myelin basic protein reactive T cell clones from patients with multiple sclerosis. Proc Natl Acad Sci USA 1991; 88:9161–9165.
19. Oksenberg JR, Panzara MA, Begovich AB, Mitchell D, Erlich HA, Murray RS, Shimonkevitz R, Sherritt M, Rothbard J, Bernard CCA, Steinman L. Selection of T-cell receptor Vβ-Dβ-Jβ gene rearrangements with specificity for a myelin basic protein peptide in brain lesions of multiple sclerosis. Nature 1993; 362:68-70.
20. Wucherpfennig KW, Ota K, Endo N, Seidman JG, Rosenzweig A, Weiner HS, Haffner DA. Shared human T cell receptor Vβ usage to immunodominant regions of myelin basic protein. Science 1990; 248:1016–1019.
21. Ben-Nun A, Liblau RS, Cohen L, Lehmann D, Tournier-Lasserve E, Rosenzweig A, Jingwu Z, Raus JCM, Bach MA. Restricted T-cell receptor Vβ gene usage by myelin basic protein-specific T-cell clones in multiple sclerosis: predominant genes vary in individuals. Proc Natl Acad Sci USA 1991; 88:2466–2470.
22. Martin R, Utz U, Coligan JE, Richert JR, Flerlage M, Robinson E, Stone R, Biddison WE, McFarlin DE, McFarland HF. Diversity in fine specificity and T cell receptor usage of the human CD4+ cytotoxic T cell response specific for the immunodominant myelin basic protein peptide 87-106. J Immunol 1992; 148:1359–1366.
23. Chou YK, Buenafe AC, Dedrick R, Morrison WJ, Bourdette DN, Whitham R, Atherton J, Lane J, Spoor E, Hashim GA, Offner H, Vandenbark AA. T cell receptor Vβ gene usage in the recognition of myelin basic protein by cerebrospinal fluid- and

blood-derived T cells from patients with multiple sclerosis. J Neurosci Res 1994; 37:169–181.
24. Bourdette DN, Whitham RH, Chou YK, Morrison WJ, Atherton J, Kenny C, Liefeld D, Hashim GA, Offner H, Vandenbark AA. Immunity to T cell receptor peptides in multiple sclerosis. I. Successful immunization of patients with synthetic Vβ5.2 and Vβ6.1 CDR2 peptides. J Immunol 1994; 152:2510–2519.

16
T-Cell-Receptor Peptide Vaccination Studies in Rheumatoid Arthritis

LOUIS W. HECK, LARRY W. MORELAND, and WILLIAM J. KOOPMAN
University of Alabama at Birmingham, Birmingham, Alabama

I. INTRODUCTION

Rheumatoid arthritis (RA) is a systemic disease characterized by chronic inflammation, abnormal humoral and cellular immune responses, and synovial hyperplasia and is defined by the pattern of joint involvement. Despite intense investigation, no etiological agent has been identified and RA remains incurable by current medical treatment (1).

RA is also considered to be an example of an autoimmune disease in which circulating autoantibodies and/or autoreactive T cells are thought to be effector mechanisms of joint tissue destruction (2). In contrast to an autoimmune response, the normal task of the human immune response is to resist foreign invading microbial agents by mounting a specific immune response to foreign antigens by discriminating self-antigens from non-self-antigens. The normal immune response to self antigens is tolerance (immune unresponsiveness). Since autoimmunity may involve aberrant humoral and/or cellular immune responses, much investigation has focused on T-cell abnormalities in autoimmune diseases since helper T cells serve pivotal roles in the regulation of immune responses to protein antigens (2,3). Furthermore, the susceptibility to several autoimmune diseases including RA is linked to particular MHC class II molecules, which could in turn regulate the T-cell immune response in at least two ways: a particular allele polymorphism may result in differences in amino acid composition allowing no or preferential binding of certain peptide antigens to T cells and MHC polypeptide chains, or the problem of autoimmunity could arise in the process of discrimination of self from nonself occurring during T-cell repertoire formation or in the maintenance of tolerance (3,4).

Although there is no exact animal model of RA, investigation of rodent models of experimental allergic encephalomyelitis (EAE), diabetes, collagen, and adjuvant arthritis have provided important insights regarding the role of the T cell as a mediator of specific tissue injury and destruction in autoimmune diseases (5,6). In EAE, which resembles multiple sclerosis, the disease can be induced by immunizing animals with myelin basic protein (MBP) or constituent peptides of this protein (5). Approximately 2–3 weeks later, the clinical symptoms become apparent and are associated with perivascular mononuclear cells and demyelination in the brain and spinal cord. In this rodent model, EAE is caused by autoreactive CD4+ T cells directed against myelin basic protein as demonstrated by adoptive transfer experiments with naïve animals. Furthermore, this experimental disease can be ameliorated or prevented by a variety of T-cell-directed immunotherapies including inhibition of MHC recognition by anti-MHC antibodies; peptides that compete for antigen, MHC, or T-cell binding; inhibition of T-cell activation with anti-T-cell-receptor antibodies or anti-CD4 antibodies; and vaccination with "pathological" T-cell-receptor peptides or whole T cells in which the recipient initiates an immune response against its own pathological T cells (6–18).

Efforts to isolate clones of specific autoreactive T cells in RA have been unsuccessful, yet the infiltration of lymphocytes and monocytes in the inflamed synovial membrane strongly suggests the presence of a cellular immune responses to an unknown, but persistent antigen necessary for maintenance of the disease (2). Most of these infiltrating cells are CD4+ T cells expressing CD29/CD45 RO antigens. Efforts to either reduce the number of circulating T cells or inhibit T-cell regulation of immune function in refractory RA patients using total nodal radiation, thoracic duct drainage, lymphophoresis, and cyclosporine as well as the amelioration of RA in AIDS patients support the notion that T cells play an important role in the pathogenesis of RA (2).

II. EAE AS A MODEL FOR T-CELL-MEDIATED AUTOIMMUNE DISEASE

As mentioned previously, EAE is an autoimmune demyelinating disease induced by immunizing mice or rats with defined peptide fragments of myelin basic protein (MBP). At least five encephalitogenic T-cell epitopes of MBP in two inbred mouse strains and one major encephalitogenic epitope in the Lewis rat have been identified (5). In these model systems, the animals are immunized with MBP

T-Cell-Receptor Peptide Vaccination 191

peptide fragments and complete Freund's adjuvant, and signs of paralysis will develop in 2–3 weeks, which is correlated with the invasion of the central nervous system with CD4+ T lymphocytes leading to demyelination. Infiltrating effector T cells may be isolated from pathological lesions and propagated in tissue culture. These cloned T cells can be transferred into naïve syngeneic animals and induce EAE. When these encephalitogenic T cells were cloned and studied by a combination of techniques including cell surface staining and Southern blot analysis, TCR sequencing limited heterogeneity in the use of Vβ or Vα gene expression was noted (19,20). In one study it was reported that 85% of MBP P1-9-specific T-cell clones in PL/J mice express Vβ 8 and all express Vα 4 (5). That this oligoclonal or biased TCR usage was not a cloning artifact was demonstrated through studies of T cells isolated from draining lymph nodes from these animals using FACS analysis. Only the CD4+/Vβ-positive cells proliferated in vitro when exposed to MBP 1-11, suggesting that this restriction is a true reflection of the in vivo immune cellular response. This limited TCR usage by encephalitogenic T cells suggested that various intervention therapies such as targeting the TCR could be developed to control the immune responses to specific antigens such as infusing antibodies to the TCR (8,9).

In a series of experiments reported by Acha-Orbea et al., EAE was prevented and reversed in mice treated with Vβ8-specific monoclonal antibodies (MAb F23.1) (9). In the first experiment, EAE was induced by infusing encephalitogenic T cells into (PL/J × SJL) F1 mice. Within 2–3 weeks paralysis developed and the mice were randomized to receive either MAb F23.1 or a monoclonal isotype control antibody. Most of the mice (13 of 16) receiving MAb F23.1 were symptom-free 10 days after starting treatment, whereas the animals receiving the control antibody did not improve. Other studies have been reported using passive administration of antibodies directed against T-cell surface proteins critical to regulating T-cell responses in various EAE model systems (8,9,13). Ben-Nun et al. were the first to demonstrate that attenuated cloned T cells reactive to MBP (pretreated with mitomycin C or irradiation) failed to induce EAE after infusion into syngenic rats whereas 18/20 rats infused with MBP-reactive T cells (untreated) acquired EAE (18). They further demonstrated that rats infused with the attenuated cloned T cells reactive to MBP followed by immunization with MBP-CFA were protected (14/40) against induction of EAE. However, pretreatment of animals with attenuated MBP-reactive T cells failed to protect against disease induced by adoptive transfer of these nontreated MBP-T cells. This problem may be one of technique since Offner et al. have reported complete protection of both active and passive (adoptive transfer) EAE by pretreatment of encephalitogenic T cells with 1300 Atm pressure for 15 min followed by irradiation with 2500 R (21).

With advances in DNA-sequencing techniques, the DNA sequences of TCR genes used by encephalitogenic T cells in both the mouse and rat models of EAE were delineated. In several studies, injection of the TCR Vβ 8.2 peptides was demonstrated to prevent and treat EAE in rats. For example, Howell et al. synthesized peptides corresponding to the TCR β chain VDJ regions and J alpha regions that were conserved among encephalitogenic T cells and demonstrated that Lewis rats previously immunized with certain peptides were protected from EAE when challenged with a guinea pig MBP (14). Offner et al. injected the TCR Vβ 39-59 peptide into rats with clinical signs of EAE and demonstrated disease amelioration in the treated animals (16). In addition to MBP, another autoantigen called proteolipid protein (PLP) can induce relapsing EAE in mice. In a series of elegant sequential experiments reported by Whitham et al., relapsing EAE was induced into SJL/J;PL/J, and (PL-SJL) F1 hybrid mice by immunizing with PLP 139-151 or PLP 43-64 (17). The draining lymph nodes were removed and stimulated with antigen to produce specific T-cell lines that were expanded and phenotyped using FACS and monoclonal antibodies. T-cell lines from the central nervous system and spinal lesions of the afflicted mice were isolated using discontinuous Percoll gradients and clonally expanded and Vβ expression analyzed with MAb and by PCR. The results indicated preferential usage of Vβ 4, Vβ 8.2, and Vβ 17a. TCR peptide sequences were synthesized and treatment of relapsing EAE in SJL/J mice with Vβ 4 and Vβ 17a peptides reduced clinical and histological disease severity (17). Thus, there is evidence in several rodent models of EAE that anti-TCR Vβ antibody, T-cell vaccination, and TCR peptide vaccination are efficacious.

Multiple sclerosis (MS) is an inflammatory disease of the brain, spinal cord, and optic nerve characterized by multiple foci of inflammation with loss of myelin leading to multiple scars (multiple sclerosis). Typically, large numbers of T cells are found in the nervous system lesions and cerebrospinal fluid. Many MS patients have evidence of MBP-reactive T cells in their blood, cerebrospinal fluid, and brain lesions, and these T cells have a limited Vβ expression suggesting a selective recruitment or expansion of cells (22–27). In a well-designed study, TCR expression was directly studied in plaque lesions from 16 autopsied MS patients (26). Seven of eight patients with HLA-DR2-Dw2 haplotype had detectable rearrangements of Vβ 5.2 and all eight had detectable rearrangements of Vβ 5.1 or 5.2 or both. This suggested that a limited Vβ usage was present in T cells in MS lesions similar to that observed in EAE. Not surprisingly, it has recently been reported that certain MBP-reactive T-cell clones from MS patients use Vβ with a high degree of sequence humology to encephalitogenic MBP-reactive T-cell clones from the mouse and rat (30).

Based on the fact that a variety of T-cell-targeted therapies were successful in preventing and treating EAE, Bourdette et al. studied the immunogenicity of synthetic 21 amino acid residues encompassing the CDR_2 region of $V\beta$ 5.2 and $V\beta$ 6.1 gene families in a phase I trial involving 11 patients with MS (28). Seven patients responded specifically in T-cell-proliferation assays to these peptides following immunization. Furthermore, three of the seven respondents demonstrated delayed-type hypersensitivity skin responses to the intradermal instilled peptides, and specific antibody was measured in two patients. No overall improvement in clinical response was observed. However, there was no evidence of immunosuppression. Data were presented in one patient showing a reduction in the peripheral blood of MBP-specific T-cell frequencies after immunization with both peptides. T-cell clones were isolated from the peripheral blood of two patients with increased frequencies of anti-TCR peptide-specific T cells (29). The phenotypes of these responding T cells were CD4+ CD8 utilizing $V\beta$ 4, $V\beta$ 6, $V\beta$ 12, and $V\beta$ 14 but not $V\beta$ 5.2 or $V\beta$ 6.1 in their TCR repertoire.

III. EXPERIENCE WITH $V\beta 17$ T-CELL-RECEPTOR PEPTIDE VACCINE IN RA

At least five studies have investigated RA synovial fluid and/or synovial tissue for restricted expression of TCR $V\alpha$ or $V\beta$ (31–35). One has reported a limited number of $V\alpha$, but not $V\beta$, gene families expressed primary in the synovial tissue in patients with early disease (32). The others have reported that the number of $V\beta$, but not $V\alpha$, gene families expressed in synovial/synovial fluid T cells is restricted. Besides the difficulties with cell isolation, cell culture, and the various polymerase chain reaction techniques, the possibility exists that the sampled cells were "bystander" cells and not of pathogenic importance. To address this issue, Howell et al. selected IL-2R+ T cells from five rheumatoid synovium samples and found a preponderance of $V\beta 3$, $V\beta 14$, and $V\beta 17$ to be expressed in the majority of samples (35). Dominant rearrangements were observed in these three $V\beta$ families (but not peripheral blood of the same patients), consistent with the interpretation of the data that there is limited oligoclonal expansion of activated T cells.

Following this observation (35) that $V\beta 17$ was overutilized in RA, we initiated a phase I open-label safety and dose-ranging study to investigate the feasibility of TCR peptide immunization as therapy for RA. The purpose of this initial study was to establish safety, optimal dose, and biological effects of immunization with varying doses of a 17-amino-acid sequence derived from the CDR2 region of the human $V\beta 17$ TCR. Fifteen patients with moderate to severe RA received an intramuscular injection of one of four doses of the peptide in incomplete Freund's adjuvant, followed by a booster injection of the dose of peptide 4 weeks later. The doses given were 10 µg (three patients), 30 µg (three patients), 100 µg (five patients), and 300 µg (four patients). All patients had active RA and were taking

≤7.5 mg/day prednisone, NSAIDS, and/or antimalarials. Patients were followed for 48 weeks postimmunization to monitor for safety. Their average age was 43.4 years and mean disease duration was 9.6 years (range 0.5–23 years). Nonblinded assessment of disease, measured by number of swollen and tender joints, indicated trends toward improvement over the 48-week follow-up period (Fig. 1). In addition to safety measurements, several studies were undertaken to determine the effect, if any, of the immunization on the immune system and the target Vβ17 T-cell population. No antibody to the immunizing peptide was observed in the patients during the 48-week follow-up. Significant T-cell proliferation (SI ≥ 3) in response to the immunizing Vβ17 peptide was observed at 6 weeks or later postimmunization in 1/3 patients in both the 10- and 30-µg dose groups and in 2/5 and 2/4 of patients in the 100- and 300-µg groups, respectively. Percentages of Vβ17 T-cells in blood that were IL-2R+(CD25+) decreased (≥20%) in 3/5 patients in the 100-µg group after initial measurement at week 2 (Fig. 2A) and in 3/4 patients in the 300-µg group after immunization (Fig. 2B). This decrease was sustained in 5/6 patients through 48 weeks. The total Vβ17+ T-cell population remained constant in the range of 3–6% of total T cells in these patients. No measurable effect of the ability of the T cells from these patients to be stimulated by PHA was noted during the course of the study. Based on these preliminary studies, TCR peptide immunization appears to be safe and well tolerated. No significant adverse events attributable to the treatment were observed. Further studies will be required to assess the significance of the biological effects observed postimmunization, and a double-blind, controlled trial will be needed to assess efficacy of this treatment approach. Studies with TCR vaccines directed against Vβ14 are also in progress.

IV. DISCUSSION

EAE in rodents is to date the best-defined experimental model of an autoimmune disease mediated by T cells. It is induced by immunizing rodents to well-defined antigens such as MBP and PLP proteins/peptides and may be transferred to unimmunized animals by CD4+ T cells from immunized animals or with MBP or PLP-specific cloned T-cell lines. Through studying this model system, new insights into T-cell specificity and pathogenicity have emerged. The provocative discovery that the encephalitogenic (pathogenic) T cells exhibit a limited or biased TCR Vβ gene expression has led to specific T-cell-directed therapy of immunizing animals with the variable regions of the B-chain of TCR, which has been successful in the prevention and treatment of EAE. The mechanism(s) of TCR vaccination therapy is controversial in that it appears to induce a population of either CD4+ or CD8+ T cells, which reduce the activity (presence) of the pathogenetic antigen-specific oligoclonal T cells (36–38).

T-Cell-Receptor Peptide Vaccination

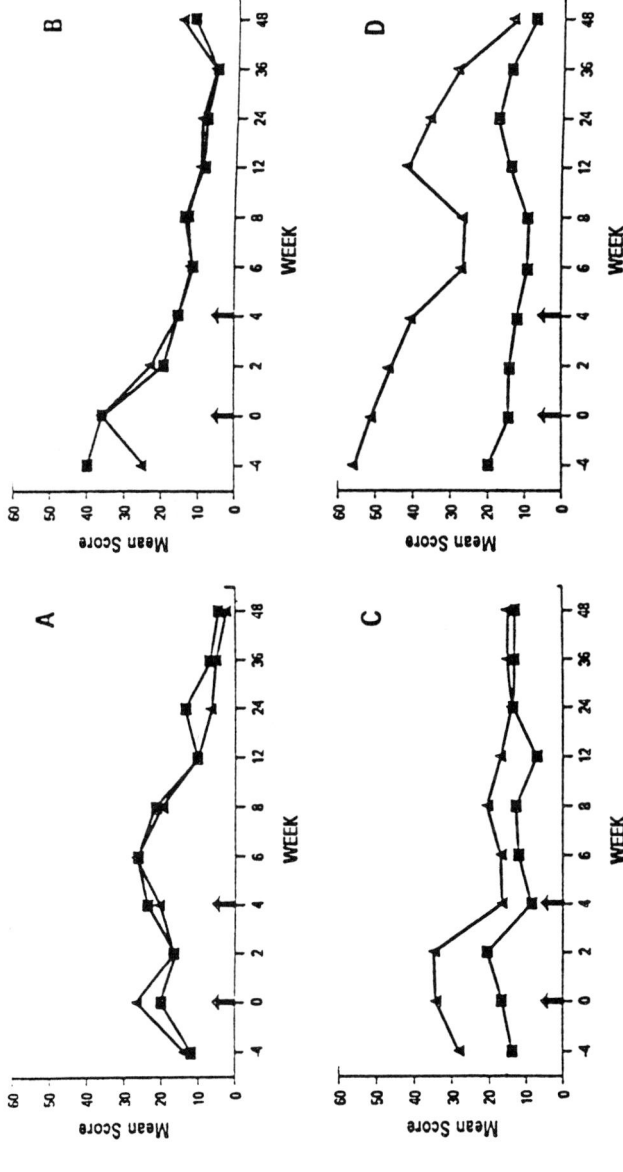

Figure 1 Mean joint scores for pain (-▲-) and swelling (-■-) in patients receiving doses of (A) 10 μg (three patients); (B) 30 μg (three patients); (C) 100 μg (five patients); and (D) 300 μg (four patients), of Vβ17 T-cell receptor peptide in incomplete Freund's adjuvant. Arrows denote times of administration of each dose.

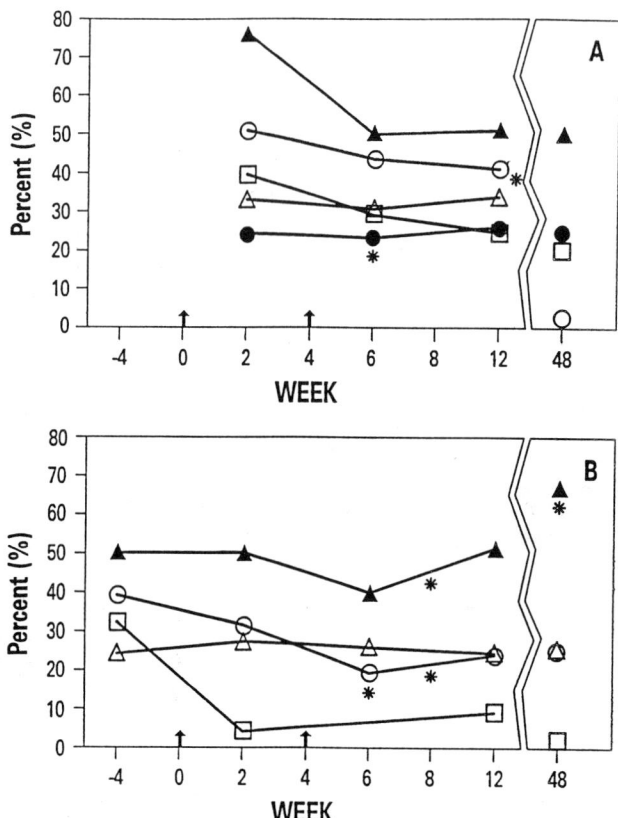

Figure 2 Percent activated Vβ17 T cells in peripheral blood of (A) five patients receiving 100 μg and (B) four patients receiving 300 μg of Vβ17 T-cell-receptor peptide in incomplete Freund's adjuvant. Scores for patient 1 (-▲-), patient 2 (-O-), patient 3 (-□-), and patient 4 (-△-) in each dose group and for patient 5 (-●-) in the 100-μg group are shown individually. *Times at which positive cell mediate immunity (stimulation index > 3.0) were measured. Arrows denote time of administration of each dose. Measurements were not determined for patient 3 at 6- or 8-week time points. Antibody for Vβ17 not available prior to the 2-week time point for the 100-μg dose group.

In MS, there is evidence for oligoclonality of TCR gene expression in MBP-specific T cells in cerebral plaque lesions. However, it is not known if these T cells play a pathogenetic role or are bystander cells. There are recent reports of a phase I trial of TCR vaccination in patients with MS without clinical improvement or adverse side effects.

Finally, the role of the T cell in the pathogenesis of RA is less secure than in MS. However, there is limited but convincing evidence for oligoclonality of activated CD4+ T cells in rheumatoid synovium. These results raise interest regarding a possible role for TCR-directed vaccine approaches in RA. In a small phase I study reported herein, no adverse reactions were noted and evidence of immunogenicity of the vaccine was obtained. We are encouraged, however, that there was no evidence for immune compromise in our patients. Expanded studies will be required to determine the efficacy of such approaches in RA.

REFERENCES

1. Harris ED. Rheumatoid arthritis. Pathophysiology and implications for therapy. N Engl J Med 1990; 322:1277–1289.
2. Koopman WJ, Gay S. Do nonimmunologically mediated pathways play a role in the pathogenesis of rheumatoid arthritis? Rheum Dis Clin North Am 1993; 19: 107–122.
3. Nepom GT, Erlich H. MHC class-II molecules and autoimmunity. Annu Rev Immunol 1991; 9:493–525.
4. Winchester R. The molecular basis of susceptibility to rheumatoid arthritis. Adv Immunol 1994; 56:389–466.
5. Steinman L. The development of rational strategies for selective immunotherapy against autoimmune demyelinating disease. Adv Immunol 1991; 49:357–379.
6. Wraight DW, McDevitt HO, Steinman L, Acha-Orbea H. T cell recognition as the target for immune intervention in autoimmune disease. Cell 1989; 57:709–715.
7. Sriram S, Steinman L. Anti I-A antibody suppresses active encephalomyelitis: treatment model for diseases linked to IR genes. J Exp Med 1983; 158:1362–1367.
8. Urban JL, Kumar V, Kono DH, Gomez C, Horvath SJ, Clayton J, Ando DG, Sercarz EE, Hood L. Restricted use of T cell receptor V genes in murine autoimmune encephalomyelitis raises possibilities for antibody therapy. Cell 1988; 54:577–592.
9. Acha-Orbea H, Mitchell DJ, Timmermann L, Wraith DC, Tausch GS, Waldor MK, Zamvil SS, McDevitt HO, Steinman L. Limited heterogeneity of T cell receptors from lymphocytes mediating autoimmune encephalomyelitis allows special immune intervention. Cell 1988; 54:263–273.
10. Wraith DC, Smiler DE, Mitchell DJ, Steinman L, McDevitt HO. Antigen recognition in autoimmune encephalomyelitis and the potential for peptide-mediated immunity. Cell 1989; 59:247–255.
11. Gaur A, Wiers B, Liu A, Rothbard J, Fathman CG. Amelioration of autoimmune encephalitis by myelin basic protein synthetic peptide-induced anergy. Science 1992; 258:1491–1494.
12. Sharma SD, Nag B, Su X, Green D, Spack E, Clark BR, Sriram S. Antigen-specific therapy of experimental allergic encephalomyelitis by soluble class II major histocompatibility complex-peptide complexes. Proc Natl Acad Sci USA 1991; 88: 11465–11469.

13. Alters SE, Sakai K, Steinman L, Oi VT. Mechanisms of anti-CD4 mediated depletion and immunotherapy: a study using a set of chimeric anti-CD4 antibodies. J Immunol 1990; 144:4587–4592.
14. Howell MD, Winters ST, Olee T, Powell HC, Carlo DJ, Brostoff SW. Vaccination against experimental allergic encephalomyelitis with T cell receptor peptides. Science 1989; 246:668–670.
15. Vandenbark AA, Hashim G, Offner H. Immunization with a synthetic T-cell receptor V-region peptide protects against experimental autoimmune encephalomyelitis. Nature 1989; 341:541–544.
16. Offner H, Hashim G, Vandenbark AA. T cell receptor peptide therapy triggers autoregulation of experimental encephalomyelitis. Science 1991; 251:430–432.
17. Whitham RH, Kotzin BL, Buenafe AC, Weinberg AD, Jones RE, Hashim GA, Hoy CM, Vandenbark AA, Offner H. Treatment of relapsing experimental autoimmune encephalomyelitis with T cell receptor peptides. J Neurosci Res 1993; 35:115–128.
18. Ben-Nun A, Wekerle H, Cohen IR. Vaccination against autoimmune encephalomyelitis with T-lymphocyte line cells reactive against myelin basic protein. Nature 1981; 292:60–61.
19. Happ MP, Kiraly A, Offner H, Vandenbark AA, Heber-Katz E. The autoreactive T cell population in EAE: TCR B chain rearrangements. J Neuroimmunol 1988; 19:191–199.
20. Burns FR, Li X, Shen N, Offner H, Chou YK, Vandenbark AA, Heber-Katz E. Both rat and mouse T cell receptors specific for the encephalitogenic determinant of myelin basic protein use similar V and Vβ chain genes even though the major histocompatibility complex and encephalitogenic determinants being recognized are different. J Exp Med 1989; 169:27–39.
21. Offner H, Jones R, Celnik B, Vandenbark AA. Lymphocyte vaccination against experimental autoimmune encephalomyelitis: evaluation of vaccination protocols. J Neuroimmunol 1989; 21:13–22.
22. Kotzin BL, Karuturi S, Chou YK, Lafferty J, Forrester JM, Better M, Nedwin GE, Offner H, Vandenbark AA. Preferential T-cell receptor B-chain variable gene use in myelin basic protein-reactive T-cell clones from patients with multiple sclerosis. Proc Natl Acad Sci USA 1991; 88:1961–1965.
23. Lee SJ, Wucherpfenning KW, Brod SA, Benjamin D, Weiner HL, Hafler DA. Common T-cell receptor Vβ usage in oligoclonal T lymphocytes derived from cerebrospinal fluid and blood of patients with multiple sclerosis. Ann Neurol 1991; 29:33–40.
24. Birnbaum G, Van Ness B. Quantitation of T-cell receptor Vβ chain expression on lymphocytes from blood, brain, and spinal fluid in patients with multiple sclerosis and other neurologic diseases. Ann Neurol 1992; 32:24–30.
25. Richert JR, Robinson ED, Johnson AH, Bergman CA, Dragovic LJ, Reinsmoen NL, Hurley CK. Heterogeneity of the T-cell receptor B gene rearrangements generated in myelin basic protein-specific T-cell clones isolated from a patient with multiple sclerosis. Ann Neurol 1991; 29:299–306.
26. Oksenberg JR, Panzara MA, Begovich AB, Mitchell D, Erlich HA, Murray RS, Shimonkevitz R, Sherritt M, Rothbard J, Bernard CCA, Steinman L. Selection for T-cell receptor Vβ-Dβ-Jβ gene rearrangements with specificity for myelin basic protein peptide in brain lesions of multiple sclerosis. Nature 1993; 362:68–70.

27. Oksenberg JR, Stuart S, Begovich AB, Bell RB, Erlich HA, Steinman L, Bernard CCA. Limited heterogeneity of rearranged T cell receptor V transcripts in brains of multiple sclerosis patients. Nature 1990; 345:344–346.
28. Bourdette DN, Whitham RH, Chou YK, Morrison WJ, Atherton J, Kenny C, Liefeld D, Hashim GA, Offner H, Vandenbark AA. Immunity to TCR peptides in multiple sclerosis. I Successful immunization of patients with synthetic Vβ 5.2 and Vβ 6.1 CDR 2 peptides. J Immunol 1994; 152:2510–2519.
29. Chou YK, Morrison WJ, Weinberg AD, Dedrick R, Whitham R, Bourdette DN, Hashim GA, Offner H, Vanderbark AA. Immunity to TCR peptides in multiple sclerosis. II. T cell recognition of Vβ 5.2 and Vβ 6.1 CDR 2 peptides. J Immunol 1994; 152:2520–2529.
30. Allegretta M, Albertini RJ, Howell MD, Smith LR, Martin R, McFarland HF, Sriram S, Brostoff S, Steinman L. Homologies between T cell receptor junctional sequences unique to multiple sclerosis and T cells mediating experimental allergic encephalomyelitis. J Clin Invest 1994; 94:105–109.
31. Paliard X, West S, Lafferty J, Clements J, Kappler J, Marrack P, Kotzin B. Evidence for the effects of a superantigen in rheumatoid arthritis. Science 1991; 253:325–329.
32. Bucht A, Oksenberg JR, Linblad S, Gronberg A, Steinman L, Klareskog L. Characterization of T-cell receptor β repertoire in synovial tissue from different temporal phases of rheumatoid arthritis. Scand J Immunol 1992; 35:159–165.
33. Jenkins RN, Nikaein A, Zimmermann A, Meek K, Lipsky PE. T cell receptor Vβ gene bias in rheumatoid arthritis. J Clin Invest 1993; 92:2688–2701.
34. Williams WV, Fang Q, Demarco D, VonFeldt J, Zurier RB, Weiner DB. Restricted heterogeneity of T cell receptor transcripts in rheumatoid synovium. J Clin Invest 1992; 90:326–333.
35. Howell MD, Diveley JP, Lundeen KA, Esty A, Winters ST, Carlo DJ, Brostoff SW. Limited T-cell receptor B-chain heterogeneity among interleukin 2 receptor-positive synovial T cells suggests a role for superantigen in rheumatoid arthritis. Proc Natl Acad Sci USA 1991; 88:10921–10925.
36. Kumar V, Sercarz EE. The involvement of T cell receptor peptide-specific regulatory CD4+ T cells in recovery from antigen-induced autoimmune disease. J Exp Med 1993; 178:909–916.
37. Gaur A, Haspal R, Mayer JP, Fathman CG. Requirement for CD8+ cells in T cell receptor peptide-induced clonal unresponsiveness. Science 1993; 259:91–94.
38. Kuhröber A, Schirmbeck R, Reimann J. Vaccination with T cell receptor peptides primes anti-receptor cytotoxic T lymphocytes (CTL) and anergizes T cells specifically recognized by these CTL. Eur J Immunol 1994; 24:1172–1180.

17
Oral Tolerance for the Treatment of Autoimmune Disease

DAVID A. HAFLER and HOWARD L. WEINER
Brigham and Women's Hospital, Boston, Massachusetts

I. OVERVIEW OF HUMAN AUTOIMMUNE DISEASE

A. Disease Initiation

There are two general hypotheses for the etiology of human autoimmune disorders. In the first instance, the organ becomes infected by a virus or other infectious agent and cells that infiltrate that organ are targeted to the infectious agent. These events would then begin the cascade leading to the organ-specific autoimmune disease. The alternative hypothesis is that the initial infiltrating cells are autoimmune and recognize the organ-specific proteins that are presented by local antigen-presenting cells in the context of MHC. At our present state of knowledge, either hypothesis is possible. However, once the autoimmune cascade begins it is likely that "epitope spreading" will occur, and T cells recognizing other organ-specific proteins will be recruited. Perhaps viruses and infectious agents acting as superantigens initially trigger or drive the autoimmune process, as opposed to being the primary target of infiltrating cells. Understanding the mechanism for initiation of autoimmune disorders and the range of antigens recognized by self-reactive T cells is critical for developing antigen-specific therapies for these disorders.

B. Organ-Specific Proteins as Targets for T Cells

What are the characteristics of a tissue-specific autoantigen that will be recognized by T cells? Approximately 0.1% of a restricting class II MHC protein must be occupied with the antigen peptide for T-cell activation to occur (1). Considering there is competition by a large variety of self and foreign peptides for binding to MHC class II proteins (2), it is possible that this requirement may be met only by peptides that are present in relatively large quantities in antigen-presenting cells, processed efficiently, and bound with a high affinity to class II molecules. Thus, the physical characteristics and processing of the antigen in the tissue may limit the number of potential autoantigens that the immune system is capable of recognizing.

Second, a T-cell autoantigen must be processed in such a way as to be presented by class II MHC by the antigen-presenting cells specific for that tissue site (Fig. 1). Using the central nervous system as an example, microglia, which are specialized central nervous system (CNS) macrophages, might phagocytose breakdown products of myelin and present peptide antigens to T cells (3,4). Alternatively, glial cells such as astrocytes, which after activation can express class II MHC and present antigen, might present CNS antigens in the context of class II MHC (5,6).

C. Epitope Mapping of Autoantigen

The CNS antigen myelin basic protein (MBP) is perhaps the most extensively characterized autoantigen and is a primary candidate autoantigen in MS because of its ability to induce EAE. We will use MBP as an example in discussing epitope mapping. There have been many investigations of MBP reactivity in MS using 7-day proliferation assays of whole peripheral blood mononuclear cells (7,8). These investigations have shown a slight increase in T-cell responses to human MBP in subjects with MS as compared to normal subjects or other neurological disease patients, but the magnitude of the difference has generally been less than convincing. T-cell cloning techniques have recently been applied to study autoreactive T cells in both the peripheral blood and spinal fluid. Antigen-specific T-cell clones are generated by repeated stimulation and culture of T cells with antigen in the presence of T-cell growth factors.

In the EAE model, T-cell clones can be derived from animals injected with MBP in adjuvant that recognize immunodominant regions of MBP that are influenced by the MHC of the animal. These clones can transfer EAE when injected into naïve animals (9–12). For example, in Lewis rats, the MBP (68–88) peptide is immunodominant and is the epitope for the majority of encephalitogenic T-cell clones (10–12). Lewis rat T-cell clones generated against MBP (residues 43–67) are only weakly able to cause disease.

Oral Tolerance for Treatment

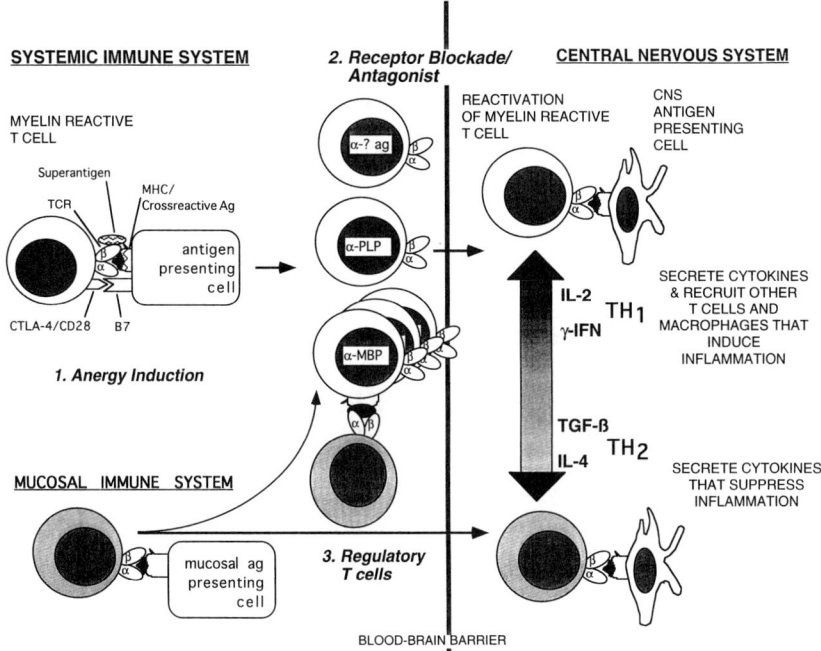

Figure 1 Overview of proposed pathophysiology for MS and specific areas where immune system can be manipulated. The minimal requirements in mammals for inducing inflammatory autoimmune disease in the CNS white matter are activated CD4+ T cells recognizing MBP or PLP. In MS, we hypothesize that some activating event, such as a viral superantigen or molecular mimicry, activates an autoreactive T cell. This allows the activate T cell to migrate into the CNS where it can now recognize CNS antigen in the context of MHC, recruiting effector cells leading to the inflammatory CNS response. The process may be blocked by (1) anergy induction, (2) by receptor blockade or antagonist, or (3) by regulatory T cells (see text).

In humans, immunodominant regions of MBP that are presented by defined MHC proteins have been defined. Immunodominant regions of MBP are found between residues 84–106 and 143–172 near the C terminus (13–17). In DR2+ individuals, a high proportion of T-cell clones reactive with MBP recognized the 84–106 region, whether or not they had MS (13). Both DR and DQ antigens were found to be restricting elements, indicating that immunodominant MBP peptides can be presented by different class II molecules. It has also been demonstrated that T cells from genetically diverse individuals responding to different epitopes of human MBP are associated with distinct MHC class II molecules (18). Of interest was the finding that DR2 possessed an unusual capacity to restrict all of the epitopes identified on MBP (15,16,18). Moreover, different types of DR2 have the capacity to present a number of different MBP epitopes (19). These data together indicate that disease-associated DR2 antigens can present a variety of MBP peptides to T cells, and this is likely to be associated with competition for peptide binding between different MBP regions. Thus, peptide competition for binding to class II molecules may in part determine immunodominant sites of an autoantigen.

D. Activation of Autoreactive T Cells

T cells recognizing MBP and PLP exist in normal humans. From investigation of the EAE model, it has been learned that for these cells to be pathogenic they must be activated in vivo. To examine whether MBP reactive T cells are activated in vivo, an hprt⁻ mutant assay was used (20). The assay is based on the observation that dividing cells acquire random mutations during DNA synthesis. Some of these mutations occur in the *hprt* gene, which results in inactivation of the hprt enzyme. Thus, mutant cells do not metabolize thioguanine to a cytotoxic metabolite, which allows a very effective selection of these mutants in culture. Eleven of 258 mutant T-cell clones cultured by mitogen from the peripheral blood of five of six multiple sclerosis (MS) patients showed strong reactivity to MBP, while none of 114 clones grown from blood of normal subjects did. However, clones were not investigated from subjects with other neurological diseases or with nonmyelin antigens. These data suggest that MBP-reactive T cells are activated in MS patients and thus are pathogenic.

We directly investigated a total of 72 subjects with definitive MS as to whether myelin-reactive T cells, which might be critical for the pathogenesis of MS, exist in a different state of activation as compared to myelin-reactive T cells cloned from the blood of normal individuals. While there were no differences in the frequencies of MBP- and PLP-reactive T cells after primary antigen stimulation, the frequency of MBP or PLP, but not tetanus toxoid–reactive T cells, generated after primary rIL-2 stimulation, was significantly higher in MS patients as compared to control individuals. Primary rIL-2-stimulated MBP-reactive T-cell lines were CD4+ and recognized MBP epitopes 84–102 and 143–168 similar to MBP-reactive T-cell lines generated with primary MBP stimulation. In the CSF of MS patients, MBP-reactive T cells generated with primary rIL-2 stimulation accounted for 7% of the IL-2-responsive cells, greater than 10-fold higher than paired blood samples, and these T cells also selectively recognized MBP peptides 84–102 and 143–168. In striking contrast, MBP-reactive T cells were not detected in cerebrospinal fluid obtained from patients with other neurological diseases. These results provide definitive in vitro evidence of an absolute difference in the activation state of myelin-reactive T cells in the CNS of patients with MS and provide evidence of a pathogenic role of autoreactive T cells in the disease.

If circulating, autoreactive T cells are present in the circulation of normal individuals, how do they become activated? (Fig. 1.) The mechanism is of particular interest in a CNS disease such as MS, since resting T cells appear unable to cross the blood-brain barrier. Possible mechanisms in the absence of autoantigen involve immune activation associated with infections, which include: (1) molecular mimicry, (2) activation by superantigens, and (3) bystander CD2 activation.

II. ORAL TOLERANCE; ANTIGEN-SPECIFIC IMMUNOTHERAPY

A. Overview of Antigen-Specific Immunotherapy

In Section I, we presented an overview of antigen-specific recognition in autoimmune diseases. In total, these experiments suggest a number of manipulations that may be used to block the immune response. It is important to note the likelihood that no in vitro experiment can prove the autoimmune hypothesis for any of the presumed autoimmune diseases. Instead, only specific manipulations in vivo where antigen-reactive cells are targeted with associated amelioration of disease activity will allow the understanding of the disease's pathogenesis. Thus, these specific manipulations of the immune response represent both attempts to treat the disease and scientific experiments to understand the disease.

There are two general approaches to antigen specific immunotherapy of autoimmune disease. *One approach is to block either the initial activation or the subsequent recognition of autoantigen by autoreactive T cells* (see Fig. 1). This approach requires the immune response against the self-antigen to be restricted in terms of either the number of epitopes recognized or the TCR repertoire that is used. This includes MHC-blocking peptides, TCR antagonists, and TCR peptides that may inhibit specific T-cell function. Using MS as an example, it appears that although there are MBP and PLP dominant epitopes and somewhat restricted use of TCRs in their recognition, the outbred human species contains too many myelin-reactive T cells with different TCR usage to successfully inhibit the total of the spreading immune response to myelin antigens.

The second approach involves the antigen-specific targeting of T cells that down-regulate immune responses to the inflammatory sites. In this instance, it is not necessary to know the inciting antigen that elucidates the immune response nor is it required that the T cells be restricted in terms of antigen recognition or TCR usage. This approach may well be physiological in regulating the normal immune response. One such method, which has attracted much attention recently, is the use of oral tolerization where the autoantigen is delivered orally with the generation of T cells secreting TGF-β, IL-4, and IL-10 that migrate to the site of inflammation and suppress the immune response. We will discuss oral tolerance below.

B. Mechanism of Oral Tolerance

Oral tolerance represents the exogenous administration of antigen to the peripheral immune system via the gut. As such, it is a form of antigen-driven peripheral immune tolerance. Immunological tolerance is not programmed into the germline but is acquired during maturation of the immune system by mechanisms that delete or inactivate antigen-reactive clones. There are three basic mechanisms to explain antigen-driven tolerance: clonal deletion, clonal anergy, and active suppression (21,22). A large number of studies have shown that one of the primary mechanisms associated with oral tolerance is the generation of active suppression (23). More recently, clonal anergy has been demonstrated (24,25). There is little evidence that orally administered antigen induces clonal deletion. During the course of our investigations we have delineated two pathways by which oral tolerization results in systemic hyporesponsiveness by either active suppression or clonal anergy (26). These pathways are described below in a schema that forms the theoretical basis for this review.

Based on recent findings in our laboratory and those of others, we believe that oral tolerance can be viewed as *anergy-driven oral tolerance* or *regulatory-cell-driven oral tolerance*. The primary factor that determines which form of peripheral tolerance develops following oral administration of antigen is the dose of antigen fed. Low doses of antigen favor the generation of active suppression or regulatory-cell-driven tolerance whereas high doses of antigen favor anergy-driven tolerance (Fig. 2). Although these forms of oral tolerance are not mutually exclusive and may occur simultaneously, they are distinct and the use of oral tolerance to treat autoimmune diseases is critically dependent on which of these two mechanisms is triggered.

The delineation of these two mechanisms of oral tolerance was based on the following: (1) investigations in our laboratory in which low doses of orally administered autoantigens were shown to suppress experimental autoimmune diseases via the generation of regulatory cells that suppressed in vitro and in vivo via the secretion of down-regulatory cytokines such as TGF-β (27); (2) investigations from other laboratories that demonstrated clonal anergy following oral administration of large doses of antigen with no evidence of active suppression (24,25); (3) a large series of investigations demonstrating transferrable suppression following oral tolerance (reviewed in 3) including work that showed two components of oral tolerance, one that was abrogated by treatment with low-dose cyclophosphamide and one that was not, a difference that was dose-dependent

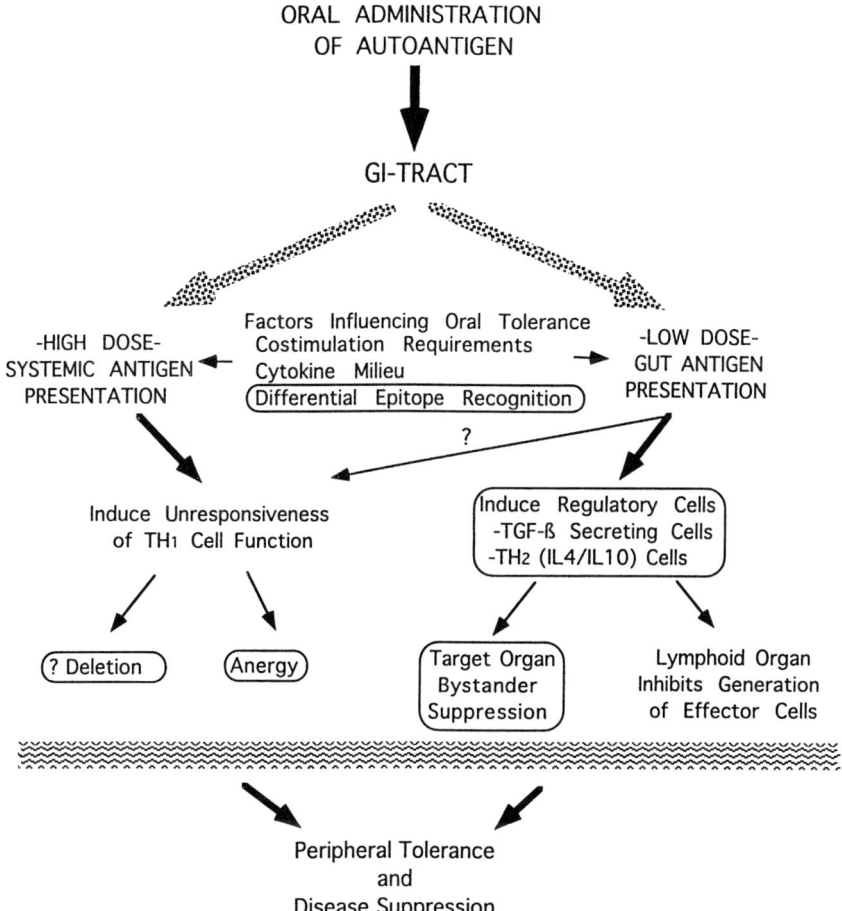

Figure 2 Mechanisms of oral tolerance. High-dose antigen administration leads to the generation of anergy, while lower-dose oral antigen induces regulatory T cells that secrete TGF-β1 and/or IL-4/IL-10, which migrate to the CNS and down-regulate the local immune response.

(28); and (4) direct comparison in our laboratory demonstrating that the two mechanisms depend on the dose (26) (Fig. 2).

As shown in Figure 2, *low doses* of antigen result in the generation of antigen-specific regulatory cells and as such involve presentation of antigen by gut-associated antigen-presenting cells. Such presentation preferentially induces regulatory cells, which, upon subsequent recognition of antigen in vivo or in vitro, secrete the suppressive cytokine TGF-β. In addition, Th$_2$ responses are preferentially generated in the gut, resulting in cells that secrete IL-4 and IL-10. These antigen-specific regulatory cells migrate to lymphoid organs and suppress immune responses by inhibiting the generation of effector cells and to the target organ and suppress disease by releasing antigen-nonspecific cytokines (bystander suppression). Several factors can affect the generation of regulatory cells, including costimulation requirements, the cytokine milieu in which the immune response is generated, and differential generation of epitopes that preferentially may trigger certain regulatory cells.

High doses of orally administered antigen result in systemic antigen presentation after antigen passes through the gut and enters the systemic circulation as either intact protein or antigen fragments. High doses of antigen induce unresponsiveness of Th$_1$ cell function, primarily via clonal anergy. The degree to which clonal anergy following high doses of antigen merely represents the direct passage of small amounts of antigen into the systemic or portal circulation or is dependent on filtration by the gut is unknown. Why there is reduced active suppression with high doses of orally administered antigen is unclear, but could relate to anergizing cells involved in the generation of active suppression. In addition, it is not known the degree to which costimulatory requirements, cytokine milieu, and differential epitope recognition may preferentially favor the generation of anergy in Th$_1$ cells.

1. Active Suppression

Many studies demonstrate that active suppression is an important mechanism for oral tolerance (23,29–34). After feeding antigens such as ovalbumin or sheep red blood cells, transferrable suppression mediated by T cells from Peyer's patches, mesenteric lymph node, and spleen has been demonstrated. Investigators have also reported initial sensitization prior to the appearance of suppression (30). Further characterization of active suppression as measured in these systems has not occurred probably owing to the difficulties in defining the biology of suppression (35,36). Nonetheless the demonstration of transferrable cellular suppression associated with oral tolerance is a recurrent theme reported by many investigators (23).

Our studies of oral tolerance in autoimmune models have found active suppression to be a primary mechanism and have identified regulatory cells generated following oral tolerance that act via the secretion to antigen-nonspecific downregulatory cytokines following triggering by the fed antigen (37). These cells have been characterized in both the rat and murine model of EAE orally tolerized to MBP. In the Lewis rat model, regulatory cells were CD8+ (38) and acted via the secretion of TGF-β following antigen-specific triggering (27). They transfer suppression in vivo and can suppress in vitro. The epitopes of guinea pig MBP triggering CD8+ regulatory cells following orally administered MBP were different than the encephalitogenic determinant (39). In addition, TGF-β-secreting regulatory cells can be found in Peyer's patches 24–48 hr after one feeding of MBP (40). Of note is the fact that cells from Peyer's patches removed after one feeding of MBP do not proliferate in response to MBP even though they release TGF-β upon in vitro stimulation. The mechanism by which these regulatory cells are induced remains unknown. Also unknown is the degree to which the generation of regulatory cells is related to unique antigen-presenting cells in the gut, the cytokine milieu, or other factors. Use of murine models of oral tolerance suggests that in addition to CD8+ TGF-β-secreting regulatory cells, CD4+ regulatory cells are also induced. These CD4+ cells secrete IL-4 and IL-10 in addition to TGF-β. Moreover, these TGF-β-secreting T cells can adoptively transfer suppression of EAE to naïve animals (41).

2. Anergy

Anergy has only recently been demonstrated as a possible mechanism for oral tolerance (24,25). Anergy is defined as a state of T-lymphocyte unresponsiveness characterized by absence of proliferation, IL-2 production, and diminished expression of IL-2R (42). Anergy may be experimentally differentiated from clonal deletion by demonstrating the presence of antigen-specific TcR clonotypes, or by release from the anergic state, which is accomplished by preculture of cells in IL-2 (43). Under these conditions, we have shown that a single feeding of 20 mg OVA induced a state of anergy in OVA-specific T lymphocytes: cells did not respond to OVA by proliferation, OVA stimulation did not induce IL-2 production or IL2R expression, and the nonresponsive state was reversed by preculture of tolerized cells in IL-2 (25). One other study has indirectly demonstrated anergy as a mechanism for oral tolerance (24). Whitacre et al. reported diminished IL-2 and IFN-γ production in rats orally fed MBP in the presence of the soybean protease inhibitor; however, anergy was not confirmed in this study by TcR analysis or by IL-2-driven release. As discussed previously, the induction of anergy depends on antigen dosage and frequency of feeding (26).

C. Treatment of Organ-Specific Autoimmune Diseases in Animals

We have examined the manner in which oral tolerance may be applied to the treatment of autoimmune conditions in both humans and animals. Thompson and Staines (44) and Nagler-Anderson et al. (45) initially described suppression of collagen-induced arthritis by feeding type II collagen. Our laboratory (46–48) and that of Whitacre (24,49,50) have studied suppression of EAE by orally administered myelin antigens. Additionally, we have investigated oral tolerance to suppress autoimmune models of uveitis (51), diabetes in the NOD mouse (52,53),

Table 1 Animal Models and Antigens Fed

Animal model	Antigen fed
EAE (MS)	MBP, PLP
Arthritis (CII, AA)	Type II collagen
Uveitis	S-antigen, IRBP
Diabetes (NOD mouse)	Insulin, GAD
Myasthenia gravis	AChR
Thyroiditis	Thyroglobulin
Transplantation	Alloantigen, MHC peptide

and adjuvant arthritis (54,55), as well as orally administered alloantigen or MHC peptide in transplantation models (56–58). Investigators have also demonstrated suppression of other autoimmune models by orally administered antigen (for review, see 59). Thus, the ability to suppress autoimmunity in animal models via oral tolerance has been established although the mechanisms responsible may differ depending on the laboratory and the models being studied.

1. EAE

Orally administered guinea pig MBP suppresses EAE in the Lewis rat model (46,49). As discussed above, this effect is mediated by both active suppression and anergy. The mechanisms were discussed in detail using the EAE model above in Section II.B.

An important question before human clinical trials were begun was whether or not ongoing disease could be suppressed. We investigated chronic relapsing EAE in the Lewis rat and strain 13 guinea pig. Disease was also suppressed by oral administration of MBP or a bovine myelin preparation that is also being administered in human clinical trials. There was no exacerbation of disease in these animals, demonstrating that orally administered antigens do not appear to prime rather than suppress in an already immunized animal. Of note in the guinea pig model is that 10 mg of bovine myelin fed 3 times per week over a 3-month period suppressed disease and histological manifestations whereas feeding 50 mg of bovine myelin did not. This may indicate that bystander suppression was responsible for the effect on chronic disease and that in some instances the oral administration of too high a dose will not suppress autoimmune models. As discussed later, loss of protection by orally administered antigen at higher doses was also seen with orally administered collagen to suppress adjuvant arthritis and orally administered insulin to suppress diabetes in the NOD mouse.

Finally, we will discuss the immunohistology associated with oral tolerization to MBP and in animals naturally recovering from EAE (56). Brains from OVA-fed animals at the peak of disease showed perivascular infiltration with activated mononuclear cells, which secreted the inflammatory cytokines IL1, IL-2, TNF-α, IFN-γ, IL-6, and IL-8. Inhibitory cytokines TGF-β and IL-4 and prostaglandin E2 (PGE$_2$) were absent. In MBP orally tolerized animals there was a marked reduction of the perivascular infiltrate and down-regulation of all inflammatory cytokines. In addition, there was up-regulation of the inhibitory cytokine TGF-β. When bacterial lipopolysaccharide (LPS) was fed in addition to MBP, protection against EAE was enhanced and was associated with elevated IL-4 and PGE$_2$ in the brain (56,60). In control recovering animals (day 18), staining for inflammatory cytokines was diminished and there was up-regulation of TGF-β and IL-4. These results suggest that the suppression of EAE, by oral tolerization and natural recovery, is related to regulatory cells that secrete inhibitory cytokines at the target organ.

2. Collagen and Adjuvant Arthritis

Previous investigators have demonstrated suppression of collagen-induced arthritis by feeding collagen type II (44,45). We have studied adjuvant arthritis (AA) in the rat, a well-characterized and more fulminant form of experimental arthritis (54). Oral administration of chicken collagen type II (CII), given at a dose of 3 µg/feeding, consistently suppressed the developed of AA. A decrease in DTH responses to CII was observed that correlated with suppression to AA. Oral administration of collagen type I also suppressed AA; only minimal effects were seen with collagen type III. Suppression was antigen-specific in that feeding collagen type II did not suppress EAE, and feeding MBP did not suppress AA. Suppression of AA could be adoptively transferred by T cells from CII-fed animals and was observed when CII was fed after disease onset. Of note is that suppression was observed at doses of 3 and 30 µg, but not at 300 or 1000 µg. These results suggest that oral collagen is suppressing AA via bystander suppression rather than clonal anergy since active suppression may be lost at higher doses. The effectiveness of such small amounts of oral collagen may be related to the fact that collagen has repeating amino acid subunits.

3. Uveitis

Oral administration of S-antigen (S-Ag), a retinal autoantigen that induces EAU, prevented or markedly diminished the clinical appearance of S-Ag-induced disease as measured by ocular inflammation (51,61–66). Furthermore, oral administration of S-Ag also markedly diminished uveitis induced by the uveitogenic M and N fragments of the S-Ag. Oral administration of S-Ag did not prevent MBP-induced EAE. In vitro studies demonstrated a significant decrease in proliferative responses to the S-Ag in lymph node cells draining the site of immunization from fed versus nonfed animals. Furthermore, the addition of splenocytes from S-Ag-fed animals to cultures of a CD4+ S-Ag-specific lymphocyte line profoundly suppressed the line's response to the S-Ag, whereas these splenocytes had no effect on a PPD-specific lymphocyte line. The antigen-specific in vitro suppression was blocked by anti-CD8 antibody, demonstrating that suppression was dependent on CD8+ T lymphocytes. As in EAE, EAU was also suppressed by feeding S-Ag-related peptides that were either uveitogenic, cross-reactive, or synthetic (62–66). Gregerson et al., using high and low doses of S-Ag peptides, also found that low doses of antigen favor suppression whereas high doses induce unresponsiveness or anergy (65).

4. Diabetes

NOD diabetic mice spontaneously develop an autoimmune form of diabetes associated with insulitis. To test oral tolerance in the NOD model, we administered porcine insulin at a dose of 1 mg orally twice a week for 5 weeks and then weekly until 1 year of age (67). The severity of lymphocytic infiltration of pancreatic islets was reduced by oral administration of insulin and there was a delay in the onset of diabetes. A decreased incidence of diabetes was seen in animals followed for 1 year. Suppression of insulitis as observed at a dose of 1 mg but not 5 mg. As expected, orally administered insulin had no metabolic affect on blood glucose levels. Furthermore, splenic T cells from animals orally treated with insulin adoptively transferred protection against diabetes, demonstrating that oral insulin generates active cellular mechanisms that suppress disease. Ongoing studies have demonstrated the ability to suppress insulitis by administering insulin peptides, the B chain of insulin, or GAD, and immunohistochemical studies have demonstrated an increase of IL-4 in the islets of insulin-fed animals. Given the mechanism of antigen-driven bystander suppression, our results do not implicate autoreactivity to insulin as a pathogenic mechanism in the NOD mouse. Indeed, initial experiments suggest that orally administered glucagon can suppress insulitis.

D. Treatment of Autoimmune Diseases in Humans

The first attempts of oral tolerization may have been utilized by Native Americans who were thought to have fed their children *Rhus* leaves to prevent them from becoming sensitized to poison ivy (68). Investigators have shown that exposure to a contact-sensitizing agent via the mucosa prior to subsequent skin challenge led to unresponsiveness in a portion of the subjects studied (69). In another study on human volunteers, serum antibodies to bovine serum albumin (BSA) were measured before and after feeding large amounts of this antigen (0.1–1.5 mg of BSA per pound per day). Those subjects that had anti-BSA antibodies prior to

Oral Tolerance for Treatment

eating BSA showed a rise in their serum anti-BSA titers. A similar response was observed when some subjects were given an injection of BSA. Subjects who did not have anti-BSA antibodies before or after the test did not respond to subsequent intradermal immunization (70). Oral desensitization has also been attempted in Rh disease (71). In an attempt at oral immunization, human volunteers were given capsules containing killed *Streptococcus mutans,* and circulating IgA-producing cells were found in some subjects (72). This suggests that some generation of a secretory immune response can occur following oral ingestion of microbial antigens. Orally administered KLH, 50 mg given daily for 2 weeks over a 3-week period, has been reported to decrease subsequent cell-mediated immune responses although antibody responses were not affected (73). In addition, preliminary experiments in subjects fed KLH have suggested that cell lines may be generated that suppress proliferative responses, although more work is needed in this area (74).

1. Multiple Sclerosis

To determine whether orally ingested autoantigens could affect the clinical course and immune responses in patients with an autoimmune disease, 15 patients with relapsing remitting multiple sclerosis were fed a capsule containing 300 mg of bovine myelin or placebo daily for 1 year (75). Results demonstrated a decrease in myelin basic protein reactive cells in the bloodstream of MS patients as compared to controls. There was no evidence of sensitization either as measured by antibody levels to MBP or PLP or by increased proliferative responses to the fed antigens at a 1-year period. Clinical responses demonstrated that 12 of 15 placebo-fed patients had major MS attacks whereas only six of 15 in the control group had attacks ($p = 0.06$). It appeared that a subgroup of patients that were either males or DR2- preferentially responded to the oral tolerization. However, the sample size was small and the degree to which this subgroup response will occur in future studies is unknown. Based on these observations, a multicenter, <500-patient, double-blind, placebo-controlled trial of bovine myelin in DR2- relapsing remitting MS patients, both male and female, has begun. Of note is that trials of myelin basic protein given by multiple subcutaneous injections resulted in prominent immune responses to the antigen (76).

2. Rheumatoid Arthritis

A 60-patient, double-blind trial of oral collagen administration to patients with rheumatoid arthritis demonstrated a decrease in joint swelling and disease index in patients fed collagen compared to placebo controls (77). Patients were those previously or currently on immunosuppressant drugs such as methotrexate who had failed such therapy. Patients were taken off these medications and treated for 3 months. In the first month they received 100 µg/day of oral collagen and in the second and third months 500 µg/day. These doses were extrapolated from the small amounts of collagen used to suppress adjuvant arthritis in the Lewis rat. There were no toxicities or evidence of sensitization to type II collagen in fed patients as measured by anticollagen antibodies. There was no linkage to either DR type or sex in the patients who responded. In patients treated with the collagen there was less need for narcotic use during the course of the study and four patients in the collagen-treated group apparently had complete remission of their rheumatoid arthritis. Given the biological effects seen, future trials will focus on other disease categories, more prolonged administration, and dosing studies. A multicenter trial is currently being planned. Given what is known of the mechanism of oral tolerization, these studies do not establish that type II collagen is a target autoantigen in the disease. Indeed, given the low doses fed, it is possible that the effect may have been mediated by regulatory cells that migrated to the joint and released anti-inflammatory cytokines such as TGF-β or IL-4. Whether patients who went into complete remission represent a separate category remains to be determined.

3. Uveitis

An open-label pilot study has been performed on two patients with uveitis, one with Behçet's disease (78) and the second with pars planitis. In this open-label trial, patients had required steroids and/or cyclosporine to maintain visual function. Patients were started on 30 mg of bovine S-antigen three times a week and then tapered from steroids and immunosuppressive mediation. A positive therapeutic response was observed in both patients in that over a 2-year period they were able to reduce their previous medication without worsening of vision and with decrease in S-antigen responsiveness. A double-blind, placebo-controlled trial of 45 patients is currently in progress.

4. Future Directions

Given the results in animal models of autoimmunity and initial studies in human disease states, it appears that orally administered autoantigens may find a place in the treatment of human organ-specific inflammatory autoimmune diseases. Such therapy would have the advantages of being orally administered, nontoxic, and antigen-specific. The mechanism of bystander suppression solves a major problem related to designing antigen or T-cell-specific therapy of inflammatory autoimmune diseases since one need not necessarily identify the target autoantigen for oral tolerance to be effective. As discussed above, it is likely that in human autoimmune disease states there are reactivities to multiple autoantigens from the target organ and multiple epitopes given that humans are an outbred population. Dosing appears to be important for stimulating the active suppression component of oral tolerance, and identification of regulatory cells in humans

Oral Tolerance for Treatment

following oral tolerization is critical for demonstrating the immunological effects of oral tolerization. Given the results in animal studies, one would predict that homologous protein and the use of synergists or enhancers, would increase the biological efficiency of oral tolerance. In this regard, recombinant human proteins and the concomitant administration of immune adjuvants to enhance generation of regulatory cells would be required.

REFERENCES

1. Harding CV, Unanue ER. Quantitation of antigen-presenting cell MHC class II/peptide complexes necessary for T-cell stimulation. Nature 1990; 346:574–576.
2. Adorini L, Muller S, Cardineaux F, Lehmann P, Falcioni F, Nagy ZA. In vivo competition between self peptides and foreign antigens in T-cell activation. Nature 1988; 334:623–625.
3. Woodroofe MN, Bellamy AS, Feldmann M, Davison AN, Cuzner ML. Immunocytochemical characterisation of the immune reaction in the central nervous system in multiple sclerosis: possible role for microglia in lesion growth. J Neurol Sci 1986; 74:135–152.
4. Hickey WF, Kimura HP. Perivascular microglial cells of the CNS are bone marrow–derived and present antigen in vivo. Science 1988; 239:290–292.
5. Fontana A, Fierz W, Wekerle H. Astrocytes present myelin basic protein to encephalitogenic T-cell lines. Nature 1984; 307:273–276.
6. Fierz W, Endler B, Reske K, Wekerle HAF. Astrocytes as antigen-presenting cells. I. Induction of Ia antigen expression on astrocytes by T cells via immune interferon and its effect on antigen presentation. J Immunol 1985; 134:3785–3793.
7. Lisak RP, Zweiman B. In vitro cell-mediated immunity of cerebrospinal-fluid lymphocytes to myelin basic protein in primary demyelinating diseases. N Engl J Med 1977; 297:850–853.
8. Johnson D, Hafler DA, Fallis RJ, Lees MB, Brady RO, Quarles RH, Weiner HL. Cell-mediated immunity to myelin-associated glycoprotein, proteolipid protein, and myelin basic protein in multiple sclerosis. J Neuroimmunol 1986; 13:99–108.
9. Schluesener HJ, Wekerle H. Autoaggressive T lymphocyte lines recognizing the encephalitogenic region of myelin basic protein: in vitro selection from unprimed rat T lymphocyte populations. J Immunol 1985; 135:3128–3133.
10. Martenson RE, Levine S, Sowniski R. The location of regions in guinea pig and bovine myelin basic proteins which induce experimental allergic encephalomyelitis in Lewis rats. J Immunol 1975; 114:592–595.
11. McFarlin DE, Blank SE, Kibler RF. Experimental allergic encephalomyelitis in the rat: response to encephalitogenic proteins and peptides. Science 1973; 179: 478–480.
12. Vandenbark AA, Hashim GA, Celnik B. Determinants of human myelin basic protein that induce encephalitogenic T cells in Lewis rats. J Immunol 1989; 143: 3512–3516.

13. Ota K, Matsui M, Milford EL, Mackin GA, Weiner HL, Hafler DA. T-cell recognition of an immunodominant myelin basic protein epitope in multiple sclerosis. Nature 1990; 346:183–187.
14. Martin R, Howell MD, Jaraquemada D, Flerlage M, Richert J, Brostoff S, Long EO, McFarlin DE, McFarland HF. A myelin basic protein peptide is recognized by cytotoxic T cells in the context of four HLA-DR types associated with multiple sclerosis. J Exp Med 1991; 173:19–24.
15. Jaraquemada D, Martin R, Rosen Bronson S, Flerlage M, McFarland H, Long EO. HLA-DR2a is the dominant restriction molecule for the cytotoxic T cell response to myelin basic protein in DR2Dw2 individuals. J Immunol 1990; 145:2880–2885.
16. Zhang JW, Chou CH, Hashim G, Medaer R, Raus JC. Preferential peptide specificity and HLA restriction of myelin basic protein-specific T cell clones derived from MS patients. Cell Immunol 1990; 129:189–198.
17. Wucherpfennig KW, Weiner HL, Hafler DA. T cell recognition of myelin basic protein. Immunol Today 1991; 12:277–282.
18. Chou YK, Vainiene M, Whitham R, Bourdette D, Chou CH, Hashim G, Offner H, Vandenbark AA. Response of human T lymphocyte lines to myelin basic protein: association of dominant epitopes with HLA class II restriction molecules. J Neurosci Res 1989; 23:207–216.
19. Pette M, Fujita K, Wilkinson D, Altmann DM, Trowsdale J, Giegerich G, Hinkkanen A, Epplen JT, Kappos L, Wekerle H. Myelin autoreactivity in multiple sclerosis: recognition of myelin basic protein in the context of HLA-DR2 products by T lymphocytes of multiple-sclerosis patients and healthy donors. Proc Natl Acad Sci USA 1990; 87:7968–7972.
20. Allegretta M, Nicklas JA, Sriram S, Albertini RJ. T cells responsive to myelin basic protein in patients with multiple sclerosis. Science 1990; 247:718–721.
21. Kroemer G, Martinez AC. Mechanisms of self tolerance. Immunol Today 1992; 13:401–404.
22. Miller JFAP, Moraham G. Peripheral T cell tolerance. Annu Rev Immunol 1992; 10:51–70.
23. Mowat AM. The regulation of immune responses to dietary protein antigens. Immunol Today 1987; 8:93–98.
24. Whitacre CC, Gienapp IE, Orosz CG, Bitar D. Oral tolerance in experimental autoimmune encephalomyelitis. III. Evidence for clonal anergy. J Immunol 1991; 147: 2155–2163.
25. Melamed D, Friedman A. Direct evidence for anergy in T lymphocytes tolerized by oral administration of ovalbumin. Eur J Immunol 1993; 23:935–942.
26. Friedman A, Weiner HL. Induction of anergy and/or active suppression in oral tolerance is determined by frequency of feeding and antigen dosage. Proc Natl Acad Sci USA 1994; 91(14):6688–6692.
27. Miller A, Lider O, Roberts AB, Sporn MB, Weiner HL. Suppressor T cells generated by oral tolerization to myelin basic protein suppress both in vitro and in vivo immune responses by the release of TGFβ following antigen specific triggering. Proc Natl Acad Sci USA 1992; 89:421–425.

28. Mowat AM, Strobel S, Drummond HE, Ferguson A. Immunological response to fed protein antigens in mice. I. Reversal of oral tolerance to ovalbumin by cyclophosphamide. Immunology 1982; 45:105–113.
29. MacDonald TT. Immunosuppression caused by antigen feeding. I. Evidence for the activation of a feedback suppressor pathway in the spleens of antigen-fed mice. Eur J Immunol 1982; 12:767–773.
30. Guatam SC, Chikkala NF, Battisto JR. Oral administration of the contact sensitizer trinitrochlorobenzene: initial sensitization and subsequent appearance of a suppressor population. Cell Immunol 1990; 125:437–438.
31. Cowdery JS, Johlin BJ. Regulation of the primary in vitro response to TNP-polymerized ovalbumin by T suppressor cells induced by ovalbumin feeding. J Immunol 1984; 132(6):2783–2789.
32. Richman LK, Chiller JM, Brown WR, Hanson DG, Vaz NM. Enterically induced immunological tolerance. I. Induction of suppressor T lymphocytes by intragastric administration of soluble proteins. J Immunol 1978; 121(6):2429–2433.
33. Strobel S, Mowat AM, Drummond HE, Pickering MG, Ferguson A. Immunological responses to fed protein antigens in mice. II. Oral tolerance for CMI is due to activation of cyclophosphamide-sensitive cells by gut-processed antigen. Immunology 1983; 49(3):451–456.
34. Miller S, Hanson D. Inhibition of specific immune responses by feeding protein antigens. IV. Evidence for tolerance and specific active suppression of cell-mediated immune responses to ovalbumin. J Immunol 1979; 123:2344.
35. Bloom BR, Modlin RL, Salgame P. Stigma variations: observations on suppressor T cells and leprosy. Annu Rev Immunol 1992; 10:453–488.
36. Sercarz E, Krzych U. The distinctive specificity of antigen-specific suppressor T cells. Immunol Today 1991; 12(4):111–118.
37. Weiner HL, Miller A, Khoury SJ, Al-Sabbagh A, Brod SA, Lider O, Higgins P, Sobel R, Matsui M, Sayegh M, Carpenter D, Eisenbarth G, Nussenblatt RB, Hafler DA. Suppression of organ-specific autoimmune diseases by oral administration of autoantigens. Progress in Immunol VIII: 8th Intl Congress of Immunol, Budapest 1992, pp. 627–634.
38. Lider O, Santos LMB, Lee CSY, Higgins PJ, Weiner HL. Suppression of experimental autoimmune encephalomyelitis by oral administration of myelin basic protein. II. Suppression of disease and in vitro immune responses is mediated by antigen-specific CD8+ T lymphocytes. J Immunol 1989; 174:791–798.
39. Miller A, Prabhu-Das M, Weiner HL. Epitopes of myelin basic protein (MBP) that trigger TGF-β release following oral tolerization to MBP are different from encephalitogenic epitopes. FASEB J 1992; 6(5):1686.
40. Santos LMB, Al-Sabbagh A, Londono A, Weiner HL. Oral tolerance to myelin basic protein induces TGF-β secreting T cells in Peyer's patches. J Immunol 1993; 150:115A.
41. Chen Y, Kuchroo V, Hafler DA, Weiner HL. Myelin basic protein specific regulatory T cell clones form orally tolerized mice suppress autoimmune encephalomyelitis. Science 1994; 265(5176):1237–1240.
42. Schwartz RH. A cell culture model for T lymphocyte clonal anergy. Science 1990; 248:1349–1356.

43. DeSilva DR, Urdahl KB, Jenkins MK. Clonal anergy is induced in vitro by T cell receptor occupancy in the absence of proliferation. J Immunol 1991; 147: 3261–3267.
44. Thompson HSG, Staines NA. Gastric administration of type II collagen delays the onset and severity of collagen-induced arthritis in rats. Clin Exp Immunol 1986; 64:581–586.
45. Nagler-Anderson C, Bober LA, Robinson ME, Siskind GW, Thorbeke FJ. Suppression of type II collagen-induced arthritis by intragastric administration of soluble type II collagen. Proc Natl Acad Sci USA 1986; 83:7443–7446.
46. Higgins PJ, Weiner HL. Suppression of experimental autoimmune encephalomyelitis by oral administration of myelin basic protein and its fragments. J Immunol 1988; 140:440–445.
47. Brod SA, Al-Sabbagh A, Sobel RA, Hafler DA, Weiner HL. Suppression of experimental autoimmune encephalomyelitis by oral administration of myelin antigens. IV. Suppression of chronic relapsing disease in the Lewis rat and strain 13 guinea pig. Ann Neurol 1992 (in press).
48. Miller A, Lider O, Al-Sabbagh A, Weiner H. Suppression of experimental autoimmune encephalomyelitis by oral administration of myelin basic protein. V. Hierarchy of suppression by myelin basic protein from different species. J Neuroimmunol 1992; 39:243–250.
49. Bitar D, Whitacre CC. Suppression of experimental autoimmune encephalomyelitis by the oral administration of myelin basic protein. Cell Immunol 1988; 112:364–370.
50. Fuller KA, Pearl D, Whitacre CC. Oral tolerance in experimental autoimmune encephalomyelitis: serum and salivary antibody responses. J Neuroimmunol 1990; 28:15–26.
51. Nussenblatt RB, Caspi RR, Mahdi R, Chan CC, Roberge F, Lider O, Weiner HL. Inhibition of S-antigen induced experimental autoimmune uveoretinitis by oral induction of tolerance with S-antigen. J Immunol 1990; 144(5):1689–1695.
52. Zhang JA, Davidson L, Eisenbarth G, Weiner HL. Suppression of diabetes in NOD mice by oral administration of porcine insulin. Proc Natl Acad Sci USA 1991; 88:10252–10256.
53. Bisaccia G, Caputo D, Landoni AM, Macchi GP. Heterogeneity of human T lymphocytes to bind sheep red blood cells in multiple sclerosis patients and controls. Boll Ist Sieroter Milan 1978; 56:603–608.
54. Zhang JZ, Lee CSY, Lider O, Weiner HL. Suppression of adjuvant arthritis in Lewis rats by oral administration of type II collagen. J Immunol 1990; 145: 2489–2493.
55. Birnbaum G, Lackovic V, Kotilinek L, Tobolt D. A comparison of regulatory cells in spinal fluid and blood in patients with multiple sclerosis and other neurologic diseases. Neurology 1990; 40:1785–1790.
56. Khoury SJ, Hancock WW, Weiner HL. Oral tolerance to myelin basic protein and natural recovery from experimental autoimmune encephalomyelitis are associated with down-regulation of inflammatory cytokines and differential upregulation of TGF-β, IL-4 and PGE expression in the brain. J Exp Med 1992; 46:1355–1364.
57. Sayegh MH, Zhang ZJ, Hancock WW, Kwok CA, Carpenter CB, Weiner HL. Down-regulation of the immune response to histocompatibility antigen and prevention of sensitization by skin allografts by orally administered alloantigen. Transplantation 1992; 53:163–166.

58. Sayegh MH, Khoury SJ, Hancock WH, Weiner HL, Carpenter CB. Induction of immunity and oral tolerance with polymorphic class II major histocompatibility complex allopeptides in the rat. Proc Natl Acad Sci USA 1992; 89:7762–7766.
59. Weiner HL, Friedman A, Miller A, Khoury SJ, Al-Sabbagh A, Santos L, Sayegh M, Nussenblatt RB, Trentham DE, Hafler DA. Oral tolerance: immunologic mechanisms and treatment of animal and human organ specific autoimmune diseases by oral administration of autoantigens. Annu Rev Immunol 1994; 12:809–837.
60. Khoury SJ, Lider O, Al-Sabbagh A, Weiner HL. Suppression of experimental autoimmune encephalomyelitis by oral administration of myelin basic protein. III. Synergistic effect of lipopolysaccharide. Cell Immunol 1990; 131:302–310.
61. Thurau SR, Caspi RR, Chan CC, Weiner HL, Nussenblatt RB. Immunological suppression of experimental autoimmune uveitis. Fort Ophthalmol 1991; 88(4): 404–407.
62. Thurau SR, Chan CC, Suh E, Nussenblatt RB. Induction of oral tolerance to S-antigen induced experimental autoimmune uveitis by a uveitogenic 20mer peptide. J Autoimmun 1991; 4(3):507–516.
63. Singh VK, Kalra HK, Yamaki K, Shinohara T. Suppression of experimental autoimmune uveitis in rats by the oral administration of the uveitopathogenic S-antigen fragment ar a cross-reactive homologous peptide. Cell Immunol 1992; 139(1):81–90.
64. Vrabec TR, Gregerson DS, Dua HS, Donoso LA. Inhibition of experimental autoimmune uveoretinitis by oral administration of s-antigen and synthetic peptides. Autoimmunity 1992; 12:175–184.
65. Gregerson D, Obritsch W, Donoso L. Suppression and clonal anergy play roles in oral tolerance and EAU. Invest Ophthalmol Vis Sci 1993; 34(Suppl)(1000-55:45):902.
66. Suh EDW, Vistica B, Chan CC, Raber JM, Gery I, Nussenblatt RB. Splenectomy abrogates the induction of oral tolerance in experimental autoimmune uveoretinitis. Curr Eye Res (in press).
67. Weiner HL, Zhang ZJ, Khoury SJ, Miller A, Al-Sabbagh A, Brod SA, Lider O, Higgins P, Sobel R, Nussenblatt RB, Hafler DA. Antigen-driven peripheral immune tolerance: suppression of organ-specific autoimmune diseases by oral administration of autoantigens. Ann NY Acad Sci 1991; 636:227–232.
68. Dakin R. Remarks on a cutaneous affection produced by certain poisonous vegetables. Am J Med Sci 1829; 4:98–100.
69. Lowney ED. Immunologic unresponsiveness to a contact sensitizer in man. J Invest Dermatol 1968; 51(6):411–417.
70. Korenblatt PE, Rothberg RM, Minden P, Farr RS. Immune responses of human adults after oral and parenteral exposure to bovine serum albumin. J Allergy 1968; 41: 226–235.
71. Gold WRJ, Queenan FT, Woody J, Sacher RA. Am J Obstet Gynecol 1983; 146:980.
72. Czerkinsky C, Prince SJ, Michaelek SM, Jackson S, Russell MW, Moldoveanu Z, McGhee JR, Mestecky J. IgA antibody-producing cells in peripheral blood after antigen ingestion: evidence for a common mucosal immune system in humans. Proc Natl Acad Sci USA 1987; 84:2449–2453.

73. Husby S, Elson CO, Moldoveanu Z, Mestecky J. Oral tolerance in humans. T cell but not B cell tolerance to a soluble protein antigen. 7th Intl Congress Mucosal Immunol, Prague, Czechoslovakia, 1992.
74. Polanski M, Matsui M, Khoury S, Weiner H. Oral tolerization to KLH in humans: generation of antigen-specific lines that suppress proliferative responses. J Immunol 1993; 150(8,II):114A.
75. Weiner HL, Mackin GA, Matsui M, Orav EJ, Khoury SJ, Dawson DM, Hafler DA. Double-blind pilot trial of oral tolerization with myelin antigens in multiple sclerosis. Science 1993; 259:1321–1324.
76. Salk RJS. A study of myelin basic protein as a therapeutic probe in patients with multiple sclerosis. In: Hallpike JF, Adams CWM, Toutellotte WW, eds. Multiple Sclerosis. University Press, 1983:621–630.
77. Trentham DE, Dynesius-Trentham RA, Orav EJ, Combitchi D, Lorenzo C, Sewell KL, Hafler DA, Weiner HL. Effects of oral administration of type II collagen on rheumatoid arthritis. Science 1993; 261:1727–1730.
78. Nussenblatt RB, Smet D, Weiner HL, Grey I. The treatment of the ocular complications of Behçet's disease with oral tolerization. In: 6th Intl Conf on Behçet's Disease. Amsterdam: Elsevier, 1993.

18
Oral Tolerance

DAVID E. TRENTHAM
Beth Israel Hospital and Harvard Medical School, Boston, Massachusetts

I. INTRODUCTION

With seeming rapidity, oral tolerance has emerged as a heuristic and potentially effective, safe, and broadly applicable approach for the treatment of autoimmune disease. This review will describe the actually considerable background of this antigen-specific intervention and indicate areas where additional testing is presently underway. A current and more comprehensive review exists (1).

II. AUTOIMMUNE DISEASE

Despite decades of research, there is no consensual explanation for the cause of autoimmune disease. Why do lymphocytes in a previously healthy individual suddenly break tolerance and initiate an attack on host tissue? It has been estimated that the body contains approximately 55,000 proteins to which it should not react (2). Each day around two million new T cells (3) and 20 million new B cells (4) form. Many T cells circulate from blood to lymph and back about once every 24 hr (5). Why do some of these cells not violate the rules and create *horror autotoxicus*? For years the explanation rested with central tolerance, enunciated by Burnet as his clonal selection theory. Stated in modern parlance, the checkmate for autoimmunity is clonal deletion occurring within the thymus; maturing thymocytes bearing high-affinity T-cell receptors (TCR) for self-antigens are killed before reaching the periphery (1,6).

But recently the notion of peripheral tolerance has come into vogue. Here the concept is one of induction of clonal anergy occurring outside the thymus. Anergy is defined as functional silencing of T cells. Unknown factors, perhaps culminating in the T cell failing to receive the complete signal set for activation, lead to a somnambulant state. The only evidence for anergy is in in vitro experiments. When cultured outside the body, anergy is proliferative unresponsiveness and is confined to the Th 1 CD4+ T-cell subset, which in turn is defined as an interleukin-2 (IL-2)-producing population that mediates delayed-type hypersensitivity (DTH). Proof that anergic T cells are there, but just asleep, is their capability of regaining reactivity in vitro by the addition of IL-2 or after resting for a period of time in the absence of antigen. Some immunologists believe that any Th1 cell that can be activated by an antigen can also be anergized; this would occur via peripheral exposure to minute amounts of autoantigen, combined with defective signaling by antigen-processing cells (APC). This scheme is envisioned to contribute to the rarity of autoimmunity.

Recently other theories for autoimmune containment have been proposed. Maybe T cells simply choose to ignore self-constituents. T-cell ignorance has been an intellectual by-product of a grouping of APC into professional and nonprofessional classes. By definition, a professional APC is one that can take antigen and activate a virgin T cell; conversely, antigen presentation by a nonprofessional APC results in T-cell ignorance. Whether these are simply semantic concepts or have a basis in reality remains to be determined.

An even more controversial hypothesis is that self-reactive lymphocytes are eliminated at thymic or extrathymic sites by veto cells. These are cells that cause inactivation or death of T cells when the T cell recognizes the veto cell. Some experiments have found evidence for veto cells within lymphocyte populations and it has been suggested that they might be natural-killer (NK) cells. According to this theory, veto cells would also be responsive to potentially autoreactive cells, explaining why the latter population does not expand.

Infectious agents as inciting factors for autoimmunity continue to be attractive candidates. Although none has been identified, lymphocytes with dual specificities, i.e., capable of recognizing both a foreign organism and a host constituent, conceivably could be triggered by an infection that expressed itself as an autoimmune disease. Acute rheumatic fever has been proposed as a process where streptococcal and heart antigens intercalate via a bifocal lymphocyte that induces the disease through corecognition.

Immunogenetic predispositions, onset during extreme emotional stress, and palliation by immunosuppressive drugs argue that autoimmunity is a violation in the immunoregulatory system. Individuals may possess self-reactive lymphocytes that can express themselves functionally by autoantibody production to a limited extent in the absence of clinically overt disease. But normally thymic and extrathymic programming curtails these cells' expansion into numbers adequate to induce autoimmunity. On occasion, control is lost; an imbalance ensues and proliferating autoaggressive lymphocytes trigger diseases that have sufficiently distinctive features to make them recognizable as clinical autoimmune syndromes. By targeting the synovium of diarthrodial joints, rheumatoid arthritis (RA) is a prototypic autoimmune disease. An almost equivalent amount of data pinpoints multiple sclerosis (MS) as a brain-sequestered autoimmune disease. Because medical subspecialists usually deal with a limited number of autoimmune diseases, an unfortunate lack of communication between researchers has often existed. Students of RA, MS, type I diabetes mellitus (DM), autoimmune uveitis (AU), and myasthenia gravis (MG) have in the past too often worked in isolation. Fortunately, in part because of fiscal contraction, this situation appears to be changing. A strategy that works in one autoimmune disease might be applicable in others.

Again, the cause of autoimmune disease is unknown. However, it is possible that these problems reflect a specific autoantigen drive. Consistent with this hypothesis is the ability of host constituents, when used as immunogens, to create morphological counterparts of many autoimmune diseases. Immunization with the native, i.e., triple helical, cartilage type of collagen (type II) induces the collagen arthritis (CA) model of RA; myelin basic protein (MBP), found in abundance in the brain, induces experimental autoimmune encephalitis (EAE), which closely resembles MS, and retinal antigen-S and the acetylcholine receptor (ACR) incite experimental autoimmune uveitis (EAU) and myasthenia gravis (EMG), respectively. Although an antigen that can induce DM in animals has not been identified, a spontaneously developing biochemical and histological DM condition develops in the nonobese diabetic (NOD) strain of mice. Passive transfer of islet-cell inflammation by T cells from diseased NOD mice indicates that the spontaneous model, like type I DM, is an autoimmune process. Another important model of RA is adjuvant arthritis (AA); here injection of rats with complete Freund's adjuvant (CFA) induces a disease characterized by the morphological and cellular autoimmune features of CA.

As a composite, the experimentally inducible models of CA, EAE, EAU, and EMG suggest that the inception of some cases of RA, MS, uveitis, and MG represents a loss of self-tolerance to a relevant self-protein, which in turn becomes a disease-specific autoantigen. Whether the established stage of human autoimmune disease (as opposed to the induction stage) has a relationship to these autoantigens is much less clear, since frequently there is no evidence of lymphocyte reactivity to these autoantigens on serological scrutiny or in vitro analysis in patients recognized as having problems of autoimmunity.

Although the pathogenetic pathways of models of autoimmunity are partially understood, many critical aspects are unexplained. For example, susceptible rats or mice immunized with CII develop brisk autoimmunity to this protein but only a proportion develop arthritis. Here there is universal recognition of collagen but not all develop an effector function culminating in arthritogenesis. In contrast to human autoimmune disease, the problem in experimentally inducible models is self-limitation. Why the inflammation ceases in the absence of additional autoantigenic stimulation is unknown. Experimentally inducible disease usually requires immunization with an oil emulsion containing the host protein and CFA. CA in the rat is the exception; it is readily inducible with oil not containing mycobacteria [incomplete Freund's adjuvant (IFA)]. CA is also unique in requiring whole native protein, as opposed to relevant protein fragments, to effectively induce disease.

III. ANTIGEN-SPECIFIC IMMUNOTHERAPY

In the past 15 years, these models have been used to develop and examine antigen-specific immunosuppressive techniques. In general, approaches that have been shown to suppress cellular responses to the inducing antigen have been capable of palliating the clinical features of the disease. Administration during the induction stage has usually been required to demonstrate effectiveness. This temporal window is similar to that required for immunosuppressive drugs, such as cyclosporine, to shown activity.

Autoantigen chemically coupled to the surface of fixed spleen cell membranes provided an interesting start to this field. In this system, subcutaneous injection of CII-coupled cells resulted in DTH but no antibody or disease production, whereas intravenous (IV) administration could suppress CA in immunized animals. The technique was also used to first show that AA was suppressible by a CII-based protocol (7). High doses of free autoantigen delivered IV were also disease suppressive but the immunological consequences of this system have not been well elucidated.

IV instillation of autoantigen-generated T-cell lines was also quite successful in EAE, CA, and AA. In CA, T-cell-line vaccination attenuated DTH without compromising autoantibody responses to CII. Presumably the mechanism involves recipient recognition of suppressing structures on the surface membrane of inoculated T cells, but the precise pathway remains elusive.

Although effective and lacking in toxicity, all of these approaches involved cumbersome preparative techniques, parenteral injection, and materials that could not be readily sterilized. Therefore, no enthusiasm could be easily engendered for human application. Nonetheless, they possess much historical significance for showing that specifically impairing an animal's capacity to mount an immune response to a host constituent abrogated the development of autoimmune disease.

IV. ORAL TOLERANCE IN ANIMALS

Perhaps receiving less attention during this period, a number of laboratories independently reported that a much more straightforward antigen-specific immunosuppressive approach, termed oral tolerance, was also effective and nontoxic in these models. In their review, Weiner et al. (1) have traced the origins of oral tolerance from purported use in the nineteenth century by native American Indian tribes to thwart the development of subsequent contact sensitivity to poison ivy by intentional feeding of this plant to children, as well as documented studies of Chase and Wells using antigen feeding to suppress sensitization in experimental animals early in this century. Several laboratories have studied the immunological consequences of antigen feeding during the past 15–30 years (cited in 8), but only in the mid-1980s was oral tolerance evaluated in models of autoimmune disease. Initial systems involved feeding MBP to abort EAE (9,10) and CII feeding to block CA (11,12). Shortly thereafter, these positive findings were replicated and extended by a number of investigators. Feeding material containing the disease-inducing epitopes suppressed CA, EAE, EMG, and EAU (13). Insulin, given orally, delayed the onset of DM in NOD mice (14). These results seem to be somewhat paradoxical, since autoimmunity to insulin is not involved in the disease process. In all studies, no side effects or evidence of toxicity was observed. There were no reports of lack of success.

Because this review focuses on RA, the experience in AA will be described briefly. As did CII-coupled cells (7), oral administration of native CII suppressed AA (15). Immunologically, there was a suppression of DTH to CII in adjuvant-injected rats but CII feeding did not alter DTH responses to the mycobacteria present in the arthritogenic CFA inoculum. Feeding the disease-irrelevant antigen, MBP, did not influence the expression of AA. These findings clearly show that the process was antigen-specific.

Of potential relevance for human use, there was a narrow dose-response range for oral CII to be effective in the AA system; i.e., daily doses of 3–30 μg were suppressive but quantities outside this window did not suppress the disease. The optimal dose actually proved to be smaller than anticipated, suggesting that human trials should commence with very low doses and perhaps include a dose-escalation format.

In conclusion, animal investigation led to a consensual opinion that oral feeding of disease-relevant antigen was a safe and effective way to down-regulate autoimmunity. Demonstrations of antigen-specific immunosuppression justified the term oral tolerance for this procedure.

V. ORAL TOLERANCE IN HUMANS

Based on the collective evidence from several models of autoimmunity and the simplicity of this method, it seemed reasonable to test oral tolerance as a therapy in human autoimmune disease. Howard Weiner, David Hafler, and their colleagues at the Brigham and Women's Hospital in Boston initiated human trials by feeding MBP, prepared from bovine brain, to patients with MS (16). The trial was randomized, placebo-controlled, extended over a year's duration, and involved 30 patients. Patient retention and compliance were excellent. In MS there are considerable vagaries in the course of the disease and it is difficult to quantify disease activity. Nonetheless, patients in the MBP group displayed a favorable trend toward improvement, compared to the placebo group. No side effects or tendency to exacerbate disease was encountered. Preliminary indications that MBP feeding reduced the number of MBP-reactive T cells in the peripheral blood were also found. Although the small number of patients and the fixed dose of antigen used in this trial make clinical judgments of efficacy inconclusive, clearly a rigorous scientific entry into human therapy had been accomplished. This effort justified larger trials in MS and was a further impetus for scrutiny in other diseases.

Collaborative discussions of the author with these neurologists formulated a plan whereby CII feeding would be tested in RA. Along with the enigma of whether CII was a "correct antigen," considerable concern existed over what dose should be utilized, given the narrow window observed in the AA studies (15). Accordingly, a minor dose-escalation format was selected for testing (17). The single issue was whether anything at all might be seen. Therefore, the initial analysis involved feeding 10 patients 0.1 mg of chicken CII daily for a month and then 0.5 mg/day for 2 additional months. To expedite the trial, it was unencumbered by a washout period from other drugs or a controlled limb. Antirheumatic drugs other than prednisone or nonsteroidal antiinflammatory drugs (NSAIDS) were stopped. Somewhat startling was that six of the 10 patients improved with CII feeding. Even more curious was the bimodal nature of the response—patients either seemed to improve to a substantial extent or appeared to be completely refractory to the intervention. Thus, there was an early suggestion of subset responsiveness to CII feeding in RA. The other intriguing outcome was the durability of the response in a few patients. Both patient and physician perception and lack of necessity for additional drug treatment indicated that some of these patients remained improved for a number of months after completing the CII treatment. When the disease reappeared, it could again be captured in the responding patients by an additional 3 months of collagen feeding. As expected, no toxicity occurred.

Although uncontrolled, the degree and duration of response observed in a few patients clearly mandated analysis of CII feeding in a larger controlled situation.

A fairly pragmatic design was selected (17). Only the initial doses would be evaluated; patients readily available because of an interest in other concurrent trials would be entered regardless of duration or severity of disease; a washout period and attendant delays would not be incorporated; and the thrust of the study would be clinical (immunological dissection was viewed as premature). The possibility of a responsive subset would be addressed by inclusion of as many patients as could be realistically recruited by a single center as well as attempted delineation of prognostic markers (HLA-DR subtyping and serum antibodies to CII). Familiarity with purifying the protein and solubility characteristics that facilitated sterilization by membrane filtration explained why chicken CII was used.

Patient recruitment proved to be relatively simple. The intended 60 patients were enrolled over a 4-month period, in part because of the described success in the 10-patient study and the apparent benignity of ingesting a natural animal protein. What was difficult to describe to participants was what might be going on—terms like oral tolerance, reverse vaccination, and so forth seemed to be only partially enlightening. Because of preselection, compliance was excellent. Only one of the 60 patients was judged to be nonevaluable at completion of the study.

Analysis of mean group data at the end of the 3-month trial showed that a trend favoring CII over placebo occurred for each disease parameter (17). This difference attained statistical significance ($p < 0.05$) for the primary clinical variables, such as the number of swollen joints and number of tender joints. As is frequent in controlled clinical trials of RA, a sizable placebo response was observed. Some patients given placebo responded to an extent that made predictions of who received CII, prior to breaking the code, frequently unreliable. No side effects were encountered and no predictors of CII responsiveness were found.

As in the initial study, a few patients (14%) in the collagen limb met clinical criteria for remission, even after it had been prospectively reformulated as disease resolution, which required that a response not be influenced by concurrent prednisone therapy (17). At present, experience with additional open-label administration is being obtained in some patients who relapsed after initial CII feeding and responded after retreatment.

Recently, direct evidence that oral tolerance can be achieved in humans has been reported (8). Keyhole limpet hemocyanin (KLH), a potent antigen to which humans are not usually exposed, was utilized in healthy test volunteers. Two brief periods of KLH feeding, followed by two subcutaneous injections of KLH to induce sensitization, extinguished the DTH skin test reaction that was observed in the unfed control group. The study was simple in design, and the outcomes were definitive and strongly suggest that oral feeding can thwart the development of DTH in humans. Although the MS (16) data suggest that a preexistent immune response in humans, i.e., to MBP, can be influenced by antigen feeding, more work is needed. For example, what happens to PPD-positive individuals fed mycobacterial particles?

VI. ORAL TOLERANCE: POTENTIAL MECHANISMS

While investigators agree that oral tolerance can be achieved in animals, no uniformity of opinion exists as to how this state is acquired. The explanation of tolerance is surprisingly enigmatic. Some phrases, often used without substantial experimental evidence, attempting to describe various central and peripheral tolerance states are shown in Table 1. Note that these processes parallel those envisioned to explain self/nonself discrimination and discussed earlier in the context of autoimmunity. Clearly many lymphocytes that begin to express receptors for self molecules as they mature within the thymus are eliminated before leaving this site. However, extrathymic scenarios that curtail self-reactivity in the periphery are much more mysterious. Anergy is a currently popular concept of peripheral tolerance that relates to a functional dormancy that can be overcome. It appears to be the net result of variant ways in which antigen stimulation is delivered to normally reactive lymphocytes. An even more recent concept is that in some situations lymphocytes exposed to antigen proliferate so vigorously that they become "exhausted" and enter an effete state of protracted functional decline.

In the case of oral tolerance, two, perhaps bifunctional, pathways have been described. Antigen feeding could lead to T-cell clonal anergy. An alternative explanation is bystander suppression, which retains antigen specificity but is orchestrated by in situ inhibitory cytokine production (chemoquiescence).

The intestinal epithelium is the largest surface in the body exposed to the external environment. It is well documented that there is a remarkable abundance of lymphoid tissue within the gut mucosal wall. Beginning in the tonsillar region and extending throughout the colon, large concentrations of lymphocytes are found. The proximal small intestine may be the most immunologically active site.

Table 1 Possible Explanations for Tolerance

Thymic censorship
Extra-thymic lymphocyte curtailment
  Deletion or purging
  Suppression
  Anergy
  Ignorance
  Exhaustion
  Chemoquiescence or bystander suppression
  Veto cell ablation

Oral Tolerance

Here, M cells, found in the lining tissue, are efficient APC. In addition, other epithelial cells (enterocytes) that line the outer membrane express IL-2 receptors and both MHC class I and II antigens, suggesting an APC function as well (18). Unlike most areas, in the proximal small intestine there is a notable phenotypic skewage of T cells toward those of the CD8 subset. Both aggregate cells and intraepithelial lymphocytes (IEL) display quite frequently the CD8 marker. Are they there to suppress dietary antigen load?

Despite the recognition of IEL for over 100 years, surprisingly little is known about gut-associated lymphoid tissue (GALT). Most evidence implicates the intestine as a second major site, after the thymus, of T lymphopoiesis. There is an age-dependent increase in the number of IEL and percentage that express the classic pan-T-cell-receptor marker $\alpha\beta$ (TCR $\alpha\beta$); both these increases appear to be maturation events dependent on a conventional diet and exposure to a germ-laden environment. In the mouse, GALT T cells are functionally more active than their splenic counterparts, as measured by production of the cytokines IL-5 and interferon-γ (IFN-γ) (19). GALT T cells are also efficient promoters of isotype switching to IgA production by B cells, a process in which IL-5 is probably involved (19,20). What else GALT cells do is, at present, solely conjecture.

Weiner and his colleagues have provided evidence for oral tolerance relating both to clonal anergy and to bystander suppression. Experiments with the antigen ovalbumin (OVA) in the guinea pig are illustrative (1). Guinea pigs fed OVA prior to immunization have lymph node cells (LNC) that fail to proliferate to OVA in vitro. This state of oral tolerance can be achieved either by infrequent feeding of relatively high doses of OVA or by repetitive feeding of low doses of antigen. The failure of LNC from guinea pigs administered high doses of OVA to respond persists in the presence of a monoclonal antibody against transforming growth factor-β (TGF-β), added to proliferation cultures. In contrast, cells from the low-dose-fed group exhibit a restoration in the lymphocyte proliferation in the in vitro presence of anti-TGF-β. In the former state of high-dose oral tolerance, presumably OVA-reactive Th-1 CD4+ cells have either been eliminated or anergized (paralyzed) and thus do not proliferate to any challenge in vitro. That the guinea pig LNC recover proliferative responsiveness when IL-2 is added to the cultures favors anergy as the mechanism for high-dose oral tolerance. The low-dose repetitive feeding schema is compatible with a state of bystander suppression (21). As mentioned earlier, antigen-proliferating CD4+ T cells can be subdivided into Th-1- and Th-2-positive populations. The former subset releases promotive cytokines during antigen stimulation, such as IL-2, IFN-γ, and tumor necrosis factor-α (TNF-α), while the latter releases inhibitory cytokines, such as TGF-β and IL-4, -5, and -10. On a molar basis, TGF-β is probably the most potent inhibitor of lymphocyte proliferation described to date. If oral tolerance resulting from low-dose frequent feeding were to be a bystander process, elimination of

TGF-β pathways by an anti-TGF-β monoclonal antibody should at least partially restore responsiveness. In the guinea pig system, this seems to be the case.

A bystander scenario might still be an antigen-specific form of tolerance. Triggering of Th-2 cells by feeding would require the specific antigen for their in vivo activation. In the guinea pig system, cells would require specific exposure to OVA to release TGF-β and thereby produce the immunosuppressive state. Chemoquiescence, i.e., mediation by cytokine, might be a more descriptive term than bystander suppression.

By inference, bystander suppression might have important implications for the capacity of oral tolerance to alter organ- or tissue-specific disease. Clonal anergy as the singular mechanism would imply that autoimmunity to MBP is a mediatory element in MS and an analogous autoaberrant response against CII would be disease-provocative in RA. But such might not be the case. Examine the NOD system, which can be palliated by oral insulin, even though epitopes on insulin do not appear to be involved in the disease process (14). Bystander suppression could work if the antigen triggering TGF-β production by Th-2 cells was found at the site of the disease, whether or not it was actually involved in inciting the disease process. Th-2 cells, primed by oral feeding, could migrate to the brain in EAE and there be activated by MBP. Analogous cells could recognize CII from the cartilage of inflamed joints. The parallel also seems reasonable for insulin in the NOD model. Thus for host antigen to be an efficient oral toleragen it might not have to be a disease-inducing material. Rather, its capacity to trigger Th-2 cells and a residence at the site of disease attack would cogovern its effectiveness.

Until recently, it was assumed that disease-inducing and oral-tolerizing epitopes were always identical. Now a report by Miller et al. (22) has shown that in some instances the stimulants may be dissociable. EAE has the advantage of being inducible by both whole MBP and certain of its fragments. Likewise, both the entire MBP molecule and some, but not all, of its fragments are effective toleragens. Some encephalitogenic fragments are not tolerogenic and some tolerogenic fragments are not encephalitogenic. Of further interest, some tolerogenic fragments stimulate TGF-β release in vitro, whereas other clinically effective portions lack this function. In this model, in vitro release of TGF-β appears to be associated with CD8+ cells. These findings are consistent with an earlier report that CD8+ cells from MBP-fed rats can passively transfer protection against EAE (23). Awareness of potential epitope differences may be important in designing molecules for oral tolerance in the future.

VII. CONCLUSIONS AND NEW DIRECTIONS

In terms of the scientific community, animal demonstrations of oral tolerance are well accepted. The only inconclusive element is the extent to which anergy versus regulatory cell (Th-2-cytokine releasing) mediation accounts for the outcome. Current animal work is focusing on ways to upgrade the potency of oral tolerance, such as the use of adjuvants or synergists. Also important are more precise delineations of the effector cells intrinsic to oral tolerance and clarification of more efficient ways to harness their properties.

The reaction of the scientific community to the claims of effectiveness of oral tolerance in human disease (16,17) has been mixed. Some of the incredulity can be traced to several factors. Perhaps most significantly, the skepticism relates to the claim itself. It is justifiably difficult to deal with such a simplistic and unsophisticated notion as oral tolerance, particularly when it is heralded as a way to benignly redirect the immune system when chemicals, irradiation in the form of total lymphoid exposure, cytokines, monoclonal antibodies, and even toxins have failed to do so in the past.

Another major problem relates to the "softness" of the evidence acquired in clinical trials to date and the propensity of the public to interpret the findings in wishful ways. The findings in the MS trial (16) were only suggestive and the p values attesting to success in the RA study (17) were often marginally significant. To date, some patients with MS or RA appear to respond to the procedure, but their numbers remain too small and the treatment periods too brief to ascertain the value of this approach. As the authors clearly state, these are *approach* papers not intended to prove to disprove efficacy. The nonclinical journal in which these papers appear connotes this opinion.

Finally, relating to the RA outcome, placebo responses are always worrisome, regardless of how expected they are to rheumatologists. If a placebo response of 20 or 30% is predicted and encountered and an active treatment response of 60% or so is reached, the study is deemed a success. However, for a clinician who might use the new therapy, is the candidate drug 60% effective or is 30% of the "treatment effect" due to the confounding placebo phenomenon? No one knows. Accordingly, it is totally unclear how much of the "collagen response" was attributable to the use of a promising new technique by enthusiastic investigators in a medical setting generally considered to be prestigious.

The considerable impact of the MS and RA studies for patients and the public also deserves comment. Clearly patients recognize that these diseases are chronic, incapacitating, and incurable. Despite this, they dislike and mistrust strong drugs that can on rare occasions produce serious side effects, particularly because of the duration with which they must be applied. Not adequately appreciated by the medical profession is the propensity of these patients to turn to alternative remedies for help. Then comes an approach with almost a homeopathic flavor, i.e., using one's own immune resources in concert with oral administration of substances that putatively are involved in the disease to remedy the state. It is naturally quite difficult for the public to understand that if a treatment involves an extract from a chicken and is, according to a newspaper headline, "something to crow about," why is the treatment not being released in a widespread and expedient manner to sufferers?

Fortunately, additional data needed to judge oral tolerance should soon be available. A study of oral retinal antigen-S is well underway at the National Eye Institute. Multicenter trials of oral insulin in type I DM should soon begin. Both the MS and RA work need enlargement into multicenter and longer-term settings as well as independent appraisals. These extensions are underway. Concerns that currently exist include microheterogenity in supplies of MBP and CII, the uncertain effectiveness of CII from species other than chickens, and the difficulty in purifying and maintaining CII in a native state. Until additional appraisals are completed, all that can be said about MBP and CII is that they are encephalogenic and arthritogenic self-immunogens in animals, and when used as oral toleragens, they arrest models of autoimmune disease in animals and appear to have the same effect in some patients with MS or RA.

REFERENCES

1. Weiner HL, Friedman A, Miller A, Khoury SJ, Al-Sabbagh A, Nussenblatt RB, Trentham DE, Hafler DA. Oral tolerance: immunologic mechanisms and treatment of animal and human organ-specific autoimmune disease by oral administration of autoantigens. Annu Rev Immunol 1994; 12:809–837.
2. Matzinger P. Tolerance, danger, and the extended family. Annu Rev Immunol 1994; 12:991–945.
3. Scollay R. Thymus migration: quantitative studies on the rate of migration of cells from the thymus to the periphery in mice. Eur J Immunol 1980; 10:210–216.
4. Osmond DG. The turnover of B-cell populations. Immunol Today 1993; 14:34–37.
5. Sprent J, Basten A. Circulating T, B lymphocytes of the mouse. II. Lifespan. Cell Immunol 1973; 7:40–59.
6. Kappler JW, Roehm N, Marrack P. T cell tolerance by clonal elimination in the thymus. Cell 1987; 4:273–279.
7. Trentham DE, Dynesius-Trentham RA. Attenuation of an adjuvant arthritis by type II collagen. J Immunol 1983; 130:2689–2692.
8. Husby S, Mestecky J, Moldoveanu Z, Holland S, Elson CO. Oral tolerance in humans: T cell but not B cell tolerance after antigen feeding. J Immunol 1994; 152:4663–4669.
9. Higgins PJ, Weiner HL. Suppression of experimental autoimmune encephalomyelitis by oral administration of myelin basic protein and its fragments. J Immunol 1988; 140:440–445.
10. Bitar DM, Whitacre CC. Suppression of experimental autoimmune encephalomyelitis by the oral administration of myelin basic protein. Cell Immunol 1988; 112:364–370.
11. Thompson HSG, Staines NA. Gastric administration of type II collagen delays the onset and severity of collagen-induced arthritis in rats. Clin Exp Immunol 1986; 64:581–586.

12. Nagler-Anderson C, Bober LA, Robinson ME, Siskind GW, Thorbecke GJ. Suppression of type II collagen-induced arthritis by intragastric administration of soluble type II collagen. Proc Natl Acad Sci USA 1986; 83:7443–7446.
13. Nussenblatt RB, Caspi RR, Mahdi R, Chan C-C, Roberge F, Lier O, Weiner HL. Inhibition of S-antigen induced autoimmune uveoretinitis by oral induction of tolerance with S-antigen. J Immunol 1990; 144:1689–1695.
14. Zhang JA, Davidson L, Eisenbarth G, Weiner HL. Suppression of diabetes in NOD mice by oral administration of porcine insulin. Proc Natl Acad Sci USA 1990; 88:10252–10256.
15. Zhang JA, Lee CSY, Lider O, Weiner HL. Suppression of adjuvant arthritis in Lewis rats by oral administration of type II collagen. J Immunol 1990; 145:2489–2493.
16. Weiner HL, Mackin GA, Matsui M, Orav EJ, Khoury SJ, Dawson DM, Hafler DA. Double-blind pilot trial of oral tolerization with myelin antigens in multiple sclerosis. Science 1993; 259:1321–1324.
17. Trentham DE, Dynesius-Trentham RA, Orav EJ, Combitchi D, Lorenzo C, Sewell KL, Hafler DA, Weiner HL. Effects of oral administration of type II collagen on rheumatoid arthritis. Science 1993; 261:1727–1730.
18. Bland P. MHC class II expression by the gut epithelium. Immunol Today 1988; 9(6):174–178.
19. Taguchi Y, McGhee JR, Coffman RL, Beagley KW, Elderidge JH, Takatsu K, Kiyono H. Analysis of the Th 1 and Th 2 cells in murine gut-associated tissue: frequencies of CD 4^+ and CD 8^+ T cells that secrete IFN-γ and IL-5. J Immunol 1990; 145:68–77.
20. Kaetzel CS, Robinson JK, Chontalachuruva KR, Vaerman J-P, Lamm ME. The polymeric immunoglobulin receptor (secretory component) mediates transport of immune complexes across epithelial cells: a local defense function for IgA. Proc Natl Acad Sci USA 1991; 88:8796–8800.
21. Miller A, Lider O, Weiner HL. Antigen-driven bystander suppression following oral administration of antigens, J Exp Med 1991; 174:791–798.
22. Miller A, Al-Sabbagh A, Santos LMB, Das MP, Weiner HL. Epitopes of myelin basic protein that trigger TGF-β release after oral tolerization are distinct from encephalitogenic epitopes and mediate epitope-drive bystander suppression. J Immunol 1993; 151:7307–7315.
23. Miller A, Lider O, Roberts AB, Sporn M, Weiner HL. Suppressor T cells generated by oral tolerization to myelin basic protein suppresses both in vitro and in vivo immune responses by the release of TGF-β following antigen specific triggering. Proc Natl Acad Sci USA 1992; 89:421–425.

# 19
Intravenous Immunoglobulin (IVIg) in the Treatment of Autoimmune Diseases

VIBEKE STRAND
Stanford University School of Medicine, Stanford, California

MARTIN L. LEE
School of Public Health, University of California, Los Angeles, Los Angeles, California

I. INTRODUCTION

The decision to use intravenously administered immunoglobulin (IVIg) in a variety of autoimmune diseases, including rheumatoid arthritis (RA), systemic lupus erythematosus (SLE), dermatomyositis/polymyositis (DM/PM), vasculitis, neurological diseases such as Guillain-Barré syndrome (GBS), multiple sclerosis, amyotrophic lateral sclerosis, and myasthenia gravis, and hematological diseases such as autoimmune hemolytic anemia was largely based on its successful empirical use in idiopathic thrombocytopenic purpura (ITP). Randomized controlled trials (RCTs) have subsequently confirmed the efficacy of IVIg therapy compared to aspirin in Kawasaki disease (KD) (1,2), to corticosteroids in ITP (3), to placebo in DM (4), and to plasmapheresis in GBS (5).

This chapter reviews the reported clinical and experimental data on the use of IVIg, as well as postulated mechanisms by which it may exert clinical effect, in autoimmune diseases. Careful, well-designed, prospective RCTs are lacking in most of these disease applications. They will clearly be necessary to accurately assess the cost-effectiveness of this potentially beneficial, but expensive, therapy.

II. TREATMENT OF IDIOPATHIC THROMBOCYTOPENIA PURPURA

In 1980, administration of IVIg to a child with ITP resulted in a rapid increase in platelet count. This anecdotal observation provided the impetus for the initiation of pilot (6) and subsequent controlled (3) clinical trials. A regimen of 400 mg/kg body weight daily for 5 consecutive days was initially used, and variations of this regimen (e.g., 2000 mg/kg as a single infusion) have been successfully employed. Although most patients respond rapidly to therapy, few can remain treatment-free for more than several months. Sustained responses can follow maintenance therapy; prolonged remissions have been reported (7). The use of IVIg is now considered standard in ITP. Its efficacy has suggested that IVIg may prove useful in other autoimmune diseases, prompting a broad variety of pilot investigations.

III. TREATMENT OF SYSTEMIC LUPUS ERYTHEMATOSUS

Treatment with IVIg has been demonstrated to reverse severe manifestations of hematological, renal, central nervous system (CNS), and pulmonary disease in SLE. It has been advocated in the treatment of recurrent pregnancy losses attributed to the presence of antiphospholipid antibodies (APL). Unfortunately, these data are gleaned from case reports and open-label clinical trials. Currently two placebo RCTs are underway to evaluate administration of IVIg to pregnant women with recurrent abortions (8). Otherwise the data are anecdotal, and by virtue of their publication, would be expected to be positive. At present, there are no data from RCT to support IVIg treatment in SLE, vasculitis, or RA.

Based on its use in treatment of ITP, IVIg has been successfully reported to treat the thrombocytopenia, leukopenia, hemolytic anemia, and pancytopenia observed in SLE (9–16). Improvement can be dramatic, but is generally transient. Sustained benefit of more than 6 months' duration as well as complete clinical resolution has been reported. No pretreatment characteristics have been identified that predict response to therapy. Similarly, IVIg administration has dramatically reversed CNS symptoms, ranging from psychosis to postinfarction neurological deficits (10,13,17,18). Refractory and severe pulmonary and pleural manifestations have also responded to IVIg therapy (18,19).

In general, severe manifestations of SLE requiring high-dose chronic corticosteroid and/or immunosuppressive therapy may be responsive to one or several courses of IVIg. The treatment of renal disease remains controversial, because of several reports of worsening renal function after administration of IVIg. Tomino et al. demonstrated that incubation of renal biopsy specimens with human sera or IVIg solubilized intraglomerular deposits of IgG immune complexes, suggesting a direct mechanisms for clinical benefit (20). Subsequently, in a series of nine patients with renal disease refractory to steroids and cytoxan, IVIg administration improved creatinine clearance and decreased proteinuria (21). In vitro incubation with IVIg resulted in dissociation of glomerular deposits. Akashi et al. also reported dramatic improvement in two patients following IVIg therapy (1).

However, in a series of six SLE patients with cytopenias accompanied by renal and/or cutaneous manifestations, demonstration of decreased C1q binding of immune complexes in vivo failed to correlate with clinical improvement after treatment with IVIg (22). Schifferli et al. reported transient and asymptomatic increases in serum creatinine in eight patients with SLE renal disease following IVIg administration; only two had renal function impairment prior to treatment (23). Barron and Sher observed worsening renal function in 3/7 patients who received 2 g/kg of IVIg over 4–5 days followed by five monthly courses of 400 mg/kg (15). Another group reported exacerbation of lupus nephritis in one of two patients after 10 months of IVIg therapy despite initial dramatic improvement (18). The etiology for worsened renal disease remains unexplained although it has been suggested to occur due to enhanced immune complex formation (24). The potential for this complication requires that IVIg usage in SLE be considered with caution.

IV. TREATMENT OF RECURRENT FETAL LOSS

Greater than 90% of pregnancies in women with APL and previous fetal loss fail to result in live births (25). Scott and colleagues first reported a patient with IgG and IgM anticardiolipin antibodies and nine consecutive fetal deaths who achieved a live birth after two courses of IVIg therapy (400 mg/kg × 5 days) at 8 and 15 weeks (26). Subsequent work has indicated that IVIg treatment decreases titers of both the lupus anticoagulant and anticardiolipin antibodies, with improvement in preeclampsia, although suppression of APL antibodies was transient (27). A variety of reports suggest improved fetal survival in patients with very bad obstetric histories associated with APL antibodies (25,28–30). At present, treatment for recurrent pregnancy loss attributed to APL antibodies remains unresolved, although IVIg may be less toxic to the mother than prednisone.

In women with recurrent abortion without associated autoantibodies, administration of IVIg may offer benefit (31), although data from the European experience remain unconvincing (8). A RCT, currently underway in the United States (31), may better determine its place in the therapeutic armamentarium.

V. TREATMENT OF RHEUMATOID ARTHRITIS/JUVENILE RHEUMATOID ARTHRITIS

Anecdotal reports in RA and JRA suggest benefit in patients with disease refractory to currently available therapies (32–34). The data are open-label and would be expected to be positive. A RCT is underway in JRA, and its results are eagerly awaited.

VI. TREATMENT OF DERMATOMYOSITIS/POLYMYOSITIS

Multiple open-label series have reported improvement in dermatomyositis/polymyositis following treatment with IVIg (35–39) and suggest it may be steroid-sparing. In others, only transient benefit has occurred (40,41). Recently Dalakas and colleagues published a double-blind RCT comparing three monthly infusions of 2 g/kg IVIg to placebo in 15 patients with biopsy-proven, treatment-resistant dermatomyositis (4). The eight patients assigned to active treatment had significant improvement in muscle strength and neuromuscular symptoms; seven receiving placebo did not. With crossover from the placebo arm, a total of 12 patients received IVIg, nine of whom improved. Repetitive muscle biopsies documented histological improvement in five and offer insight into the mechanisms of action of IVIg. Clinical effect was short-lived, with mean duration of 6 weeks; long-term benefit required repeated treatment courses. RCTs are currently underway to evaluate IVIg therapy in polymyositis and inclusion body myositis.

VII. TREATMENT OF KAWASAKI DISEASE

Although the mechanisms of action leading to beneficial effects of IVIg treatment in autoimmune diseases are incompletely understood, its use has been derived from the positive experience in KD and ITP. Over the past decade, the treatment of acute, severe KD has evolved to include intravenously administered IVIg (1,2,42–44). Despite its expense, and requirement for intravenous (IV) administration, patients have less incidence of coronary artery aneurysms as well as more rapid resolution of symptoms and disease. A recently published economic analysis demonstrated cost-effectiveness compared to aspirin and superiority for the high-dose regimen (single dose of 2000 mg/kg vs. 4 days of 400 mg/kg), particularly as it was usually associated with shorter hospital stays (45). Comparison of the 4-day 400 mg/kg/day regimen with a single lower dose of 100 mg/kg has shown no significant difference in outcome (42), allowing for further economic benefit. IVIg is considered accepted level of care in the United States, Canada, and Europe for KD (46,47) and has been recommended by the American Academy of Pediatrics.

VIII. TREATMENT OF VASCULITIS

Jayne and Lockwood published the first series of seven patients with systemic vasculitis treated with 5-day courses of 400 mg/kg/day of IVIg (48). All improved, with associated decreases in antineutrophil cytoplasmic antibodies (ANCA) and C-reactive protein (CRP). Subsequent reports extended this open-label experience to 16, with 15 improved (49), and 26, with 18 improved (50). Fourteen patients had Wegener's granulomatosis (WG), 11 had microscopic polyarteritis, and one had rheumatoid vasculitis. Mild disease flares were observed in many patients (number not stated) immediately after treatment, associated with increases in ANCA titers, which peaked at 10 days. Benefit was maintained in 18 for 1 year from treatment, although many patients continued to receive concomitant immunosuppressive therapy. No deterioration in renal function occurred. A smaller series published by Gross and colleagues described significant but transient improvement in five of nine patients, eight with WG and one with pANCA-associated vasculitis (51). Reductions in ANCA titer and CRP occurred in three of nine. Rostoker and colleagues recently published an open-label series of 11 patients with severe IgA nephropathy and Henoch-Schönlein purpura, treated with three monthly courses of IVIg (2 g/kg/month) followed by six monthly IM treatments (52). Posttreatment glomerular filtration rate improved or stabilized in nine patients with deteriorating renal function; this improvement was maintained in four for as long as 19 months.

In summary, several cases and open-label series have reported improvement in patients with small to medium-sized vessel vasculitis, associated with ANCA, IgA, or IgE (53) treated with IVIg. Prospective RCTs will be required to confirm these positive data.

IX. TREATMENT OF AUTOIMMUNE HEMATOLOGICAL DISEASES

A. Autoimmune Neutropenia/Hemolytic Anemia

Anecdotal studies have shown that high-dose IVIg (300–500 mg/kg for 3–5 days) produces a rapid increase in neutrophils in subjects with autoimmune neutropenia (54–61). On the other hand, in the treatment of autoimmune hemolytic anemia (warm antibody type), the data have been mixed (62–68). Better results occur in those patients for whom the anemia is secondary to an underlying lymphoproliferative disease. A recent meta-analysis of the reported data suggested that the presence of hepatomegaly and low pretreatment hemoglobin may be good predictors of response (69). It has been hypothesized that IVIg may exacerbate the anemia by promoting the clearance of autoantibody-coated red blood cells.

B. Factor VIII Inhibitors

Antibodies to the factor VIII molecule may occur spontaneously postpartum, in autoimmune diseases and in malignancies, or may be acquired (alloimmune). Treatment with high-dose IVIg (400 mg/kg × 5 days) appears to ameliorate the problem in patients with autoimmune factor VIII inhibitors (70–73) but is not always successful (74). Approximately 10–15% of patients with hemophilia receiving regular infusions of replacement factor develop alloantibodies to factor VIII. These types of factor VIII inhibitors typically require more aggressive therapy utilizing a combination of IVIg treatment and cyclophosphamide, prednisone, and factor VIII infusions (75,76).

C. Other Diseases

As previously noted, IVIg has proven quite useful in treating platelet autoimmunization, such as ITP and the thrombocytopenia associated with HIV disease. However, its use in the treatment of platelet alloimmunization, secondary to repetitive platelet transfusions, has not been successful (77). Together, the data suggest that alloimmunization resulting from transfusions may be more resistant to IVIg administration than cytopenias resulting from autoimmune processes.

X. TREATMENT OF AUTOIMMUNE NEUROLOGICAL DISEASES

A. Myasthenia Gravis

Myasthenia gravis (MG) is an autoimmune disease characterized by antibodies to the acetylcholinesterase receptor (AChR) in the myoneural junction, although a direct correlation between antibody levels and disease severity is often lacking. Treatment for MG has included cholinesterase inhibitors, thymectomy, corticosteroids, and plasmapheresis. Following anecdotal reports that high-dose IVIg (400 mg/kg × 5 days) ameliorated motor weakness and reduced AChR antibody titers (78,79). Asura and colleagues reported its beneficial effects in a cohort of 11 patients (80). Similar findings have been reported in other small open-label series (81,82).

All reports have been positive; only one study showed equivocal results (82), using half the dose employed in the other studies. The effect of IVIg appears to be palliative and relatively short-lived with a maximum effect occurring by day 10 posttherapy. Its benefit may be in the ability to taper or eliminate corticosteroids and for emergency use such as prior to surgery. Controlled clinical studies are clearly needed to accurately evaluate its role in the therapy of MG.

B. Guillain-Barré Syndrome

Acute GBS is a neurological disease whose pathogenesis may be linked to the development of autoantibodies to peripheral nerve myelin (83). The disease is characterized by a progressive muscle weakness, which in its most severe form can lead to impaired respiration due to intercostal muscle paralysis. The use of IVIg has been studied extensively in GBS and includes one of the only, large RCT in an autoimmune disease. In a comparison of IVIg with plasmapheresis, van der Meche and colleagues demonstrated greater improvements in motor function (on a graded scale) at 4 weeks posttreatment after IVIg administration, as well as a shorter time to achieve that improvement (84).

C. Chronic Inflammatory Demyelinating Polyneuropathy

Chronic inflammatory demyelinating polyneuropathy (CIDP), sometimes referred to as chronic Guillain-Barré syndrome, is also characterized by a progressive weakening of the muscles, occurring over a much longer period than acute GBS (typically 6–12 months). A similar pattern of autoimmunity has been demonstrated (85). Several uncontrolled studies indicated that high-dose IVIg (400 mg/kg × 5 consecutive days) may be useful in the treatment of CIDP as measured by an improvement in the ranking scale of functional disability (86,87).

This was followed by a small placebo (albumin)-controlled crossover study of seven patients, which again showed improvement following IVIg treatment (87). However, a larger RCT was unable to show significant treatment effect over placebo (88). An analysis of the results suggested that only a subset of CIDP patients may benefit: those with disease duration of less than 1 year, and a nerve conduction velocity median of <80% normal. Results may have been confounded in that some patients studied may have had a neuropathy other than CIDP. Other RCTs with a well-defined diagnosed subset of patients will be needed to better define the role of IVIg therapy in CIDP.

D. Multiple Sclerosis

Multiple sclerosis (MS) is an autoimmune disease with autoantibodies directed against the perineural myelin sheaths in the CNS (89). There have been limited attempts to establish the effectiveness of IVIg in this setting. Initial work utilized 5000-mg doses regardless of body weight; clinical improvement compared to a nonrandomized control group was reported (90).

Recent studies have employed higher doses of IVIg. An Israeli study of 20 MS subjects randomized to either IVIg (400 mg/kg × 5 days initially with 400 mg/kg booster doses bimonthly) or no treatment suggested a reduction in disease exacerbation rate and decreased disability status in the IVIg groups (91). A second study comparing 14 patients receiving IVIg (500–2000 mg/kg on a monthly basis), to methylprednisolone pulse treatment failed to demonstrate a benefit on disease exacerbations or steroid sparing (92). The results to date in MS are equivocal and require a large RCT study for confirmation.

E. Other Diseases

Intractable childhood epilepsy may have an immunologically mediated etiology. The use of IVIg in this condition has met with mixed results (93–97) and the lack of controlled studies makes interpretation of the findings difficult. IVIg has also been reported useful in treatment of the Lambert-Eaton syndrome, a myasthenia-like disease (94,95) and in a limited series of patients with amyotrophic lateral sclerosis (96,97). With regard to the former, a double-blind, crossover trial in 10 patients indicated an objective improvement in muscle strength, drinking time (150 ml of water), and vital capacity by the use of IVIg (1000 mg/kg × 2 days).

XI. TREATMENT OF OTHER AUTOIMMUNE DISEASES

IVIg administration has proven to be a possibly useful therapy in a number of other diseases including idiopathic ulcerative or Crohn's colitis (98), Graves' ophthalmopathy (99,100), type I diabetes (101–104), and uveoretinitis (105) in an animal model. In insulin-dependent diabetes, data suggest that early intervention with IVIg may slow the initial phases of metabolic deterioration. Nonetheless, all reports are based on limited, nonrandomized studies and must be viewed with caution.

XII. POTENTIAL MECHANISMS OF ACTION

The potential mechanisms of action for IVIg in KD are multiple; the data do not definitively identify a predominant process. Work by Leung et al. has demonstrated increased numbers of circulating activated T cells as well as increased spontaneous IgG and IgM secretion by peripheral blood mononuclear cells (pbmcs) in patients with KD, which decreased after treatment in the first controlled trial of IVIg (106). Increased production of IL-1, TNF-α, IFN-γ, and IL-6 occurs during the acute phase of the disease, with activation of vascular endothelial cells associated with increased expression of E-selectin (107,108). Furukawa et al. reported that levels of soluble (or circulating) ICAM-1 were elevated during active disease and correlated with the presence of coronary artery disease (108a); flow cytometry indicated that activated T cells were temporarily removed from the peripheral circulation (108b). Taken together, data suggest that

IVIg reduces cytokine secretion, adhesion molecule up-regulation, and thus endothelial cell activation, thereby preventing further vascular damage.

Finberg et al. studied patients with KD and seizure disorders before and after IVIg treatment and showed a significant increase in the number of peripheral circulating NK cells, suggesting that IVIg inhibits margination of NK cells, presumably by its effects on endothelial cells (109). But as Saulsbury (and others) have pointed out, the rapidity of improvement in treated patients with KD argues for mechanisms other than direct modulation of cell activation markers (110) and suggests a role for microbial superantigens in mediation of the disease (111).

IVIg treatment of ITP also results in rapid onset of benefit. The most frequently cited hypothesis for its effect is a reversibly mediated blockade of Fc receptors present on reticulendothelial (RE) cells and B cells (112–115). Direct interference of antibody binding to platelets may occur because of competitive binding, leading to Fc receptor (FcR) blockade, through decreased synthesis of autoantibodies, or via small immune complexes contained in the preparation that may act to down-regulate macrophage Fc receptor expression. Three studies suggest diminished FcR-mediated clearance of antibody-coated cells: administration of anti-Fcγ receptor antibodies resulted in improvement in patients with refractory ITP (116); infused red blood cells (rbcs) coated with anti-RhD antibodies survived longer in patients following IVIg therapy (117); and in vitro incubation with IVIg resulted in decreased phagocytosis of anti-RhD-coated rbcs by adherent pbmcs (118). Over time FcRs on the cell surface are regenerated; therefore, this mechanism does not explain the long-term effects that may be observed after treatment.

Recent data suggest that IVIg administration may function more specifically through anti-idiotype suppression, thereby decreasing titers of disease-associated autoantibodies. Very small amounts of autoantibodies (autoAbs) and anti-idiotypes (anti-ids) to these antibodies exist in normal serum; concentrated preparations pooled from 10,000–15,000 donors would be expected to contain a vast assortment of autoAb idiotypes (id) and anti-ids. Furthermore, as a consequence of variable region complementarity between immunoglobulins in the pool, IgG dimers are plentiful in IVIg preparations. Id anti-id interactions may function in several ways: An autoantibody may be neutralized by the formation of an id anti-id dimer, thereby preventing its binding to antigen. RE system clearance may be facilitated when the Fc portion of the dimer binds to FcR on phagocytic cells. AutoAb production may be inhibited following binding to FcR on the surface of B cells. T cells recognize either anti-id antibodies or the id anti-id dimer; although cell activation may not be inhibited, its consequent effects may be down-regulated by this interaction.

Work by Dietrich, Kaveri, Kazatchkine, and others supports the hypothesis that id-anti id regulatory mechanisms underlie many of the beneficial effects of IVIg administration. F(ab')₂ fragments of IVIg neutralize and/or inhibit Ag binding to anti-factor VIII, anti-gpIIb/III Abs, anti-DNA Abs, antithyroglobulin Abs, and ANCA Abs, among others (119,120). Affinity columns of F(ab')₂ fragments of IVIg coupled to sepharose bind these autoantibodies (autoAbs) often in higher amounts than are measured in patients' sera (121). Further, autoAbs in autoimmune diseases often express cross-reactive idiotypes (122–126). Thus infusion of IVIg may not only passively transfer autoantibody anti-idiotypes; but actively alter endogenous regulation of the expressed autoAb repertoire in treated patients through id anti-id interactions. Pooled preparations from older donors with broader expression of public idiotypes are thought to be more effective.

Other mechanisms may explain the effects of IVIg in treatment of autoimmune diseases. In vitro data from the early use of IVIg suggested its capability to solubilize existing immune complexes, be they tissue-bound or circulating (20,21,24). More recent data suggest little influence (127). Frank et al. postulate that interference with complement-mediated damage may be particularly important in the reversal of observed thrombocytopenia as well as hemolytic anemia and leukopenia (128). Fluid IgG inhibits C3b and C4b binding without effect on C1q deposition. C3 and the Fd portion of IgG form complexes that may act as "superopsonins," sponging up activated complement. Modulation of cytokine secretion (decreased IFN-γ, IL-1, TNF-α) may occur as demonstrated in KD, as either a primary or secondary process, although data to support actions in vivo are necessarily inferred (115,129–132). IVIg preparations contain antibodies to the CD5 cell surface Ag, which may also function to regulate B-cell production of autoAbs (132). Finally, soluble HLA class I and II molecules are present in IVIg preparations; they and other molecules (including soluble CD4) may act directly to down-modulate specific T-B-cell interactions (133–135).

Recent published work suggests that infusions of IVIg may have direct effects on cell activation and proliferation. VanSchaik et al. described a dose-dependent inhibition upon antigen (Ag) (MLR: mixed lymphocyte reactions) and mitogen (PHA, r IL-2) stimulated proliferation of pbmcs and EBV transformed cell lines by supraphysiological concentrations of IVIg; levels attainable with the current high-dose IV regimens (136). Others have shown inhibition of MLR and/or mitogen-stimulated proliferation of pbmcs after incubation with IVIg at concentrations corresponding to serum levels following doses of 250–500 mg/kg body weight (137,138). Incubation with IVIg in vitro acts directly on B cells and indirectly on monocytes to reduce total immunoglobulin secretion in nonspecific fashion (139).

The hypothesized mechanisms of action underlying the efficacy of IVIg in KD may be similar in SLE and vasculitis; recent data suggest anti-DNA and ANCA id anti-id interactions are immunomodulatory. Evans and Abdou showed that pooled IVIg contained anti-id Abs against anti-DNA, which specifically inhibited the binding of SLE area to DNA but did not affect the binding of anti-tetanus toxoid to tetanus toxoid (140). The anti-id Ab was eluted from anti-DNA affinity columns after depletion of anti-DNA, anti-Fc, and anti-F(ab')$_2$ of normal IgG. In vitro incubation of pbmcs from SLE patients with this anti-id Ab partially inhibited spontaneous secretion of anti-DNA Abs, and was specific for anti-DNA, having no effect on polyclonal IgG secretion (141). Pooled IVIg preparations also contain anti-id Abs to ANCA (142), and inhibition of ANCA binding in vitro by IVIg (F(ab')$_2$ fragments correlates with decreases in ANCA titers in patients with vasculitis following IVIg treatment (49,143). A transient rise in ANCA binding is frequently observed during and after IVIg infusion, followed by sustained decreases in ANCA titers 10–40 days later, which is attributed by Jayne and Lockwood to id anti-id regulation.

Passive neutralization of APL Abs may result from id anti-id interactions. IVIg can neutralize lupus anticoagulant activity in vitro (144,145). In experimental antiphospholipid syndrome, induced by passive transfer of anticardiolipin antibodies to naïve mice, IVIg treatment resulted in significantly less fetal resorptions than in the untreated cohorts (146).

In the Dalakas study of patients with dermatomyositis, repetitive muscle biopsies were performed blinded to treatment, but only in those who demonstrated improvement in muscle strength, therefore only in patients who received IVIg (4). Posttreatment biopsies revealed abundant deposits of IgG around muscle fibers and capillaries, suggesting the Fc portion of the administered IgG bound directly to FcR on the vessel walls. CD8+ T-cell endomysial inflammatory infiltrates had largely resolved; MHC-I, prominent on the periphery of muscle fascicles pretreatment, was barely detectable; and ICAM-I, strongly expressed on endothelial cells and lymphocytic infiltrates pretreatment, was weakly expressed. Deposits of the complement membrane attack complex, previously prominent on capillaries and necrotic muscle fibers, were absent after treatment. Regenerating muscle fibers, prominent before treatment, became sparse, and the mean number of capillaries increased. Together these observations suggest down-regulation of T-cell activation, cytokine secretion, vascular endothelial cell activation, adhesion molecule expression, and resultant lymphocytic infiltrate. Administration of IVIg appears also to stop complement-mediated immune damage to endothelial cells and necrosis of muscle fibers. These data indicate that many of the hypothesized mechanisms of action for IVIg occur in the reversal of an ongoing autoimmune process.

XIII. ADVERSE EFFECTS

It is safe to say that most patients tolerate IV infusions of pooled IgG very well. There are at present at least 17 preparations of IVIg in use worldwide, all with similar profile of adverse effects (147). Often manifestations of headache, flushing, low back pain, nausea, and wheezing can be resolved by slowing the infusion rate. Severe headaches, fever, and chills have responded to symptomatic treatment (4,52,148). Rarely, anaphylaxis may occur, particularly in IgA-immunodeficient patients with anti-IgA antibodies. Risk for such reactions has been substantially reduced in those IVIg preparations with low IgA content (149). The rare incidence of Coombs-positive hemolysis is due to cross-reactivity with the patient's rbcs (150). Deterioration of renal function (151–154) has been reported in a small number of patients with SLE, ITP, and neurological diseases following IVIg treatment, without preexisting renal disease. Transient neutropenia (155) has occasionally been reported.

Recently, Sekul and colleagues reported the incidence of aseptic meningitis in 54 patients with neurological and neuromuscular diseases consecutively treated with IVIg (148). Signs and symptoms developed within 24 hr of completion of treatment in six, or 11%, and cleared without sequelae within 3–5 days. Although there was no correlation with age or underlying disease, increased susceptibility was associated with family history or previous incidence of migraine. Despite readministration of a different preparation of IVIg, aseptic meningitis recurred in susceptible patients. In response to this report, Scribner et al. reviewed previously published reports as well as those received as part of the FDA MEDWatch program (156). It appears that this complication is most likely to occur in patients receiving high-dose IVIg (1000 mg/kg) with presumed preexisting compromise of the blood-brain barrier due to underlying disease or prior history of migraine.

In the 1980s several batches of IVIg were associated with outbreaks of non-A, non-B hepatitis (probable hepatitis C). Detergent and other methods of viral inactivation, as well as exclusion of HCV+ donors, has significantly decreased this risk. Nonetheless, occasional reports of hepatitis C seroconversion have been associated with various IVIg preparations (157), but their frequency has been reduced through the utilization of these new methods of viral inactivation.

XIV. CONCLUSIONS

Clinical and experimental data suggest administration of IVIg to be beneficial in some autoimmune diseases. However, its use may be limited by its considerable expense. Well-designed, prospective, RCTs will be necessary for an accurate cost-benefit analysis of its use in diseases such as SLE, DM/PM, RA, and vasculitis; neurological diseases such as MS and MG; and autoimmune cytopenias.

REFERENCES

1. Newburger J, Takahashi M, Burns J. The treatment of Kawasaki syndrome with intravenous immune globulin. N Engl J Med 1986; 315:341–347.
2. Newburger J, Takahashi M, Beiser A. A single intravenous infusion of gamma globulin as compared with four infusions in the treatment of acute Kawasaki syndrome. N Engl J Med 1991; 324:1633–1639.
3. Imbach P, Wagner H, Berchtold W, Gaedicke G, Hirt A, Joller P. Intravenous immunoglobulin versus oral corticosteroids in acute immune thrombocytopenic purpura in childhood. Lancet 1985; 2:464–468.
4. Dalakas M, Illa I, Dambrosia J, Soueidan S, Stein D, Otero C, Dinsmore S, McCrosky S. A controlled trial of high dose intravenous immune globulin infusions as treatment for dermatomyositis. N Engl J Med 1993; 329:1993–2000.
5. vanderMeche FGA, Schmitz PIM, Group tDG-BS. A randomized controlled trial comparing intravenous immune globuline and plasma exchange in Guillain Barré syndrome. N Engl J Med 1992; 326:1123–1129.
6. Imbach P, Barandun S, d'Apuzzo V, Baumgarten C, Hirt A, Morell A. High dose intravenous gammaglobulin for idiopathic thrombocytopenic purpura in childhood. Lancet 1981; 1:1228–1231.
7. Bussel JB, Pham LC, Aledort L, Nachman R. Maintenance treatment of adults with chronic refractory immune thrombocytopenia purpura using repeated infusions of gammaglobulin. Blood 1988; 72:121–127.
8. Coulam CB, Coulam CH. Update on immunotherapy for recurrent pregnancy loss. Am J Repro Immunol 1992; 27:124–127.
9. Gaedicke G, Teller W, Kohne E, Dopfer R, Niethammer D. IgG therapy in systemic lupus erythematosus—two case reports. Blut 1984; 48:387–390.
10. Corvetta A, BittA RD, Gabrielli A, Spaeth P, Danieli G. Use of high dose intravenous immunoglobulin in SLE: report of 3 cases. Clin Exp Rheum 1989; 7:295–299.
11. Akashi K, Nagasawa K, Mayumi T, Yokota E, OOchi N, Kusaba T. Successful treatment of refractory SE with intravenous immunoglobulins. J Rheumatol 1990; 17:375–379.
12. Maier W, Gordon D, Howard R, Saleh M, Miller S, Lieberman J, Woodlee P. Intravenous immunoglobulin therapy in SLE associated thrombocytopenia. Arthritis Rheum 1990; 33:1233–1239.
13. Sturfelt G, Mousa F, Jonsson H, Nived O, Thysell H, Wollheim F. Recurrent cerebral infarction and the antiphospholipid syndrome: effect of intravenous gammaglobulin in a patient with SLE. Ann Rheum Dis 1990; 9:939–941.
14. terBorg E, Kallenberg C. Treatment of severe thrombocytopenia in systemic lupus erythematosus with intravenous gammaglobulin. Ann Rheum Dis 1992; 51: 1149–1151.
15. Barron K, Sher M. Intravenous immunoglobulin therapy: magic or black magic. J Rheumatol 1992; 19:94–97.
16. Ilan Y, Naparstek Y. Pure red cell aplasia associated with systemic lupus erythematosus: remission after a single course of intravenous immunoglobulin. Acta Haematol 1993; 89:152–154.

17. Tomer Y, Schoenfeld Y. Successful treatment of psychosis secondary to SLE with high dose intravenous immunoglobulin. Clin Exp Rheumatol 1992; 10: 391–393.
18. Winder A, Molad Y, Ostfeld I, Kenet G, Pinkhas J, Sidi Y. Treatment of systemic lupus erythematosus by prolonged administration of high dose intravenous immunoglobulin: report of 2 cases. J Rheumatol 1993; 20:495–498.
19. Ben-Chetrit E, Putterman C, Naparstek Y. Lupus refractory pleural effusion: transient response to intravenous immunoglobulins. J Rheumatol 1991; 18:1635–1637.
20. Tomino Y, Sakai H, Takaya M, Miura M, Suga T, Endoh M, Nomoto Y. Solubilization of intraglomerular deposits of IgG immune complexes by human sera or gamma globulin in patients with lupus nephritis. Clin Exp Immunol 1984; 58:42–48.
21. Lin C-V, Hsu H-C, Chiang H. Improvement of histological and immunological change in steroid and immunosuppressive drug resistant lupus nephritis by high dose intravenous gamma globulin. Nephron 1989; 53:303–310.
22. Lin R, Racis S. In vivo reduction of circulating C1q binding immune complexes by intravenous gammaglobulin administration. Int Arch Allergy Appl Immunol 1986; 79:286–290.
23. Schifferli J, Leski M, Favre H, Imbach P, Nydegger U, Davies K. High dose intravenous IgG treatment and renal function. Lancet 1991; 337:457–458.
24. Jordon S. Intravenous g globulin therapy in SLE and immune complex disease. Clin Immunol Immunopath 1989; 53:S164–169.
25. Parke A. The role of intravenous immune globulin in the management of patients with antiphospholipid antibodies and recurrent pregnancy losses. Clin Rev Allergy 1992; 10:105–118.
26. Scott J, Branch D, Kochenour N, Ward K. Intravenous immunoglobulin treatment of pregnant patients with recurrent pregnancy loss caused by antiphospholipid antibodies and Rh immunization. Am J Obstet Gynecol 1988; 159:1055–1056.
27. Katz V, Thorp J, Watson W. Human immunoglobulin therapy for preeclampsia associated with lupus anticoagulant and anticardiolipin antibody. Obstet Gynecol 1990; 76:986–988.
28. Orvieto R, Achiron A, Ben-Rafael Z, Achiron R. Intravenous immune globulin treatment for recurrent abortions caused by antiphospholipid antibodies. Fertil Steril 1991; 56:1013–1020.
29. Christiansen O, Mathiesen O, Lauritsen J. Intravenous immune globulin treatment of women with multiple miscarriages. Hum Reprod 1992; 7:718–722.
30. Kaaja R, Julkunen H, Ammala P, Palosuo T, Kurki P. Intravenous immune globulin treatment of pregnant patients with recurrent pregnancy losses associated with antiphospholipid antibodies. Acta Obstet Gynecol Scand 1993; 72:63–66.
31. Coulam CB, Stern JJ, Bustillo M. Ultrasonographic findings of pregnancy losses after treatment for recurrent pregnancy loss: intravenous immunoglobulin vs. placebo. Fertil Steril 1994; 61:248–251.
32. Groothoff J, vanLeeuwen E. High dose IV gamma globulin in chronic systemic juvenile arthritis. Br J Med 1988; 296.
33. Silverman E, Laxer R, Greenwald M, Gelfand E, Shore A, Stein L, Roifman C. IV gamma globulin therapy in systemic JRA. Arthritis Rheum 1990; 1015–1022.

34. Tumiati B, Casoli P, Veneziani M, Rinaldi G. High dose immunoglobulin therapy as an immunomodulatory treatment of RA. Arthritis Rheum 1992; 35: 1126–1134.
35. Roifman C, Schaffer F, Wachsmuth S, Murphy G, Gelfand E. Reversal of chronic polymyositis following immune serum globulin therapy. JAMA 1987; 258:513–515.
36. Lang B, Laxer R, Murphy G, Silverman E, Roifman C. Treatment of dermatomyositis with intravenous immune globulin. Am J Med 1991; 91:169–172.
37. Cherin P, Herson S, Wechsler B, Piette J, Bletry O, Coutellier A, Ziza J, Godeau P. Efficacy of intravenous immune globulin in chronic refractory polymyositis and dermatomyositis: an open study with 20 patients. Am J Med 1991; 91:162–166.
38. Hanslik T, Jaccard A, Guillon J, Vernant J, Wechsler B, Godeau P. Polymyositis and chronic GvHD: efficacy of intravenous immune globulin. J Am Acad Dermatol 1993; 28:492–493.
39. Soueidan S, Dalakas M. Treatment of inclusion body myositis with high dose intravenous immunoglobulin. Neurology 1993; 43:876–879.
40. Cherin P, Piette J, Wechsler B, Piette J, Bletry O, Ziza J, Laraki R, Godeau P, Herson S. Intravenous immune globulin as first line therapy in polymyositis and dermatomyositis: an open study in 11 patients. J Rheumatol 1994; 21:1092–1097.
41. Reimold A, Weinblatt M. Tachyphylaxis of intravenous immunoglobulin in refractory inflammatory myopathy. J Rheumatol 1994; 21:1144–1146.
42. Barron K, Murphy D, Silverman E, Ruttenberg H, Wright G, Franklin W, Goldberg S, Higashino S, Cox D, Lee M. Treatment of Kawasaki syndrome: comparison of two dosage regimens of intravenously administered immune globulin. J Pediatrics 1990; 117:638–644.
43. Akagi T, Rose V, Benson L. Newman A, Freedom R. Outcome of coronary artery aneurysms after Kawasaki disease. J Pediatrics 1992; 121:689–694.
44. Rowley A, Shulman S. The clinical efficacy of IVGG in Kawasaki disease: In: Ballow M, ed. IVIG Therapy Today. Humana, 1992:81–91.
45. Klassen T, Rowe P, Gafni A. Economic evaluation of intravenous immune globulin for Kawasaki syndrome. J Pediatrics 1993; 123:538–542.
46. NIH Concensus Conference. Intravenous immunoglobulin: prevention and treatment of disease. JAMA 1990; 264:3189–3193.
47. Dajani A, Taubert K, Gerber M, Shulman S, Ferrieri P, Freed M, Takahashi M, Bierman F, Karchmer A, Wilson W, Rahimtoola S, Durack D, Peter G. Diagnosis and therapy of Kawasaki disease in children. Circulation 1993; 87:1776–1780.
48. Jayne D, Davies M, Fox C, Black C, Lockwood C. Treatment of systemic vasculitis with pooled intravenous immune globulin. Lancet 1991; 337:1137–1139.
49. Jayne D, Esnault V, Lockwood C. ANCA anti-idiotype antibodies and the treatment of systemic vasculitis with intravenous immunoglobulin. J Autoimmun 1993; 6:207–219.
50. Jayne J, Lockwood C. Pooled intravenous immune globulin in the management of systemic vasculitis. In: Gross W, ed. ANCA-Associated Vasculitides: Immunological and Clinical Aspects. New York: Plenum Press, 1993:469–472.
51. Richter C, Schnabel A, Csernok E, Reinhold-Keller E, Gross W. Treatment of Wegener's granulomatosis with intravenous immune globulin. In: Gross W, ed.

ANCA-Associated Vasculitides: Immunological and Clinical Aspects. New York: Plenum Press, 1993:487–489.
52. Rostoker G, Desvaux-Belghiti D, Pilatte Y. High dose immunoglobulin therapy for severe IgA nephropathy and Henoch Schönlein purpura. Ann Intern Med 1994; 120:476–484.
53. Hamilos D, Christensen J. Treatment of Churg-Strauss syndrome with high dose intravenous immune globulin. J Allergy Clin Immunol 1991; 823–824.
54. Pollack S, Cunnningham-Rundles C, Smithwick EM, Barandun S, Good RA. High dose intravenous gammaglobulin in autoimmune neutropenia. N Engl J Med 1982; 307:252.
55. Bussel J. Lalezari P, Hilgartner M, Partin J, Fikrig S, O'Malley J, Barandun S. Reversal of neutropenia of infancy. Blood 1983; 62:398–400.
56. Hilgartner MW, Bussel J. Use of intravenous gamma globulin for the treatment of autoimmune neutropenia of childhood and autoimmune hemolytic anemia. Am J Med 1987; 83(Suppl 4A):25–29.
57. Bussell J, Lalezari P, Fikrig S. Intravenous treatment with gammaglobulin of autoimmune neutropenia of infancy. J Pediatrics 1988; 112:298–301.
58. Kuretzberg J, Friedman HS, Chaffee S, Falletta JM, Kinney TR, Kurlander R, Matthews TJ, Schwartz RS. Efficacy of intravenous gamma globulin in autoimmune-mediated pediatric blood dyserasias. Am J Med 1987; 83(Suppl 4A):4–9.
59. Ahern M, Harkness J, Maddison P, Forskitt S. High-dose immunoglobulin in Felty's syndrome. Ann Rheum Dis 1983; 42:476–477 (letter).
60. Breedveld FC, Brand A, VanAken WG. High dose intravenous gamma globulin for Felty's syndrome. J Rheumatol 1985; 12:700–702.
61. Murakami H, Kikuchi M, Toyama K, Ikeda Y. A case of autoimmune neutropenia and thrombocytopenia: effect of high-dose intravenous immunoglobulin. Keio J Med 1985; 34:227–232.
62. Mueller-Eckhardt C, Salama A, Mahn I, Kiefel V, Neuzner J, Graubner M. Lack of efficacy of high dose intravenous immunoglobulin in autoimmune haemolytic anaemia: a clue to its mechanism. Scand J Haematol 1985; 34: 3941–4000.
63. Richmond GW, Ray I, Korenblitt A. Initial stabilization preceding enhanced hemolysis in autoimmune hemolytic anemia treated with intravenous immunoglobulin. J Pediatrics 1987; 110:917–919.
64. Oda H, Honda A, Sugita K, Nakamura A, Nakajima H. High dose intravenous intact IgG infusion in refractory autoimmune hemolytic anemia (Evans syndrome). J Pediatrics 1985; 107:744–746.
65. Besa EC. Rapid transient reversal of anemia and long-term effects of maintenance intravenous immunoglobulin for autoimmune hemolytic anemia in patients with lymphoproliferative disorders. Am J Med 1988; 84:691–698.
66. MacIntyre EA, Linch DC, Macey MG, Newland AC. Successful response to intravenous immunoglobulin in autoimmune haemolytic anaemia. Br J Haematol 1985; 60:387–388.
67. Ritch PS, Anderson T. Reversal of autoimmune hemolytic anemia associated with chronic lymphocytic leukemia following high-dose immunoglobulin. Cancer 1987; 60:2637–2640.

68. Sasaki H, Akutagawa H, Kuwadado K, Uemura M, Emi I. High-dose intravenous IgG therapy in a seven-week-old infant with chronic autoimmune hemolytic anemia. Am J Hematol 1987; 25:215–218.
69. Flores G, Cunningham-Rundles C, Newland AC, Bussel JB. Efficacy of intravenous immunoglobulin on the treatment of autoimmune hemolytic anemia: results in 73 patients. Am J Hematol 1993; 44:237–242.
70. Sultan T, Kazatchkine MD, Malsonneuve P, Nydegger UE. Anti-idiotypic suppression of autoantibodies to factor VIII (antihemophilic factor) by high-dose intravenous gammaglobulin. Lancet 1984; 2:765–768.
71. Glanella-Borradori A, Hirt A, Luthy A, Wagner HP, Imbach P. Haemophilia due to factor VIII inhibitors in a patient suffering from an autoimmune disease: treatment with intravenous immunoglobulin. Blut 1984; 48:403–407.
72. Green D, Dwaan HC. An acquired factor VIII inhibitor responsive to high-dose gammaglobulin. Thromb Haemost 1987; 58:1005–1007.
73. Zimmerman R, Kommerell B, Harenberg J, Elch W, Rother K, Schimpf K. Intravenous IgG for patients with spontaneous inhibitor to factor VIII. Lancet 1985; 1:273–274.
74. Heyman MR, Chakravarthy A, Edelman BB, Needleman SW, Schiffer CA. Failure of high-dose IV gammaglobulin in the treatment of spontaneously acquired factor VIII inhibitors. Am J Hematol 1988; 28:191–194.
75. Nilsson IM, Berntorp E, Zettervall O. Induction of immune tolerance in patients with hemophilia and antibodies to factor VIII by combined treatment with intravenous IgG, cyclophosphamide and factor VIII. N Engl J Med 1988; 315:947–950.
76. Nilsson IM, Berntorp E. Induction of immune tolerance in hemophiliacs with inhibitors, by combined treatment with IVIG, cyclophosphamide and factor VIII or IX—the Malmo model. In: Imbach P, ed. Immunotherapy with Intravenous Immunoglobulins. London: Academic Press, 1991:333–344.
77. Kickler T, Braine HG, Plantadosl S, Ness PM, Herman JH, Rothko K. A randomized, placebo-controlled trial of intravenous gammaglobulin in alloimmunized thrombocytopenic patients. Blood 1990; 75:313–316.
78. Gajdos Ph, Outin H, Elkhanat D. High dose intravenous gammaglobulin for myasthenia gravis. Lancet 1984; 1:406–407 (letter).
79. Fateh-Maghadam A, Besinger U, Geursen RG. High dose intravenous immunoglobulin in the management of myasthenia gravis. Lancet 1984; 1:848–849 (letter).
80. Asura EL, Bick A, Brunner NG, Namba T, Grob D. High-dose intravenous immunoglobulin in the management of myasthenia gravis. Arch Intern Med 1986; 146:1365–1368.
81. Ibars IB, Ponsetl J, Espanol T, Matias-Gulu J, Cordina-Pulggros A. High-dose intravenous gamma-globulin therapy for myasthenia gravis. J Neurol 1987; 234–363.
82. Uchiyama M. Ichikawa Y, Takaya M, Moriuchi J, Shimizu H, Arimori S. High-dose gammaglobulin therapy of generalized myasthenia gravis. Ann NY Acad Sci 1987; 505:868–871.
83. Koski CL, Gratz E, Sutherland J, Mayer RF. Clinical correlation with anti-peripheral-nerve myelin antibodies in Guillain-Barré syndrome. Ann Neurol 1986; 19:573–577.

84. van der Meche FGA, Schmitz PIM, Dutch Guillain-Barré Study Group, et al. A randomized trial comparing intravenous immune globulin and plasma exchange in Guillain-Barré syndrome. N Engl J Med 1992; 326:1123–1129.
85. Hartung HP, Helninger K, Schafer B, Flerz W, Toyka NV. Immune mechanisms in inflammatory polyneuropathy. Ann NY Acad Sci 1988; 540:122–161.
86. Vermeulen M, van der Meche FGA, Speciman JD, Weber A, Busch HFM. Plasma and gammaglobulin infusion in chronic inflammatory polyneuropathy. J Neurol Sci 1985; 70:317–326.
87. van Dooran PA, Brand A, Strengers PFW, Meulstee J, Vermeulen M. High-dose intravenous immunoglobulin in chronic inflammatory demyelinating polyneuropathy: a double-blind, placebo-controlled, crossover study. Neurology 1990; 40:209–212.
88. Vermeulen M, van Dooran PA, Brand A, Strengers PF, Jennekesn FG, Busch HF. Intravenous immunoglobulin treatment in patients with chronic inflammatory demyelinating polyneuropathy: a double-blind, placebo controlled study. J Neurol Neurosurg Psychiatry 1993; 56:36–39.
89. Schwartz SA. Intravenous immunoglobulin (IVIG) for the therapy of autoimmune disorders. J Clin Immunol 1990; 10:81–89.
90. Schuller E, Govaerts A. First results of immunotherapy with immunoglobulin G in multiple sclerosis patients. Eur Neurol 1993; 22:305–212.
91. Achiron A, Pras E, Gilad R, Ziv I, Mandel M, Gordon CR, Noy S, Sacrova-Pinkes I, Melaned E. Open controlled therapeutic trial of intravenous immune globulin in relapsing-remitting multiple sclerosis. Arch Neurol 1992; 49:1233–1236.
92. Cook SD, Troiano R, Rokowsky-Kochin C, et al. Intravenous gammaglobulin in progressive MS. Acta Neurol Scand 1992; 86:171–175.
93. Ariizumi M, Shiihara H, Hibio S, Ryo S, Baba K, Ogawa K, Suzuki Y, Momoki T. High-dose gammaglobulin for intractable childhood epilepsy. Lancet 1993; 2:162–163.
94. Barn P, Elrington G, Goodger E, Misbah S, Panegyres P, Macpherson K, Chapel H, Newson-Davis J. A randomized double-blind controlled study of IVIG in the Lambert-Eaton myasthenic syndrome. Association of British Neurologists Annual meeting, May 1994.
95. Bird SJ, Clinical and electrophysiologic improvement in Lambert-Eaton syndrome with intravenous immunoglobulin therapy. Neurology 1992; 42:1422–1423.
96. Dalakas MC, Stern DP, Otero C, Sekul E, Cupler SE, McCrosky S. Effect of high-dose (IVIG) on amyotrophic lateral sclerosis and multifocal motor neuropathy. Anch Neurol 1994 (in press).
97. Savery F, Hang LM. Immunodeficiency associated with motor neuron disease treated with intravenous immunoglobulin. Clin Ther 1986; 8:700–702.
98. Levin DS, Fischer SH, Christle DL, Haggitt RC, Ochs HD. Intravenous immunoglobulin therapy for active, extensive, and medically refractory idiopathic ulcerative or Crohn's colitis. Am J Gastroenterol 1992; 87:91–100.
99. Antonell A, Saracino A, Alberti B, et al. High-dose intravenous immunoglobulin treatment in Graves' ophthalmopathy. Acta Endocrinol 1992; 126:13–23.

100. Dwyer JM, Benson EM, Currine JN, O'Day J. Intravenously administered IgG for the treatment of thyroid eye disease. In: Imbach P, ed. Immunotherapy with Intravenous Immunoglobulins. London: Academic Press, 1991:387–394.
101. Heinze E, Thon A, Vetter U, Gaedicke G, Zuppinger K. Gammaglobulin therapy in 6 newly diagnosed diabetic children. Acta Pediatr Scand 1985; 74:605–606.
102. Pocecco M, Campo CD, Cantoni L, Tedesco F, Panizon F. Effect of high dose intravenous IgG in newly diagnosed diabetic children. Helv Paediatr Acta 1987; 42:289–295.
103. Panto F, Giordano C, Amato MP, et al. The influence of high-dose intravenous immunoglobulin on immunological and metabolic pattern in newly diagnosed type I diabetic patients. J Autoimmun 1990; 3:587–592.
104. Leong GM, Tharper Z, Antony G, et al. High-dose intravenous immunoglobulin therapy for insulin-dependent diabetes melitis. In: Imbach P, ed. Immunotherapy with Intravenous Immunoglobulins. London: Academic Press, 1991:269–282.
105. Saoudi A, Hurez V, deKozak Y, et al. Human immunoglobulin preparations for intravenous use prevent experimental autoimmune uveoretinitis. Int Immunol 1993; 5:1559–1567.
106. Leung D, Burns J, Newburger J, Geha R. Reversal of lymphocyte activation in vivo in the Kawasaki syndrome by intravenous gammaglobulin. J Clin Invest 1987; 79:468–472.
107. Leung D, Kurt-Jones E, Newburger J, Cotran R, Burns J, Pober J. Endothelial cell activation and high interleukin-1 secretion in the pathogenesis of acute Kawasaki disease. Lancet 1989; 1298–1302.
108. Leung D. The immunoregulatory effects of intravenous immune globulin in Kawasaki disease and other autoimmune diseases. In: Ballow M, ed. IVIg Therapy Today. 1992:93–104.
108a. Furukawa S, Khozoh I, Matsubara T, Yone K, Yachi A, Okumura K, Yabuta K. Increased levels of circulating ICAM-1 in Kawasaki disease. Arth Rheum 1992; 35:672–677.
108b. Furukawa S, Matsubara T, Tsuji K, Okumura K, Yabuta K. Transient depletion of T cells with bright Cd11a/CD18 expression from peripheral circulation during acute Kawasaki disease. Scand J Rheumatol 1993; 31:377–380.
109. Finberg R, Newburger J, Mikati M, Heller A, Burns J. Effect of high doses of intravenously administered immune globulin on natural killer cell activity in peripheral blood. J Pediatr 1992; 120:376–380.
110. Salsbury F. The effect of intravenous immune globulin on lymphocyte populations in children with Kawasaki syndrome. Clin Exp Rheumtol 1992; 10:617–620.
111. Rich R. Intravenous IgG: supertherapy for superantigens? J Clin Invest 1993; 91:378 (editorial; comment).
112. Ballow M. Mechanisms of action of intravenous immunoglobulin therapy and potential use in autoimmune connective tissue diseases. Cancer 1991; 68:1430–1436.
113. Gelfand E. Treatment of autoimmune diseases with intravenous immune globulin. Semin Hematol 1992; 29:127–133.
114. Dwyer J. Manipulating the immune system with immune globulin. N Engl J Med 1992; 326:107–116.

115. Rosen F. Putative mechanisms of the effect of intravenous gamma-globulin: putative mechanisms of the effect of intravenous gamma-globulin. 1993; 67: S41–43.
116. Clarkson S, Bussel J, Kimberly R, Valinsky J, Nachman R, Unkeless J. Treatment of refractory ITP with anti Fc gamma receptor antibody. N Engl J Med 1986; 314:1236–1239.
117. Fehr J, Hofmann V. Transient reversal of thrombocytopenia in ITP by high dose intravenous gamma globulin. N Engl J Med 1982; 306:1254–1258.
118. Sheth K, Al-Sedairy S, Lee J. Effectiveness of intravenous immune globulin preparations in prevention of phagocytosis of anti-Rh-D-coated erythrocytes by mononuclear phagocytes. Vox Sang 1993; 65:190–193.
119. Kaveri S, Dietrich G, Hurez V, Kazatchkine M. Intravenous immune globulin in the treatment of autoimmune diseases. Clin Exp Immunol 1991; 86:192–198.
120. Kaveri S, Dietrich G, Kazatchkine M. Can intravenous immunoglobulin treatment regulate autoimmune responses? Semin Hematol 1992; 29(3 Suppl 2): 64–71.
121. Dietrich G, Kaveri S, Kazatchkine M. A V region connected autoreactive subfraction of normal human immunoglobulin G. Eur J Immunol 1992; 22:1701–1706.
122. Dietrich G, Kazatchkine M. Normal immunoglobulin G(IgG) for therapeutic use (intravenous Ig) contain antiidiotypic specificities against an immunodominant, disease-associated, cross-reactive idiotype of human anti-thyroglobulin autoantibodies. Clin Invest 1990; 85:620–626.
123. Dietrich G, Kaveri S, Kazatchkine M. Modulation of autoimmunity by intravenous immune globulin through interaction with the function of the immune/idiotypic network. Clin Immunol Immunopathol 1992; 62:S73–81.
124. Kaveri S, Want H, Rowen D, Kazatchkine M, Kohler H. Monoclonal anti-idiotypic antibodies against human anti-thyroglobulin autoantibodies recognize idiotopes shared by disease associated and natural anti-thyroglobulin autoantibodies. Clin Immunol Immunopathol 1993; 69:333–340.
125. Ronda N, Haury M, Nobrega A, Coutinho A, Kazatchkine M. Selectivity of recognition of variable (V) regions of autoantibodies by intravenous immunoglobulin. Clin Immunol Immunopathol 1994; 70:124–128.
126. Barbouche M, Guilbert B, Makni S, Gorgi Y, Ayed K, Avrameas S. Common idiotypes expressed on human, monoclonal, abnormal immunoglobulins and cryoglobulins with polyreactive autoantibody activities. Clin Exp Immunol 1993; 91: 196–201.
127. Halma C, Daha M, vanderMeer J, Cohen A, vanFurth R, Breedveld F, Evers-Schouten J. Pauwels E, vanEs L. Effect of monomeric IgG on the clearance of soluble aggregates of IgG in man. J Clin Lab Immunol 1991; 35:9–15.
128. Frank M, Basta M, Fries L. The effects of intravenous immune globulin on complement-dependent immune damage of cells and tissues. Clin Immunol Immunopathol 1992; 62:S82–86.
129. RuizDeSouza V, Kaveri S, Kazatchkine M. Intravenous immune globulin in the treatment of autoimmune and inflammatory diseases. Clin Exp Rheumatol 1993; 11:S33–S36.

130. Ling Z, Yeoh E, Webb B, Farrell K, Doucette J, Matheson D. Intravenous immunoglobulin induces interferon-gamma and interleukin-6 in vivo. J Clin Immunol 1993; 13:302–309.
131. Achiron A, Margalit R, Hershkoviz R, Markovits D, Reshef T, Melamed E, Cohen I, Lider O. Intravenous immunoglobulin treatment of experimental T cell–mediated autoimmune disease: upregulation of T cell proliferation and downregulation of tumor necrosis factor alpha secretion. J Clin Invest 1994; 93:600–605.
132. Vassilev T, Gelin C, Kaveri S, Zilber M, Boumsell L, Kazatchkine M. Antibodies to the CD5 molecule in normal human immunoglobulins for therapeutic use. Clin Exp Immunol 1993; 92:369–372.
133. Grosse-Wilde H, Blasczyk R, Westhoff U. Soluble HLA class I and II concentrations in commercial immunoglobulin preparations. Tissue Antigens 1992; 39:74–77.
134. Blasczyk R, Westhoff U, Grosse-Wilde H. Soluble CD4, CD8 and HLA molecules in commercial immunoglobulin preparations. Lancet 1993; 341:789–790.
135. Santoso S. Quantitation of soluble HLA class I antigen in human albumin and immunoglobulin preparations for IV use by solid phase immunoassay. Vox Sang 1992; 62:29–33.
136. vanSchaik I, Lundkvist I, Vermeulen M, Brand A. Polyvalent immunoglobulin for intravenous use interferes with cell proliferation in vitro. J Clin Immunol 1992; 12:325–334.
137. Klaesson S, Ringden O, Markling L, Remberger M, Lundkvist I. Immune modulatory effects of immunoglobulins on cell-mediated immune responses in vitro. Scand J Immunol 1993; 38:477–484.
138. Sbrana S, Ruocco L, Vanacore R, Azzara A, Ambrogi F. In vitro effects of an immunoglobulin preparation for intravenous use on T-cell activation. Allerg Immunol (Paris) 1993; 25:35–37.
139. Kondo N, Ozawa T, Musihake K, Motoyoshi F, Kameyama T, Kasahara K, Kaneko H, Yamashina M, Kato Y, Orri T. Suppression of immunoglobulin production of lymphocytes by intravenous immunoglobulin. J Clin Immunol 1991; 11:152–158.
140. Evans M. Detection and purification of antiidiotypic antibody against anti-DNA in intravenous immune globulin. J Clin Immunol 1991; 11:291–295.
141. Evans M, Abdou N. In vitro modulation of anti-DNA secreting peripheral blood mononuclear cells of lupus patients by anti-idiotypic antibody of pooled human intravenous immune globulin. Lupus 1993; 2:371–375.
142. Rossi F, Jayne D, Lockwood C, Kazatchkine M. Anti-idiotypes against anti-neutrophil cytoplasmic antigen autoantibodies in normal human polyspecific IgG for therapeutic use and in the remission sera of patients with systemic vasculitis. Clin Exp Immunol 1991; 83:298–303.
143. Jayne D, Lockwood C. Pooled intravenous immunoglobulin in the management of systemic vasculitis. Adv Exp Med Biol 1993; 336:469–472.
144. Said P, Martinuzzo M, Carreras L. Neutralization of lupus anticoagulant activity by human immunoglobulin in vitro. Nouv Rev Fr Hematol 1992; 34:37–42.
145. Matsuda J, Gohchi K, Kawasugi K, Tsukamoto M, Saitoh N, Kinoshita T. In vitro lupus anticoagulant neutralizing activity of intravenous immunoglobulin. Thromb Res 1993; 70:190–110 (letter).

146. Bakimer R, Guilburd B, Zurgil N, Shoenfeld Y. The effect of intravenous gammaglobulin on the induction of experimental antiphospholipid syndrome. Clin Immunol Immunopathol 1993; 69:97–102.
147. Misbah S, Chapel H. Adverse effects of intravenous immunoglobulin. Drug Saf 1993; 9:254–262.
164. Sekul E, Cupler E, Dalakas M. Aseptic meningitis associated with high dose intravenous immunoglobulin therapy: frequency and risk factors. Ann Intern Med 1994; 121:259–262.
149. Apfelzweig R, Piskiewicz D, Hooper JA. Immunoglobulin A concentrations in commercial immune globulins. J Clin Immunol 1987; 7:46–50.
150. Copelan EA, Strohm PL, Kennedy MS, Tutschka PJ. Hemolysis following intravenous immune globulin therapy. Transfusion 1986; 26:410–412.
151. Rault R, Piraino B, Johnston JR, Oral A. Pulmonary and renal toxicity of intravenous immunoglobulin. Clin Nephrol 1991; 36:83–86.
152. Stewart R, Winney R, Cash J. Renal toxicity of intravenous immunoglobulin. Vox Sang 1993; 65:244 (letter).
153. Ruggeri M, Castaman G, Nardi GD, Rodeghiero F. Acute renal failure after high dose intravenous immune globulin in a patient with idiopathic thrombocytopenia purpura. Haematologica 1993; 78:338–339.
154. Tan E, Hajinazarian M, Bay W, Neff J, Mendell JR. Acute renal failure resulting from intravenous immunoglobulin therapy. Arch Neurol 1993; 50:137–139.
155. Ben-Chetrit E, Putterman C. Transient neutropenia induced by intravenous immune globulin. N Engl J Med 1992; 326:271.
156. Scribner C, Kapit R, Phillips E, Rickles N. Aseptic meningitis and intravenous immunoglobulin therapy. Ann Intern Med 1994; 121:305–306.
157. Yap PL, McOmish F, Webster ADB, Hammerstrom L, Smith CIE, Bjorkander J, Ochs HD, Fischer S, Quionti I, Simmonds P. Hepatitis C virus transmission by intravenous immunoglobulin. J Hepatol 1994; Sep; 21(3):455–60.

VI
Purine and Pyrimidine Synthesis Inhibitors

When lymphocytes are stimulated, the enzymes responsible for increased de novo synthesis of purines and pyrimidines are induced. These nucleotides act as substrates for increased DNA and RNA synthesis, and glycosylation of intracellular proteins. Before an activated lymphocyte can replicate it must repair all damage to its DNA. If an imbalance of purines or pyrimidines exists, the DNA cannot be correctly repaired and the cell undergoes apoptosis. Lymphocytes may be uniquely sensitive to such drug effects because a "salvage" pathway for nucleotide synthesis is lacking.

Agents that inhibit de novo purine synthesis include azaribine, mycophenolate mofetil, and mizoribine; those that inhibit pyrimidine synthesis include brequinar and leflunomide (Chapters 20 and 21). Clinical trials with these agents have been conducted in RA and transplantation; several are promising and deserve further study.

20

Inhibitors of De Novo Nucleotide Synthesis in the Treatment of Rheumatoid Arthritis

ROBERT I. FOX
Scripps Clinic and Research Foundation, La Jolla, California

RANDALL E. MORRIS
Stanford University School of Medicine, Stanford, California

I. INTRODUCTION

Therapeutic control of the immune system is the goal in a vast array of autoimmune diseases, which differ in their organ-specific involvement, pathogenetic cofactors, response to treatment, and prognosis. They range from diseases with "spontaneous" onset such as rheumatoid arthritis (RA) to rejection reactions after allograft organ transplantation. Nevertheless, a relatively similar group of medications has been used to control the immune response (1). This chapter will discuss a group of medications that may act through partial or complete inhibition of de novo synthesis of purine and/or pyrimidine nucleotides.

At first appearance, it may appear paradoxical that agents that influence de novo synthesis of nucleotides might have a "selective" effect on immune function, since all dividing cells must replicate DNA. Multiple factors influence the ability of different cell types to survive or replicate in the presence of inhibitors of de novo nucleotide inhibitors, as shown schematically in Figure 1. For example, dividing cells may replete their pool of available nucleotides through either salvage or de novo synthesis pathways (described below). This is dependent on levels of circulating nucleosides (or their precursors) and the ability of the cell to transport and enzymatically transform them into nucleotides for DNA and RNA synthesis. Both vary significantly among different cell types. Within a given cell type, relevant enzymes may be induced or suppressed by local microenvironmental

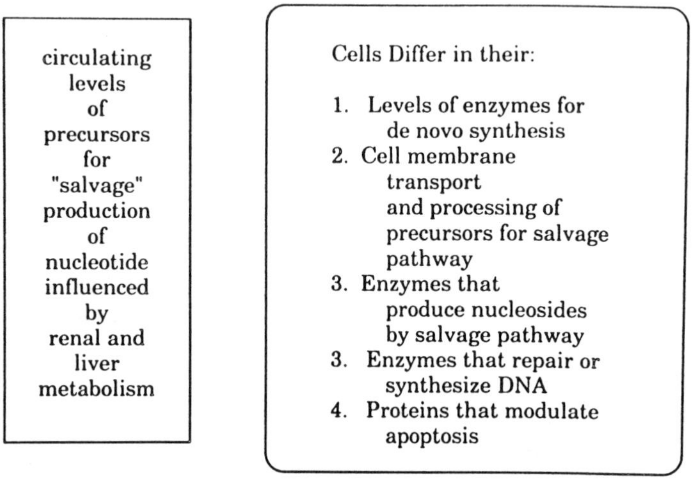

circulating levels of precursors for "salvage" production of nucleotide influenced by renal and liver metabolism	Cells Differ in their: 1. Levels of enzymes for de novo synthesis 2. Cell membrane transport and processing of precursors for salvage pathway 3. Enzymes that produce nucleosides by salvage pathway 3. Enzymes that repair or synthesize DNA 4. Proteins that modulate apoptosis

Figure 1 Cells fulfill their needs for nucleotides either by de novo synthesis or by salvage pathways.

conditions. Thus, a particular cell may be more resistant to cell death than other cells when either purine or pyrimidine nucleotide synthesis or turnover is inhibited.

When lymphocytes are stimulated, those enzymes responsible for increased de novo synthesis of purines and pyrimidines are induced, leading to increased intracellular pools of nucleotides (2,3). These nucleotides act as substrates for RNA and DNA synthesis, sources of energy, phosphate donors for intracellular signaling, and intermediate metabolites for the transfer of sugars to proteins (4). Although enzymes of the salvage pathway for synthesis for nucleotides are also activated, sensitivity of activated cells to agents that inhibit de novo synthesis (described below) indicates the relative importance of the de novo pathway.

The regulatory points for control of DNA synthesis in the cell cycle are illustrated in Figure 2. After stimulatory signals are received from the cell membrane, early protein synthesis includes cyclins C and D in early G1 and cyclin A in late G1 phase. During S phase, proliferating cell nuclear antigen (pcna) is synthesized and cell-cycle-arresting oncogenes may lead the cell toward apoptosis rather than replication. The proposed site of action of proposed antirheumatic drugs is shown in Figure 3, from a "therapeutic" point of view. The proposed block by agents such as leflunomide, brequinar, azaribine, mycophenalic mofetil, and mizoribine would be either during the late stages of phase G1 or during early S1 phase.

Inhibitors of De Novo Nucleotide Synthesis

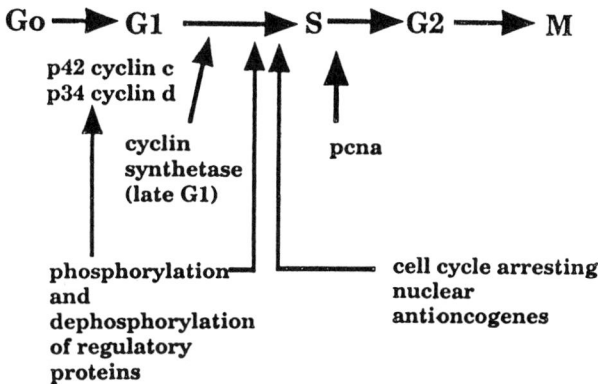

Figure 2 Regulation of DNA synthesis progression through cell cycle.

Recent studies have emphasized the role of de novo nucleotide synthesis in the process of cell death by necrosis and apoptosis. For example, purine synthesis (which induces ATP) is required to prevent the free radical (thiol) damage, which influences transcription factors such as c-*jun*, c-*fos*, growth arrest and DNA damage-induced proteins (GADD), and p53 (5). Inhibition of pyrimidine synthesis leads to DNA fragmentation, requiring new mRNA and protein synthesis and leading to apoptosis (6,7). Of importance, the antioncogene p53 serves as a checkpoint in cell cycle phase G1 for the adequacy of DNA repair, triggering cell death if the genome is not intact (8). Thus, drugs that interfere with de novo nucleotide synthesis can be considered to deliver a specific genotoxic stress. If the DNA cannot be adequately repaired, then apoptotic pathways are triggered and the damaged cell is removed. This is shown schematically in Figure 4. However, not all drugs that interfere with de novo purine or pyrimidine synthesis should be

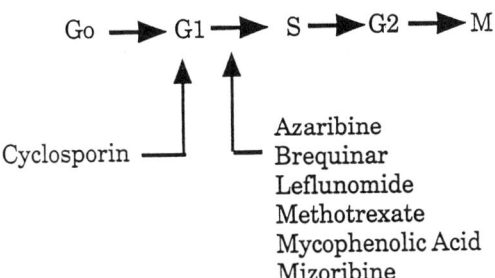

Figure 3 Therapeutic agents affect cell cycle.

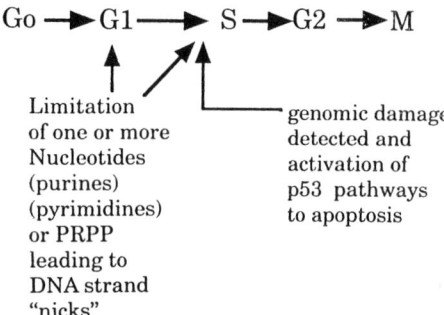

Figure 4 Imbalance of nucleotides during S phase leads to apoptosis.

assumed to activate cell death by the same apoptotic pathway. For example, 5-azacytidine and arabinosyl guanine induced DNA fragmentation in thymocytes, but methotrexate and 6-mercaptopurine failed to do so at in vitro concentrations likely to be achieved in vivo (7).

In this chapter, the widely used antirheumatic drugs azathioprine and methotrexate will be briefly reviewed. Then new candidates for immune modulation including mycophenolic mofetil, mizoribine, leflunomide, brequinar, and azaribine will be discussed. These agents were first isolated as part of drug discovery programs for their antibiotic, anti-inflammatory, and/or antitumor activity. Each mediates its effect on the immune system at least partly through inhibition of de novo synthesis of purine or pyrimidine nucleotides. It must be emphasized that their putative antirheumatic mechanisms are based largely on activities present in in vitro assays utilizing cultured cell lines and activated lymphocytes. Correlation of these in vitro data in vivo immunomodulatory or immunosuppressive effects appear likely but have not been rigorously demonstrated.

II. PATHWAYS OF DE NOVO SYNTHESIS OF PURINES AND PYRIMIDINES

The requirement for pyrimidines and purines in living organisms is ubiquitous and can be fulfilled by two synthetic pathways: de novo synthesis and/or through a salvage pathway (9–11).

For pyrimidine synthesis, uridine-5'-monophosphate (UMP) is a common product of both pathways. The de novo pathway consists of six enzymes: carbamyl phosphate synthetase II (CPSII); L-aspartate transcarboxylase (ATCase); L-dihydroorotase (DHOase); L-dihydroorotate dehydrogenase

Table 1 Inhibitors of De Novo Nucleotide Synthesis Potentially Useful in Autoimmune Disorders

Inhibitor[a]	Target[b]
Pyrimidines	
Methotrexate	TS
Azathioprine	
Leflunomide	DHO-DH
Brequinar	DHO-DH
Azaribine	OMP-DC
Pyrazofurin	OMP-DC
Purines	
Mycophenolic acid	IMP-DH
Mizoribine	IMP-DH

[a]The inhibitor or a metabolite of the inhibition may inhibit the target enzyme.
[b]Other target enzymes may also play a key role.
TS, thymidylate synthetase; DHO-DH, L-dihydroorotate dehydrogenase; OMP-DC, orotidine-5′-monophosphate decarboxylase; IMP-DH, inosine 5′-monophosphate dehydrogenase.

(DHO-DH); orotate phosphoribosyl transferase (OPRTase); and orotidine-5′-monophosphate decarboxylase (OMP-DC). These are shown schematically in Figure 5A and 5B.

The intracellular location of these six enzymes provides both efficient response to physiological stimuli and sites for regulation. CPSII, ATCase, and DHOase exist as one cytosolic, multienzyme complex (i.e., assembly line), which allows efficient transfer of products from one active site to the next without the influence of dilution or diffusion into the ambient cytoplasmic milieu. In comparison, the enzyme DHO-DH is sequestered on the outer face of the inner mitochondrial membrane (Fig. 6) at a site that may exert regulatory control over the rate of pyrimidine ring assembly since both substrate and product must diffuse across the mitochondrial membrane. Another cytosolic, multienzyme complex, composed of OPRTase and OMP-DC, affects the last two steps of pyrimidine biosynthesis, producing UMP as the product for its further transformation into cytidine or thymidine nucleotides (Fig. 5B). Enzymes responsible for de novo synthesis of pyrimidines are allosterically regulated by the levels of their precursor molecules or their products (9). This helps ensure a "balance" of final pyrimidine nucleotides and avoids apoptotic pathways (described above) associated with an imbalance of nucleotides. This allosteric regulation of enzyme activity by precursor/products

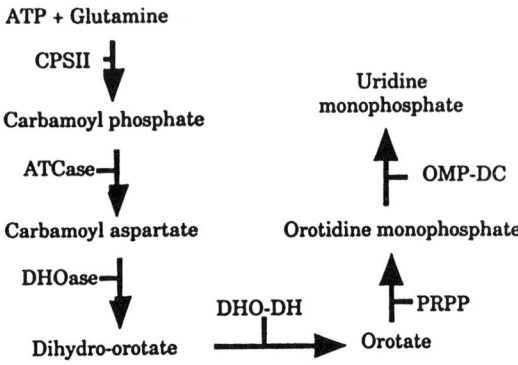

(A)

Figure 5 Pyrimidine biosynthetic pathway. The specific activities of the enzymes of de novo pyrimidine biosynthesis are quite variable in mammalian cells. However, the relative rates of these enzymes, under optimal substrate concentrations, show considerable homology.

also will act to modulate the activity of any therapeutic agent that interferes with enzyme(s) in the de novo production of pyrimidines.

For purine synthesis, phosphoribosylpyrophosphate (PRPP) and glutamine are converted to inosine 5'-monophosphate (IMP) (Fig. 7). Intracellular levels of IMP are carefully regulated by feedback inhibition (Fig. 7B). Elevated levels of ADP or GDP lead to decreased enzymatic conversion of PRPP to 5-phosphoribosylamine, an early stage in de novo purine synthesis. The enzymes of de novo purine synthesis also are carefully regulated to produce purines in the correct relative proportion to pyrimidines (9); orotate and orotate monophosphate (Fig. 5B) play an important bridge in coordinating the synthesis of the correct relative proportions of purines and pyrimidines (9). From a therapeutic point of view, critical levels of GMP and AMP are required for cellular activation of lymphoid cells (Fig. 7B), and subtle decreases in purine biosynthesis after administration of pharmaceutical agents may significantly influence immune response.

PRPP synthetase produces the sugar portion of the purine nucleosides. Since this enzyme is allosterically activated by guanine nucleosides, less PRPP is produced when guanosine levels are low. Ribonucleotide reductase then converts

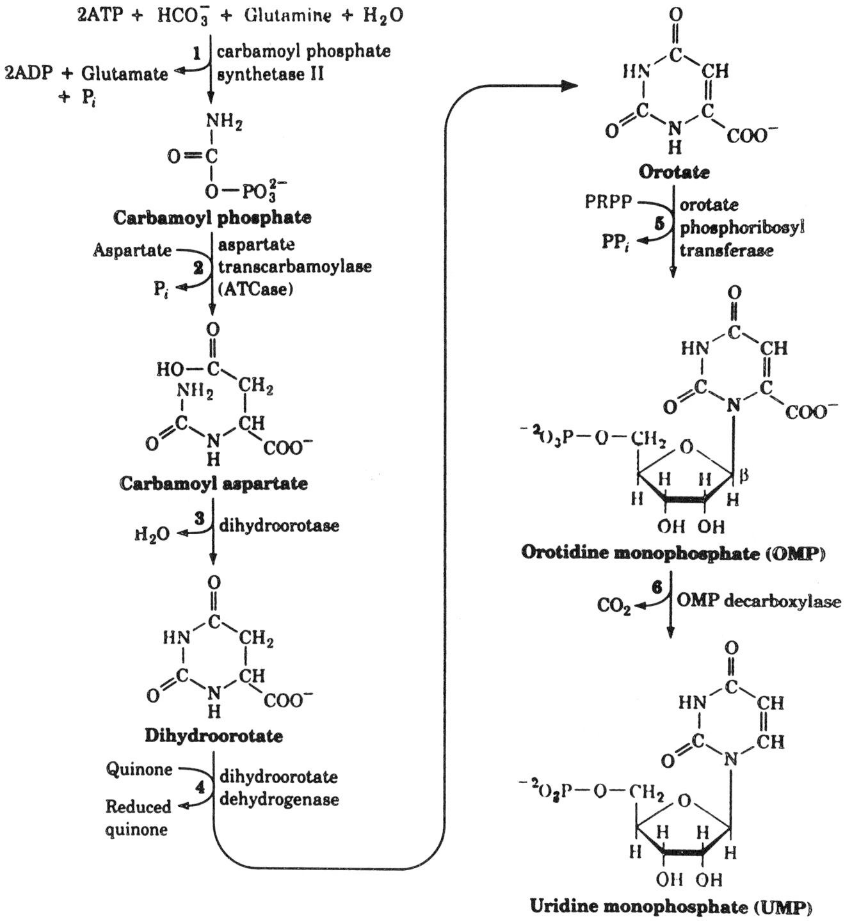

Inhibitors of De Novo Nucleotide Synthesis 263

(B)

Figure 5 (Continued)

purine ribonucleotides into deoxyribonucleotides; it is also allosterically activated by guanosine nucleotides (Fig. 7B). Therefore, depletion of guanosine nucleotides by therapeutic agents also results in decreased ribonucleotide reductase activity and decreased production of deoxyribonucleotides.

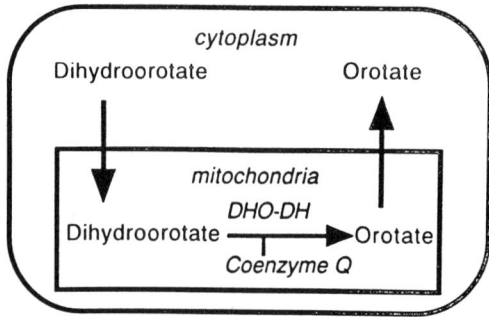

Figure 6 Dihydroorotate dehydrogenase (DHO-DH).

An important clinical side effect of therapeutic agents with antimetabolite properties is "mucositis" involving the oral and intestinal epithelial cells. Epithelial cells undergo continual replication and would therefore need to continuously replenish their nucleotide pools. These epithelial cells could fulfill their needs by either salvage pathway or de novo pathway synthesis. Although certain epithelial cell lines exhibit a dependence on de novo synthesis when cultured in vitro, they may fulfill their nucleotides for nucleotides in vivo by salvage pathway. This is possible since intestinal organisms can provide high levels of nucleotides and their precursors for salvage pathway (9). Thus, dietary intake and antibiotics that alter microbial organisms can predispose to "mucositis" associated with therapeutic agents.

III. METHOTREXATE

The treatment of RA has been changed by the relative safety and efficacy of "low-dose" oral weekly methotrexate (MTX) (12). Methotrexate is a folic acid analog, which is variably absorbed from the gastrointestinal tract with oral absorption ranging from 23 to 95% of the administered dose (13). However, individual variation in the absorption of oral methotrexate occurs and may be influenced by medical conditions such as inflammatory bowel syndrome or psoriasis (13–16). Absorption, protein binding, intracellular transport, and renal excretion of methotrexate can be altered by coadministration of other drugs including NSAIDS and probenecid (17). After absorption, methotrexate is transported intracellularly and metabolized to polyglutamates, which are retained within cells longer than unglutamated methotrexate (16).

The mechanism of the antirheumatic effects of methotrexate remains unknown. It has been hypothesized that myelomonocytic precursors of type A synoviocytes are unusually sensitive to MTX-induced alterations in purine synthesis and undergo apoptosis as a result of their imbalance in nucleotide pools. This would result in the delayed clinical onset of action of MTX in RA since the type A synoviocytes already resident in the RA joint have an estimated half-life of 4–8 weeks (18,19). In this model, the actual site of MTX action is the development of type A synoviocytes from myelomonocytic cells in the bone marrow, where methotrexate polyglutamate inhibits the enzyme thymidylate synthetase (20). This enzyme catalyzes the conversion of $2'$-deoxyuridylate (dUMP) to thymidylate (dTMP) and concomitant transfer of the one-carbon unit of 5,10-methylene tetralydrofolate (Fig. 8). Since thymidylate synthetase is the sole de novo source of thymidine, it appears to be a pivotal enzyme for the synthesis of DNA (9). However, the effect of MTX on cell replication is not simply due to depletion of thymidine or folic-acid pools, since replacement of thymidine or folic acid to cultures in vitro does not completely rescue cells from the block in their replication caused by methotrexate polyglutamate. This paradox may be explained partly by the effect of methotrexate polyglutamate on AICAR production (discussed below). Thus, it is possible for low doses of folic acid to ameliorate the potential marrow suppression of MTX but yet not overcome a block in purine synthesis in "would-be" activated lymphocytes or type A synoviocytes (21).

It has been proposed that methotrexate polyglutamate promotes intracellular accumulation of 5-aminoimidazole-4-carboxamide ribonucleotide (AICAR) by inhibiting enzyme AICAR transformylase (22). This model proposes a complex mechanism whereby transmethylation reactions result in locally increased levels of adenosine, which binds to specific (A2) receptors and provides an anti-inflammatory signal (23). In addition, AICAR inhibits the synthesis of guanine, and this could influence de novo purine synthesis through feedback inhibition at the level of PRPP synthesis as described above (Fig. 7B) (9). Increased AICAR levels have been demonstrated in RA patients treated with low-dose MTX (22).

In addition, methotrexate exhibits anti-inflammatory activity in the type II delayed hypersensitivity reactions and in assays of granulocyte function in vivo. These anti-inflammatory effects of MTX may result from effects on altered leukotriene B4 metabolism (24), adenosine secretion due to AICAR production that effects "wound repair" and neovascularization (25), and/or decreased metalloproteinase synthesis by fibroblasts in response to interleukin-1 (26).

Because of its relatively rapid onset of action and efficacy in RA, methotrexate has become the "gold standard" for future drug development. For example, 64% of RA patients enrolled in a prospective study remained on methotrexate 5 years later (12). Its relative ease of administration and efficacy have led many non-rheumatologists to assume that the treatment of RA may no longer pose a

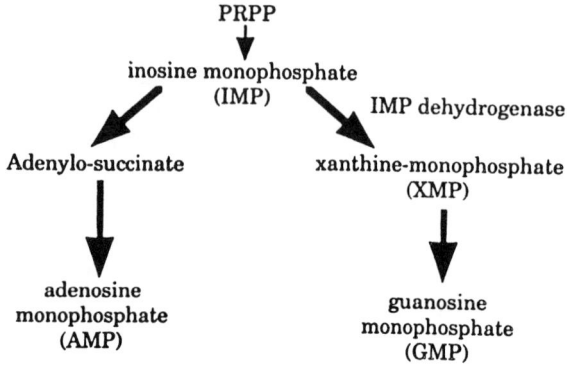

(A)

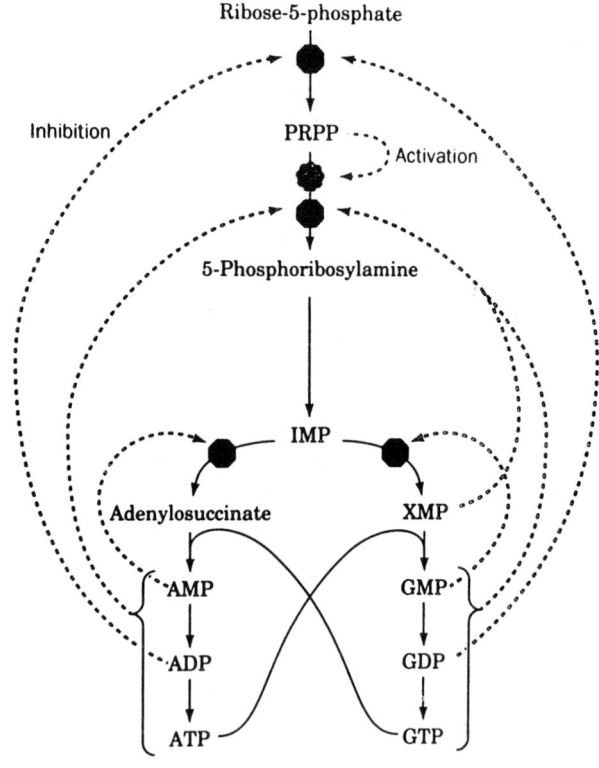

(B)

significant clinical problem. Several points need emphasis: 36% of RA patients initially enrolled in the protocol were considered "failures" due to lack of efficacy and/or toxicity; in rheumatology practice, many patients have exhausted the other therapeutic alternatives and are maintained on this drug for lack of other therapeutic options. Further, patients included in the multicenter, long-term methotrexate study were excluded if they had a history of lung or liver dysfunction (12), yet many RA patients have these problems, and the "prospective" studies (12,27) will underestimate the frequency of adverse events in "real world" RA patients (28).

Late onset of liver and lung problems as well as hematopoietic toxicity (29) must be considered in patients receiving long-term methotrexate therapy (12). Prior long-term studies in psoriasis patients recognized the risk of increased hepatic fibrosis only after prolonged methotrexate treatment (30). A meta-analysis demonstrated light microscopic evidence of liver fibrosis in 29% of RA patients receiving methotrexate who had no other risk factors (e.g., alcoholism, hepatitis) (31). Electron microscopic evidence of fibrosis manifested by increased sinusoidal collagen occurred in approximately 50% of RA patients receiving methotrexate, representing a dramatic increase over the frequency of fibrosis in patients not receiving methotrexate (32). However, the clinical significance of this finding remains controversial, since significant loss of liver function is rare in methotrexate-treated RA patients (30). Nevertheless, in view of the rapidly increasing numbers of RA patients receiving this therapy (31–33), the possibility that a subset of RA patients may develop progressive hepatic fibrosis is extremely disturbing.

A recent study showed that the addition of cyclosporine A (2.5–5 mg/kg/day) could yield significant clinical improvement (34) in patients who have had an inadequate response to MTX. This finding of potential synergy was attributed to the effect of MTX on the type A synoviocyte in addition to cyclosporine A influence on the T-cell function (34). However, cyclosporine may alter the renal excretion of MTX and thus predispose to adverse effects. Also, both cyclosporine A and methotrexate have been associated with Epstein-Barr virus lymphoproliferative disease in RA patients (35,36). Thus, RA patients on this combination of drugs must be closely observed for accelerated liver toxicity and for EBV-associated lymphomas.

Figure 7 De novo purine synthesis. (A) Pathway of purine biosynthesis showing the regulatory site of PRPP and nucleotides. AMP = adenosine monophosphate; GMP = guanosine monophosphate; IMP = inosine 5′-monophosphate; PRA = phosphoribosylamine; PRPP = phosphoribosylpyrophosphate. (B) Specific chemical conversions for synthesis of IMP.

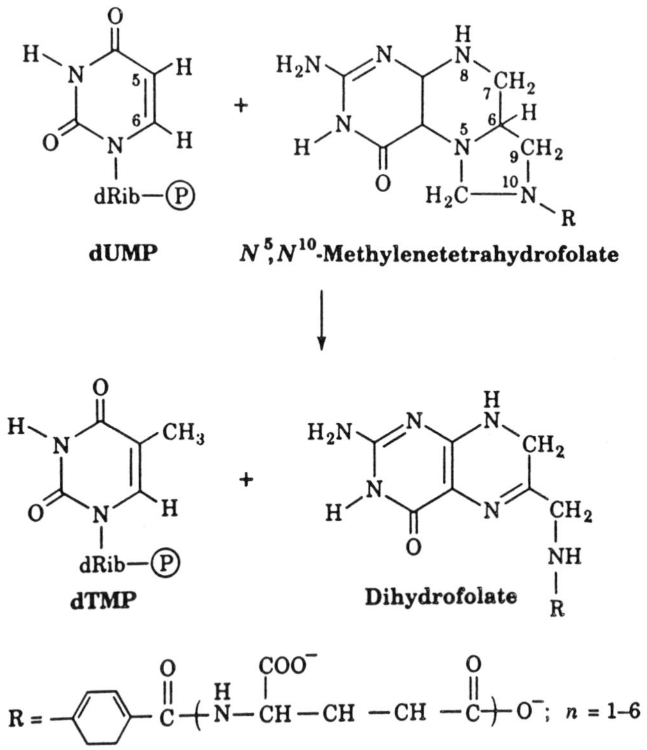

Figure 8 Synthesis of dTMP by thymidylate synthetase.

IV. AZATHIOPRINE

In the early 1960s, 6-mercaptopurine (6-MP) was reported effective in the treatment of RA, SLE, and other autoimmune disorders (37,38). Efforts to protect 6-MP from rapid metabolism led to the development of azathioprine, converted in vivo to 6-MP. Subsequent reports described the benefit of azathioprine in open-label studies in RA and SLE (39); controlled studies and long-term follow-up confirmed its efficacy in RA (40–42), although its role in SLE, polymyositis, and psoriatic arthritis remains in debate (41,43–45).

Following absorption, approximately 30% of azathioprine is protein-bound. In patients with normal renal and hepatic function, its half-life is about 3 hr (46). Azathioprine is metabolized to 6-MP by the liver; multiple metabolites have been identified, including 6-thio-inosinic acid and 6-thio-guanylic acid (47). These derivative thio-purine analogs can be incorporated into mRNA and may interfere with multiple cellular functions (9,48).

6-Thio-inosine acid, formed by the action of hypoxanthine phosphoryibosyl transferase on 6-MP, suppresses several steps in the synthesis of adenosine and guanosine; the decrease in these nucleotides subsequently leads to feedback inhibition of IMP synthesis (Fig. 7B) (49). 6-MP can also lead to cell toxicity following its conversion to 6-thioguanine (48), which is incorporated into both RNA and DNA, thereby causing faulty protein synthesis and DNA replication (50).

6-MP is converted to the same IMP analog (6-thio-MP) as mycophenolic mofetil (discussed below). However, guanosine is not able to reverse the action of 6-MP on in vitro cell growth or antibody production, in contrast to mycophenolic mofetil (discussed below) (51). Surprisingly, reversal of growth inhibition by 6-MP was achieved with addition of adenine (51). This did not occur with azathioprine (AZA), suggesting that AZA and 6-MP do not work through a common pathway (51).

Approximately 30% of RA patients discontinue azathioprine treatment because of adverse effects, usually gastrointestinal (GI) side effects (52). Hematopoietic toxicity (usually modest leukopenia) is relatively common and is generally reversible. Low levels of thiopurine methyltransferase (TPMT) have been associated with the increased incidence of serious bone marrow toxicity in a variety of conditions including RA (53). Recently, bone marrow toxicity in RA patients receiving azathioprine also has been associated with polymorphism of several other enzymes in purine biosynthesis (48).

V. MYCOPHENOLIC MOFETIL (MPA)

MPA is a prodrug that is hydrolyzed by plasma and liver esterases to produce the active metabolite (mycophenolic acid) (Fig. 9). MPA was originally derived from cultures of *Penicillium* (fungal) species in 1896. It was developed for the treatment of cancer at Eli Lilly in the late 1960s. During the 1970s, attempts to modify its immunosuppression and increase its anticancer properties were pursued (54). After initial clinical trials failed in solid tumors, it was studied for the treatment of mild to severe psoriasis using 3 g/day, based on the hypothesis that it was active against proliferating keratinocytes. In one long-term study, MPA was administered to 85 psoriatic patients for as long as 13 years with continued efficacy, without dosage escalation (55). Although 12% of patients developed herpes zoster, other evidence of immunosuppression such as opportunistic infections or lymphomas was not noted (55). This increased incidence of herpes infection led to its discontinuance in psoriasis patients but emphasized its potential action as an immunosuppressive drug (56).

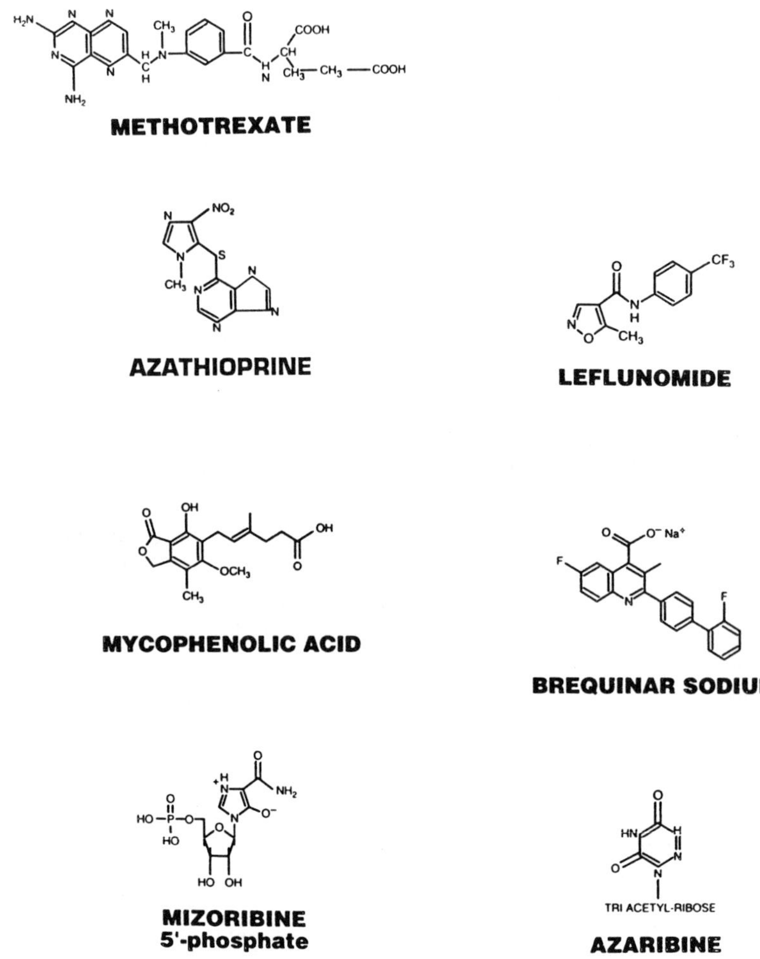
METHOTREXATE
AZATHIOPRINE
LEFLUNOMIDE
MYCOPHENOLIC ACID
BREQUINAR SODIUM
MIZORIBINE
5'-phosphate
TRI ACETYL-RIBOSE
AZARIBINE

Figure 9 Structure of several drugs that inhibit de novo nucleotide synthesis.

Mycophenolic acid is a noncompetitive, reversible inhibitor of the enzyme IMP-dehydrogenase (IMP-DH) (Fig. 10). This enzyme is the rate-limiting step for the de novo synthesis of purines during lymphocyte activation (57). Different isoforms of IMP-DH have been isolated; lymphocytes predominantly have type II, which appears to be more sensitive to inhibition by MPA (58). Regardless, observations about selective isoform inhibition remain controversial (3).

Inhibitors of De Novo Nucleotide Synthesis

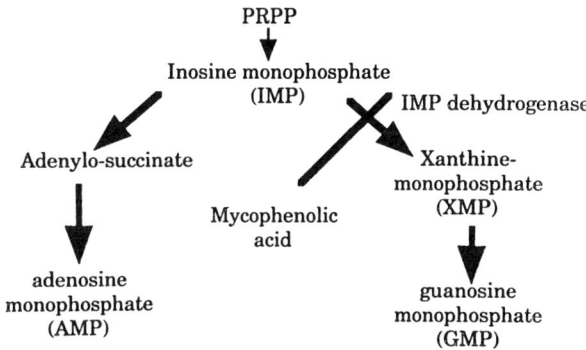

Figure 10 Mycophenolic acid action.

Resistance to MPA has been associated with up-regulation of the enzyme IMP-DH (59), indicating a potential mechanism for "escape" during chronic therapy.

Inhibition of IMP-DH by mycophenolic acid may cause (56,60–62):

1. Decreased levels of guanosine and adenosine nucleotides necessary for production of DNA and RNA
2. Altered levels of the enzyme PRPP synthetase, which is allosterically activated by guanosine nucleotides
3. Decreased levels of ribonucleotide reductase, which is allosterically activated by guanine nucleotides
4. Decreased glycosylation of molecules including cell surface adhesion molecules (63)
5. Elevated levels of calcium-binding proteins MRP8 and MRP14, which may arrest the growth and differentiation of myelomonocytic cells (64)

In vitro studies indicate that MPA can inhibit lymphocyte proliferation in vitro in response to T-cell and B-cell mitogens even when added 2 hr after stimulation. Inhibition of cell replication can be reversed by addition of guanosine (62). Inhibition of lymphocytes and myelomonocytic cell replication by MPA occurs at significantly lower drug levels than for fibroblasts or endothelial cells (65). Further, differentiation of myelomonocytic precursors into "mature" monocytic cells is dependent on critical intracellular guanosine levels and is therefore inhibited by MPA (66). MPA does not inhibit IL-2, IL-4, IL-10, or IL-13 production by stimulated lymphocytes (60,67,68) or IL-1, IL-6, or TNF-α production by monocytes (60). At higher in vitro concentrations, MPA depleted GTP and ATP, thereby leading to altered activity of a family of ATP and GTP proteins including g-*ras* (61), proto-oncogenes that play an important role in lymphocyte activation and division.

Clinical studies with MPA showed benefit in patients with severe psoriasis, but its use was limited by hematological toxicity and viral infections (69–71). Recently, a multicenter clinical trial of MPA for RA appeared promising (72). Treatment resulted in improvement in joint scores, reduction of rheumatoid factor titers, decreased immunoglobulin IgG levels, and decreased numbers of activated T cells (CD2) in the peripheral circulation. In vitro lymphocyte mitogen responses were inhibited and delayed-type hypersensitivity skin reactions decreased (72). Nonetheless, the sponsoring pharmaceutical company terminated clinical trials in RA after MPA was recently approved for use in transplant rejection.

VI. MIZORIBINE

Mizoribine (Fig. 9) was isolated in 1974 in Japan by screening culture filtrates of the soil fungus *Eupenicillum brefaldianum* and was approved for treatment of transplantation rejection in 1984. In Japan, mizoribine has largely supplanted azathioprine since it is more immunosuppressive and less myelotoxic (73,74,75). The incidence of severe leukopenia with mizoribine was 1% in comparison to 4% in patients receiving azathioprine in European studies of transplant recipients (76–79).

Mizoribine is a prodrug that must be phosphorylated to 5′-monophosphate by adenosine kinase to be active. It is a reversible inhibitor of the enzyme IMP-DH (the same enzyme inhibited by MPA). It is not incorporated into DNA or RNA and is dephosphorylated intracellularly by 5′-nucleotidase. Pharmacokinetic studies in rats showed a peak level 1.5 hr after oral dosing, with 60% excreted unchanged in urine at 4 hr and 87% at 24 hr.

Mizoribine inhibits DTH in mice, suppresses acute graft-versus-host reactions, and does not have anti-inflammatory activity in models developed to screen NSAIDS. It suppresses primary antibody responses as well as the generation of memory T-helper and memory B cells (74). Studies in dogs showed that mizoribine with cyclosporine increased long-term survival of heart, renal, and pancreas grafts (73,74). However, its therapeutic index in dogs was close to the toxic index, similar to the ideosyncratic reaction observed with other inhibitors of de novo nucleotide synthesis in dogs.

Clinical studies with mizoribine have focused on renal transplantation (76–79). However, mizoribine has also shown benefit in patients with SLE (80) and children with nephrotic syndrome (81). Mizoribine is also used as a single therapy or in combination with methotrexate in Japanese patients with severe RA (Saito I, personal communication). However, double-blind studies using mizoribine in RA have not yet been published.

VII. LEFLUNOMIDE

Leflunomide is a synthetic isoxazole derivative (Fig. 11), which is efficiently absorbed by the intestinal mucosa and converted in the GI tract and plasma to the active metabolite (A77-1726). The mechanisms of the antiproliferative, antiinflammatory, analgesic, and immunosuppressive actions of leflunomide are not well understood in vitro, let alone in vivo.

Two mechanisms have emerged to explain the antiproliferative actions of A77 1726 on T cells in vitro: (1) inhibition of pyrimidine biosynthesis (83–86); and (2) inhibition of Src family–mediated protein tyrosine phosphorylation (82,87). These two mechanisms of action are not mutually exclusive. The dominance of each mechanism for the antiproliferative effects of A77 1726 may depend on several variables including the cell type, the species, and the concentration of A77 1726 to which cells are exposed.

After conversion of leflunomide to A77-1726, the latter compound can block the replication of stimulated lymphocytes in vitro at drug levels that could be obtained in vivo. The addition of pyrimidine nucleosides, but not purine nucleosides, completely antagonizes the antiproliferative effects of A77 1726 for Jurkat T cells in vitro (83–86). Recent work has identified dihydroorotate dehydrogenase (DHO-DH), an enzyme in the pyrimidine biosynthetic pathway, as a target for A77 1726 in lymphocytes (83–86). This is the same enzyme proposed as a target for the effect of brequinar (discussed below).

An additional action of leflunomide may involve the inhibition of protein tyrosine kinase in Jurkat T cells. A77 1726 inhibits the following early tyrosine

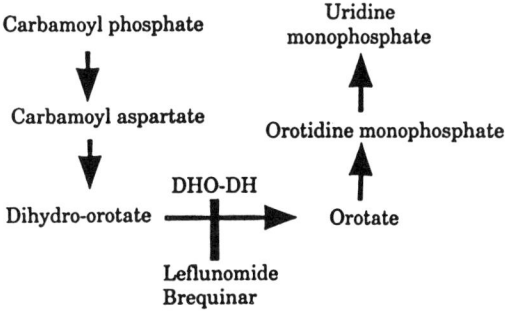

Figure 11 Leflunomide and brequinar action.

kinase–dependent events: total tyrosine phosphorylation, including phosphorylation of the zeta chain of the CD3 complex, the gamma chain of phospholipase C, $p59^{fyn}$ and $p56^{ick}$ tyrosine kinase activities. The IC_{50} required to achieve these inhibitions of tyrosine kinase activity by A77-1726 was 150 µM in vitro (82). The IC_{50} for inhibition of Jurkat T-cell proliferation by A77 1726 in vitro is more than an order of magnitude *less* than the IC_{50} for inhibition of tyrosine kinase activity. This suggests that the primary mechanism for A77-1726 is not inhibition of tyrosine kinase activity and supports the hypothesis that drug action occurs through inhibition of the enzyme DHO-DH (discussed above). Further, A77-1726 is very strongly protein bound in vivo and it is unclear whether A77 1726 levels of 150 µM could be safely obtained in patients. Other lines of evidence indicate that inhibition of tyrosine kinases is not the predominant action of A77-1726 for blocking T-cell proliferation in vitro (84–86). A77 1726 had little effect on IL-2 production in the time period immediately after T-cell stimulation through the CD3 receptor. If A77-1726 had its predominant effect on tyrosine kinase activity, a decrease in IL-2 and IL-2 promoter activity would have been expected; a decrease in early IL-2 production or IL-2 promoter activity was not observed (84–86). A decrease in IL-2 production was observed at later times after stimulation (i.e., 24–48 hr) but probably resulted from an indirect effect of the A77-1726. For example, the inhibition of cell proliferation due to inhibition of DHO-DH would lead to reduced cell numbers and a corresponding reduction in total IL-2. The undiminished inhibition of T-cell proliferation by A77 1726 added to cells 24 hr after stimulation also supports the contention that A77 1726 mediates its antiproliferative effects at a site or sites late in the biochemical cascade leading to T-cell DNA synthesis.

Other actions of leflunomide have been described that cannot be easily attributed to either inhibition of tyrosine kinases or DHO-DH. Leflunomide weakly inhibits the generation of 5-HETE and LTB4 from exogenous arachidonic acid (88,89) but does not interfere with prostaglandin production by the constitutive cyclooxygenase (COX-1) pathway.

Leflunomide prevents and reverses allograft rejection in rats and is synergistic when used with cyclosporine (90,91,107). Leflunomide administration was beneficial in the treatment of adjuvant arthritis in Lewis rats (93) and prevented disease progression in murine SLE (MRL/lpr) (94). MRL/lpr mice showed decreased lymphoproliferation and autoantibody formation, which persisted for several weeks after treatment was stopped, in contrast to cyclosporine A where rapid disease rebound occurred (94). Leflunomide has also shown activity in animal models of proteoglycan-induced spondylitis, allergic encephalomyelitis, interstitial nephritis (95), uveitis (96), and graft-versus-host reaction (97).

Initial clinical trials in RA included patients treated with 5, 10, or 25 mg of leflunomide daily. Results of a double-blind, placebo-controlled trial of 402 patients in Yugoslavia using 5, 10, or 25 mg/day showed statistically significant improvement compared to placebo in both clinical and laboratory parameters of active RA in the 10- and 25-mg dose groups. In continued open-label treatment, 146 patient were still receiving study drug at 24 months. Adverse effects included gastrointestinal symptoms, rash, allergic reactions, and elevated liver function tests, and were dose-related. Anemia, leukopenia, and thrombocytopenia were not significant problems (98).

VIII. BREQUINAR

Brequinar is a substituted 4-quinoline carboxylic acid analog (Fig. 9) synthesized for potential anticancer activity (74). It blocks the de novo synthesis of pyrimidines by reversibly inhibiting the enzyme DHO-DH, the same proposed site of action as leflunomide. Structural analogs of brequinar have shown the importance of the C2 position for hydrophobic binding and the C4 carboxylic acid for ionic binding. Although brequinar has recently been dropped from clinical trials owing to its high frequency of adverse side effects, the potential similarities to leflunomide must be carefully evaluated since the latter drug is currently in clinical trials.

In vitro proliferation of T and B cells to mitogens and stimulation with anti-CD2 or anti-CD3 are inhibited by brequinar. It blocks the progression of stimulated lymphocytes from "G1" to "S" phase, which can be reversed by the addition of uridine (99,100). The addition of cytidine potentiates the immunosuppressive effects of brequinar, which may result from an imbalance of nucleotides, i.e., decreased uridine and relatively increased cytidine levels, or an effect on the salvage pathway.

Brequinar is efficiently absorbed from the GI tract and is highly protein-bound in the plasma. The drug accumulates in the tissues, even after blood levels have fallen. A high affinity for the enzyme DHO-DH may partially explain tissue accumulation and the long half-life (99). Adverse effects include headache, thrombocytopenia, leukopenia, and stomatitis and appear to be dose-related. Interindividual variation of brequinar clearance requires close monitoring to avoid toxicity. Owing to these effects, the clinical development of brequinar as an immunosuppressive agent has been discontinued.

Although both brequinar and leflunomide may inhibit the enzyme DHO-DH, there are significant differences between these agents. (1) Therapeutic doses of brequinar have been associated with thrombocytopenia, stomatitis, and mucositis, whereas only transient thrombocytopenia has been reported with leflunomide (101). (2) In animal models of transplantation, brequinar was less effective in blocking vascular graft rejection and smooth muscle proliferation than leflunomide (102–104). (3) the K_i for brequinar (105) is at least an order of magnitude greater for human DHO-DH than is the K_i for A77 1726 (Ruuth E and Williamson RA, personal communication, 1995). (4) Administration of brequinar causes a reduction in serum uridine levels (106), but no reduction is seen after treatment with leflunomide (Horn W, personal communication). (5) The antiproliferative effects of A77 1726 on cells in culture are readily reversed by washing the cells, but washing does not reverse the antiproliferative effects of brequinar (Bartlett R, personal communication, 1995). (6) A77 1726 inhibits the proliferation of smooth muscle cells in vitro (107) but brequinar does not (103). The difference in the therapeutic index for leflunomide and brequinar emphasizes our incomplete understanding of the potential site(s) of action of these drugs.

Based on the available data, one could conjecture that brequinar must depend entirely on its high affinity for and substantial inhibition of DHO-DH for its therapeutic effect, whereas leflunomide's relatively weak inhibition of DHO-DH may be compensated by its weak additional effects on tyrosine kinase pathways at the drug levels achieved in patients. In addition, brequinar may have additional targets to which it is strongly bound and these latter inhibitions may result in the observed toxicity. Leflunomide may also have similar additional targets but may bind with a lower affinity and not lead to similar toxicity.

IX. AZARIBINE (TRIACETYL 6-AZAURIDINE)

During the late 1950s and early 1960s, 6-azauridine was found to have antineoplastic activity in animal models. In particular, 6-azauridine was found effective in mycosis fungoides (108) which given intravenously. However, central nervous system toxicity occurred when 6-azauridine was given orally owing to the breakdown to 6-azauracil by intestinal organisms (109). To overcome this problem, the triacetyl derivative of 6-azauridine (azaribine) was synthesized. When azaribine was administered orally, it was efficiently absorbed and converted in the bloodstream to 6-azauridine. 6-Azauridine is converted intracellularly by uridine kinase to the ribonucleotide, i.e., 6-azauridine 5'-monophosphate, which is the biologically active form (110,111). The primary mode of action of this metabolite is reversible inhibition of the enzyme decarboxylation of the ribonucleotide of orotic acid, i.e., orotidylic acid, to uridine 5'-monophosphate, thus inhibiting the de novo biosynthesis of pyrimidines (112) (Fig. 12). The ability of uridine to reverse the block by azaribine to cellular proliferation strongly supports this proposed mechanism (unpublished observations).

Azaribine has previously been shown effective in refractory psoriasis (113) and psoriatic arthritis (114). Indeed, azaribine was described as the "wonder drug" for severe psoriasis since patients who had previously failed both methotrexate and mycophenolic mofetil therapy responded to azaribine treatment (113,115–117). After its approval in 1975 by the Food and Drug Administration for treatment of

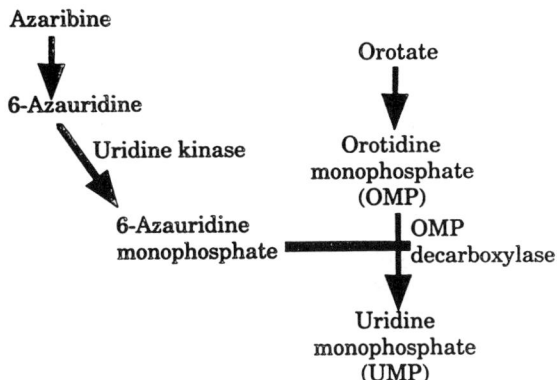

Figure 12 Azaribine action.

severe psoriasis, more than 1000 psoriatic patients were treated with azaribine (117). An increased frequency of arterial and venous thromboses (in an estimated 3–5% of treated patients) was reported, leading to its voluntary withdrawal from the market [reviewed in (118)]. These thrombotic events generally occurred within 6 weeks of initiation of therapy. Subsequently, it was recognized that a breakdown product of azaribine inhibited the breakdown metabolism of homocysteine (117), a compound that predisposes to vascular thrombosis (119–122). Of importance, approximately 5% of the general population have genetically inherited defects in homocysteine metabolism and are at risk for vascular occlusive disease (120). These individuals, as well as patients treated with azaribine, respond to oral supplementation of vitamin B, vitamin B_{12}, and folic acid with lowered homocysteine levels. Vitamins B_6 and B_{12} serve to activate the enzyme homocysteine lysase to catabolize homocysteine, and folic acid facilitates the transmethylation of homocysteine to methionine (117). By activating both pathways, lower homocysteine levels are expected to lead to lower risk for vascular occlusive disease (123). Owing to the efficacy of azaribine in the treatment of refractory psoriasis and/or psoriatic arthritis, it has continued in use during the past 20 years on a compassionate basis by Dr. Charles McDonald of Brown University (personal communication). Patients on these long-term compassionate treatment programs for psoriasis as well as RA patients (enrolled in a phase II pharmacokinetic study at Scripps Clinic) currently receive oral supplementation with pyridoxine 150 mg/day, folic acid 2 mg/day, and vitamin B_{12} 100 μg/day.

Azaribine was shown effective in the treatment of psoriatic arthritis at doses ranging from 100 to 400 mg/kg/day (124). Therapy was successful ($p < 0.01$) in reducing average tender joint count and pain, ring size, morning stiffness, and in increasing grip strength. Responses in the 32 patients varied from total remission in five to no improvement in seven. This study confirmed earlier case reports (113,125–127).

Recently in Bucharest, Romania, azaribine was administered to 25 patients with refractory RA at 20–25 mg/kg/day t.i.d. for 3 months, followed by an increased dose of 100 mg/kg/day t.i.d. for 2 months (127) in an open-label study with concomitant pyridoxine HCl 150 mg/day (manuscript in preparation). A significant decrease in clinical parameters of arthritis activity as well as improvement in acute-phase reactants occurred. Significant decreases ($p < 0.05$) in mean ESR (56 to 40), joint counts (172 to 153), and patient assessment of morning stiffness and pain occurred. When utilizing this lower initial dose and the combination with pyridoxine, no thromboembolic or central nervous system complications were observed. A prospective trial of azaribine plus pyridoxine to assess efficacy and safety in RA is in progress in the United States.

X. SUMMARY

Despite an intensive search for improved treatment for RA in the past decade, the arsenal of medications routinely used to treat RA (gold, methotrexate, azathioprine, etc.) has remained relatively unchanged. Agents demonstrated useful in other therapeutic indications such as organ allograft rejection are therefore be considered. Among such drugs, those inhibiting de novo nucleotide synthesis appear promising for the treatment of RA as single agents or in combination. This chapter reviews azathioprine, mycophenolic mofetil, mizoribine, brequinar, and azaribine. It appears that lymphocytes and myelomonocytes have a particular susceptibility to the agents based on their dependence on de novo nucleotide synthesis. Although data to confirm which mechanisms occur in vivo remain unclear, it is possible that an imbalance of nucleotide levels in activated cells leads to apoptosis cell death. Variability of enzyme levels responsible for the de novo synthesis or salvage nucleotide pathway may occur in different individuals and different disease states. It is hoped that better understanding of the mechanisms of action of proposed therapeutic agents will allow rational combination therapy and avoidance of severe drug toxicity in patients with RA.

ACKNOWLEDGMENTS

This work was supported by grants from the National Institutes of Health (M01-RR-000833), the Department of Academic Affairs of Scripps Clinic (AF-3205), and the Scripps-Stedham, Florence, Ramsdell, and Thornton Foundations (RIF) and Ralph and Marian Falk Foundation (REM). We appreciate the

Inhibitors of De Novo Nucleotide Synthesis 279

helpful suggestions and critical review of this manuscript by Drs. William Drell, Charles Carrera, and Ken Pischel.

REFERENCES

1. Morris RE. New small molecule immunosuppression for transplantation: review of essential concepts. J Heart Lung Transplant 1993; 12:S275–286.
2. Chambers D, Martin DW, Weinstein Y. The effect of cyclic nucleotides on purine biosynthesis and the induction of PRPP synthetase during lymphocyte activation. Cell 1974; 3:375–380.
3. Dayton J, Lindsten T, Thompson CB, Mitchell BS. Effects of human T lymphocyte activation on inosine monophosphate dehydrogenase expression. J Immunol 1994; 152:984–991.
4. Allison A, Kowalski WJ, Muller CJ, Waters RV, Eugui EM. Mycophenolic acid and brequinar, inhibitors of purine synthesis, block the glycosylation of adhesion molecules. Transplant Proc 1993; 25:67–70.
5. Powell SN, Abraham EH. The biology of radioresistance: similarities, differences and interactions with drug resistance. Cytotechnology 1993; 12:325–345.
6. Dessi F, Pollard H, Moreau J, Ben-Ari Y, Charriaut-Marlangue C, Cytosine arabinoside induces apoptosis in cerebellar neurons in culture. J Neurochem 1995; 64:1980–1987.
7. Kizaki H, Ohnishi Y, Azuma Y, Mizuno Y, Ohsaka F. 1-Beta-D-arabinosylcytosine and 5-azacytidine induce internucleosomal DNA fragmentation and cell death in thymocytes. Immunopharmacology 1993; 25:19–27.
8. Smith ML, Chen IT, Zhan Q, O'Connor PM, Fornace AJ Jr. Involvement of the p53 tumor suppressor in repair of u.v.-type DNA damage. Oncogene 1995; 10:1053–1059.
9. Voet D, Voet JG. Biochemistry. New York: Wiley, 1990:748–763.
10. Jones ME. Pyrimidine nucleotide biosynthesis in animals: genes, enzymes, and regulation of UMP biosynthesis. Annu Rev Biochem 1980; 49:253–279.
11. Cory JG. Purine and pyrimidine nucleotide metabolism. In: Devlin TM, ed. Textbook of Biochemistry with Clinical Correlations. New York: Wiley-Liss, 1992:529–566.
12. Kremer JM, Phelps CT. Long-term prospective study of the use of methotrexate in the treatment of rheumatoid arthritis: update after a mean of 90 months. Arthritis Rheum 1992; 35:138–145.
13. Jolivet J. The pharmacology and clinical use of methotrexate. N Engl J Med 1983; 309:1094–1104.
14. Hendel L, Hendel J, Johnson A, Gudmand-Hoyer E. Intestinal function and methotrexate absorption in psoriatic patients. Clin Exp Dermatol 1982; 7:491–497.
15. Pinkerton CR, Glasgow JFT, Welshman SG, Bridges JM. Can food influence the absorption of methotrexate in children with acute lymphoblastic leukaemia? Lancet 1980; 2:8197–8209.
16. Edelman MB, Biggs DF, Jamali F, Russell AS. Low-dose methotrexate kinetics in arthritis. Clin Pharmacol Ther 1984; 35:382–386.

17. Evans WE, Christensen ML. Drug interactions with methotrexate. J Rheumatol 1985; 12:15–20.
18. Fox RI, Robinson CA, Williams GW, Curd JG, Colwell CW Jr. Future approaches to the diagnosis and therapy of rheumatoid arthritis. In: Furst DE, Weinblatt ME, eds. Immodulators in the Rheumatic Diseases. New York: Marcel Dekker, 1990: 223–256.
19. Fox RI, Kang H-I. Structure and function of synoviocytes. In: McCarthy DJ, Koopman WJ, eds. Arthritis and Allied Conditions. Philadelphia: Lea & Febiger, 1993:263–278.
20. Kremer JM, Galivan J, Streckfuss A, Kamen B. Methotrexate metabolism analysis in blood and liver of rheumatoid arthritis patients: association with hepatic folate deficiency and formation of polyglutamates. Arthritis Rheum 1986; 29:832–835.
21. James SJ, Basnakian AG, Miller BJ. In vitro folate deficiency induces deoxynucleotide pool imbalance, apoptosis, and mutagenesis in Chinese hamster ovary cells. Cancer Res 1994; 54:5075–5080.
22. Cronstein B, Naime D, Ostad E. The anti-inflammatory mechanism of methotrexate. J Clin Invest 1993; 92:2675–2680.
23. Balzarini J, Karlsson A, Wang L, et al. Eicar (5-ethynyl-1-beta-D-ribofuanosyl-imidazole-4-carboxamide): a novel potent inhibitor of inosinate dehydrogenase activity and guanylate biosynthesis. J Biol Chem 1993; 268:24591-24598.
24. Sperling R, Coblyn J, Austin K, Weinblatt M. Inhibition of leukotriene B4 synthesis in neutrophil by methotrexate. Arthritis Rheum 1990; 33:1149–1155.
25. Hirata S, Matsubara T, Saura R, Tateishi H, Hirohata K. Inhibition of in vitro vascular endothelial cell proliferation and in vivo neovascularization by low-dose methotrexate. Arthritis Rheum 1989; 32:1065–1073.
26. Firestein GS, Paine MM, Boyle DL. Mechanisms of methotrexate action in rheumatoid arthritis: selective decrease in synovial collagenase gene expression. Arthritis Rheum 1994; 37:193–200.
27. Weinblatt ME, Kaplan H, Germain BF, et al. Methotrexate in rheumatoid arthritis: effects on disease activity in a multicenter prospective study. J Rheumatol 1991; 18:334–338.
28. Fries JF, Bloch DA, Sharp JT, et al. Assessment of radiologic progression in rheumatoid arthritis: a randomized, controlled trial. Arthritis Rheum 1986; 29:1–9.
29. Alarcon GS, Tracy IC, Blackburn WD Jr. Methotrexate in rheumatoid arthritis: toxic effects as the major factor in limiting long-term treatment. Arthritis Rheum 1989; 32:671–776.
30. Whiting-O'Keefe O, Fye K, Sack K. Methotrexate and histologic abnormalities. Am J Med 1991; 90:711–716.
31. Phillips CA, Cera PJ, Mangan TF, Newman ED. Clinical liver disease in patients with rheumatoid arthritis taking methotrexate. J Rheumatol 1992; 19:229–233.
32. Bjorkman P, Boschert M, Tolman K, Clegg D, Ward J. The effect of long-term methotrexate therapy and hepatic fibrosis in RA. Arthritis Rheum 1993; 12:1697–1701.
33. Walker AM, Funch D, Dreyer NA, et al. Determinants of serious liver disease among patients receiving low-dose methotrexate for rheumatoid arthritis. Arthritis Rheum 1993; 36:329–335.

Inhibitors of De Novo Nucleotide Synthesis 281

34. Tugwell P, Pincus T, Yocum D, et al. Combination therapy with cyclosporine and methotrexate in severe rheumatoid arthritis. N Engl J Med 1995; 333:137–141.
35. Kamel OW, van de Rijn M, Hanasono MM, Warnke RA. Immunosuppression-associated lymphoproliferative disorders in rheumatic patients. Leuk Lymphoma 1995; 16:363–368.
36. Ferraccioli GF, Casatta L, Bartoli E, et al. Epstein-Barr virus–associated Hodgkin's lymphoma in a rheumatoid arthritis patient treated with methotrexate and cyclosporin A. Arthritis Rheum 1995; 38:867–868.
37. Dameshek W, Schwartz R. Treatment of certain "autoimmune" diseases with antimetabolites; a preliminary report. Trans Assoc Am Physicians 1960; 73: 113–127.
38. Myles AB. 6-Mercaptopurine (6 M.P.) in the treatment of rheumatoid arthritis and related conditions. Ann Rheum Dis 1965; 24:179–180.
39. Corley CC, Lessner HE, Larsen WE. Azathioprine therapy of "autoimmune" diseases. Am J Med 1966; 41:404–412.
40. Hunter T. Azathioprine in rheumatoid arthritis: a long-term follow-up study. Arthritis Rheum 1975; 18:15–20.
41. Pinals RS. Azathioprine in the treatment of chronic polyarthritis: long-term results and adverse effects in 25 patients. J Rheumatol 1976; 3:140–144.
42. DeSilva M, Hazleman BL. Long-term azathioprine in rheumatoid arthritis: a double-blind study. Ann Rheum Dis 1981; 40:560–563.
43. Levy J. A double-blind controlled evaluation of azathioprine treatment in rheumatoid arthritis and psoriatic arthritis. Arthritis Rheum 1972; 15:116–117.
44. Klippel HJ. Studies in the treatment of lupus nephritis. In: Systemic lupus Erythematosus: evolving concepts. Moderated by J.L. Dekker. Ann Intern Med 1979; 91:587–604.
45. Dinant HJ. Alternative modes of cyclophosphamide and azathioprine therapy in lupus nephritis. Ann Intern Med 1982; 96:728–736.
46. Huskisson EC. Azathioprine. Clin Rheum Dis 1984; 10:325–332.
47. Elion GB. Biochemistry and pharmacology of purine analogues. Fed Proc 1967; 26:898–904.
48. Kerstens PJSM, Stolk JN, De Abreu RA, Lambooy LHJ, van de Putte LBA, Boerbooms AAMT. Azathioprine-related bone marrow toxicity and low activities of purine enzymes in patients with rheumatoid arthritis. Arthritis Rheum 1995; 38: 142–145.
49. Bertino JR. Chemical action and pharmacology of methotrexate, azathioprine and cyclophosphamide in man. Arthritis Rheum 1973; 16:79–83.
50. LePage GA. Incorporation of 6-thioguanine into nucleic acids. Cancer Res 1960; 20:403–408.
51. Dayton JS, Turka LA, Thompson CB, Mitchell BS. Comparison of the effects of mizoribine with those of azathioprine, 6-mercaptopurine, and mycophenolic acid on T lymphocyte proliferation and purine ribonucleotide metabolism. Mol Pharmacol 1992; 41:671–676.
52. Currey HLF. Comparison of azathioprine, cyclophosphamide, and gold in treatment of rheumatoid arthritis. Br Med J 1974; 3:763–766.

53. Lennard L, Van Loon JA, Weinshilboum RM. Pharmacogenetics of acute azathioprine toxicity: relationship to thiopurine methyltransferase genetic polymorphism. Clin Pharmacol Ther 1989; 46:149–154.
54. Carter S, Franklin TJ, Jones DF, et al. Mycophenolic acid: an anticancer compound with unusual properties. Nature 1969; 223:848–851.
55. Epinette WW, Parker CM, Jones EL, Greist MC. Mycophenolic acid for psoriasis: a review of pharmacology, long-term efficacy, and safety. J Am Acad Dermatol 1987; 17:962–971.
56. Allison AC, Eugui EM. Preferential suppression of lymphocyte proliferation by mycophenolic acid and predicted long-term effects of mycophenolate mofetil in transplantation. Transplant Proc 1994; 26:3205–3210.
57. Langman LJ, LeGatt DF, Yatscoff RW. Pharmacodynamic assessment of mycophenolic acid-induced immunosuppression by measuring IMP dehydrogenase activity. Clin Chem 1995; 41:295–299.
58. Natsumeda Y, Carr SF. Human type I and II IMP dehydrogenases as drug targets. Ann NY Acad Sci 1993; 696:88–93.
59. Lightfoot T, Snyder FF. Gene amplification and dual point mutations of mouse IMP dehydrogenase associated with cellular resistance to mycophenolic acid. Biochim Biophys Acta 1994; 1217:156–162.
60. Nagy S, Andersson JP, Andersson UG. Effect of mycophenolate mofetil (RS-61443) on cytokine production: inhibition of superantigen-induced cytokines. Immunopharmacology 1993; 26:11–20.
61. Franklin TJ, Morris WP. Pharmacodynamics of the inhibition of GTP in vivo by mycophenolic acid. Adv Enzyme Regul 1994; 34:107–117.
62. Stet EH, De Abreu RA, Bokkerink JP, Vogels-Mentink TM, Keizer-Garritsen JJ, Trijbels FJ. Inhibition of IMP dehydrogenase by mycophenolic acid in Molt F4 human malignant lymphoblasts. Ann Clin Biochem 1994; 31:174–180.
63. Allison AC, Kowalski WJ, Muller CJ, Waters RV, Eugui EM. Mycophenolic acid and brequinar, inhibitors of purine and pyrimidine synthesis, block the glycosylation of adhesion molecules. Transplant Proc 1993; 25:67–70.
64. Warner-Bartnicki AL, Murao S, Collart FR, Huberman E. Regulated expression of the MRP8 and MRP14 genes in human promyelocyticleukemic HL-60 cells treated with the differentiation-inducing agents mycophenolic acid and 1-alpha,25-dihydroxyvitamin D3. Exp Cell Res 1993; 204:241–246.
65. Eugui EM, Almquist SJ, Muller CD, Allison AC. Lymphocyte-selective cytostatic and immunosuppressive effects of mycophenolic acid in vitro: role of deoxyguanosine nucleotide depletion. Scand J Immunol 1991; 33:161–173.
66. Kamano H, Tanaka T, Yamaji Y, et al. *E. coli gpt* gene expression effects on K562 human leukemia cell proliferation and erythroid differentiation altered by mycophenolic acid. Biochem Int 1992; 26:537–543.
67. Chang C-C, Aversa G, Punnonen J, Yssel H, de Vries JE. Brequinar sodium, mycophenolic acid, and cyclosporin A inhibit different stages of IL-4 or IL-13-induced human IgG4 and IgE production in vitro. Ann NY Acad Sci 1993; 696: 108–122.

68. Lemster B, Woo J, Wang SC, Todo S, Starzle TE, Thomson AW. Cytokine gene expression in murine lymphocytes activated in the presence of FK506, bredinin, mycophenolic acid, or brequinar sodium. Transplant Proc 1992; 24:2845–2846.
69. Gomez EC, Menendez L, Frost P. Efficacy of mycophenolic acid for the treatment of psoriasis. J Am Acad Dermatol 1979; 1:531–537.
70. Lynch WS, Roenigk HH Jr. Mycophenolic acid for psoriasis. Arch Dermatol 1977; 113:1203–1208.
71. Marinari R, Fleischmajer R, Schragger AH, Rosenthal AL. Mycophenolic acid in the treatment of psoriasis: long-term administration. Arch Dermatol 1977; 113:930–932.
72. Goldblum R. Therapy of rheumatoid arthritis with mycophenolate mofetil. Clin Exp Rheumatol 1993; 11:S117–S119.
73. St. Georgiev V. Enzymes of the purine metabolism: inhibition and therapeutic potential. Ann NY Acad Sci 1993; 685:207–216.
74. Allison A, Eugui EM. Inhibitors of de novo purine and pyrimidine synthesis as immunosuppressive drugs. Transplant Proc 1993; 25:8–18.
75. Mitchell B, Dayton JS, Turka LA, Thompson CB. IMP dehydrogenase inhibitors as immunomodulators. Ann NY Acad Sci 1993; 685:217–224.
76. Zaoui P, Serre-Debeauvais F, Bayle F, et al. Clinical use of mizoribine (Bredinin) and pharmacologic monitoring assessment in renal transplantation. Transplant Proc 1995; 27(1):1064–1065.
77. Lee HA, Slapak M, Raman GV, Mason JC, Digard N, Wise M. Mizoribine as an alternative to azathioprine in triple therapy immunosuppressant regimens in cadaveric renal transplantation: two successive studies. Transplant Proc 1995; 27(1):1050–1051.
78. Teraoka S, Toma H, Hihei H, et al. Current status of renal replacement therapy in Japan. Am J Kidney Dis 1995; 25(1):151–164.
79. Lee H, Slapak M, Venkatraman G, Mason J, Digard N, Wise M. Mizoribine as an alternative to azathioprine in triple-therapy immunosuppressant regimens in cadaveric renal transplantation. Transplant Proc 1994; 25:2699–2700.
80. Iwasaki T, Hamano T, Aizawa K, Kobayashi K, Kakishita E. A case of systemic lupus erythematosus (SLE) successfully treated with mizoribine (Bredinin). Ryumachi 1994; 34(5):885–889.
81. Igarashi Y, Moro Y, Kondo Y, Inoue CN. Steroid-sparing effect of mizoribine in long-term nephrotic. Pediatr Nephrol 1994; 8(3):396–397.
82. Xu, X, Williams JW, Bremer EG, Finnegan A, Chong AS-F. Inhibition of protein tyrosine phosphorylation in T cells by a novel immunosuppressive agent, leflunomide. J Biol Chem 1995; 270:12398–12403.
83. Zelinski, T, Seiter D, Bartlett RR. Leflunomide, a reversible inhibitor of pyrimidine biosynthesis? Inflammation Res 1995 (in press).
84. Cao W, Kao P, Xu JC, Chao A, Gardener P, Morris R. Molecular mechanism of lymphocyte-specific anti-proliferative action of leflunomide, a novel and effective immunosuppressant. FASEB J 1995; 9:A1370 (abstract).
85. Williamson RA, Yea CM, Roboson PA, et al. Dihydroorotate dehydrogenase, a high affinity binding protein for A77 1726 and mediator of a range of biological effects of the immunomodulatory compound. J Biol Chem 1995; 270(38):22467–22472.

86. Cao W, Kao P, Xu JC, Chao A, Gardener P, Morris R. Mechanism of antiproliferative action of leflunomide. A77 1726, the active metabolite of leflunomide, does not block T cell receptor-mediated signal transduction but its antiproliferative effects are antagonized by pyrimidine nucleosides. J Heart Lung Transplant 1995 (in press).
87. Nikcevich DA, Finnegan A, Chong AS, William JW, Bremer EG. Inhibition of interleukin 2 (IL-2)-stimulated tyrosine kinase activity by leflunomide. Agents Actions 1994; 41:C279–C282.
88. Thoenesm GE, Sitter T, Langer KH, Bartlett RR, Schleyerbach R. Leflunomide (HWA 486) inhibits experimental autoimmune tubulointerstitial nephritis in rats. Int J Immunopharmacol 1989; 11:921–929.
89. Zelinski T, Zeitter D, Mullner S, Bartlett RR. Leflunomide, a reversible inhibitor of pyrimidine biosynthesis? Agents Actions 1995 (in press).
90. Morris RE, Huang XF, Shorthouse R, Reichenspurner H, Adams B, Berry G. Studies in experimental models of chronic rejection: use of rapamycin (sirolimus) and isoxazole derivatives (leflunomide and its analogues) for the suppression of graft vascular disease and obliterative bronchiolitis. Transplant Proc 1995; 27:2068–2069.
91. Morris RE, Huang XF, Shorthouse R, Reichenspurner H, Adams B, Berry G. Use of cyclosporine (CsA), mycophenolic acid (MPA), rapamycin (RPM), leflunomide (LFM) or deoxyspergualin (DSG) for prevention and treatment of obliterative airway disease (OAD) in new animal models. J Heart Lung Transplant 1995; 14:S65.
92. Morris R, Huang X, Gregory CR, et al. Studies in experimental models of chronic rejection: use of rapamycin (sirolimus) and isoxazole derivatives (leflunomide and its analogue) for the suppression of graft vascular disease and obliterative bronchiolitis. Transplant Proc 1995; 27:2068–2069.
93. Bartlett RR, Schleyerbach R. Immunopharmacological profile of a novel isoxazol derivative, HWA486 with potential antirheumatic activity. I. Disease modifying action on adjuvant arthritis of the rat. Int J Immunopharmacol 1985; 7:7–18.
94. Bartlett RR, Anagnostopulos H, Zielinski T, Mattar T, Schleyerbach R. Effects of leflunomide on immune responses and models of inflammation. Springer Semin Immunopathol 1993; 14:381–394.
95. Thoenes GH, Sitter T, Langer KH, Bartlett RR, Schleyerbach R. Leflunomide (HWA 486) inhibits experimental autoimmune tubulointerstitial nephritis in rats. Agents Actions 1989; 11:921–929.
96. Smith-Lang L, Glaser RL, Miller ST, et al. Efficacy of novel immunomodulators leflunomide and rapamycin in autoimmune uveitis. FASEB J 1992; (Part I):A1048.
97. Popovic S, Bartlett RR. The use of the murine chronic graft versus host (CGvH) disease, a model for systemic lupus erythematosus (SLE), for drug discovery. Agents Actions 1987; 21:284–286.
98. Mladenovic V, Domljam Z, Rozman B, et al. Safety and effectiveness of leflunomide in the treatment of patients with active rheumatoid arthritis—results of a randomized double blind placebo controlled Phase II trial. Arth Rheum 1995; 38:1595–1603.
99. Forrest T, Ware RE, Howard T, Jaffee BD, Denning SM. Novel mechanisms of brequinar sodium immunosuppression of T cell activation. Transplantation 1994; 58:920–926.

100. Makowka L, Sher L, Cramer DV. The development of brequinar as an immunosuppressive drug for transplantation. Immunol Rev 1993; 136:51–70.
101. de Forni M, Chabot GG, Armand J-P, et al. Phase I and pharmacokinetic study of brequinar (DUP 785; NSC 368390) in cancer patients. Eur J Cancer 1993; 29A:983–988.
102. Steele DM, Hullett DA, Bechstein WO, et al. Effects of immunosuppressive therapy on the rat aortic allograft model. Transplant Proc 1993; 25:754–755.
103. Allison AC, Eugui EM. Inhibitors of de novo purine and pyrimidine synthesis as immunosuppressive drugs. Transplant Proc 1993; 25:8–18.
104. Sher LS, Eiras-Hreha G, Kornhauser DM, et al. Safety and pharmacokinetics (PK) of brequinar sodium (BQR) in liver allograft recipients on cyclosporine (CYA) and steroids. Hepatology 1993; 18:746 (abstract).
105. Simon P, Townsend RM, Harris EA, Jones EA, Jaffee BD. Brequinar sodium: inhibition of dihydroorotic acid dehydrogenase, depletion of pyrimidine pools, and consequent inhibition of immune functions in vivo. Transplant Proc 1993; 25:77–80.
106. Peters GJ, Schwartsmann G, Nadal JC, et al. In vivo inhibition of the pyrimidine de novo enzyme dihydroorotic acid dehydrogenase by brequinar sodium (DUP-785; NSC 368390) in mice and patients. Cancer Res 1990; 50:4644–4649.
107. Nair RV, Cao WW, Morris RE. Inhibition of murine smooth muscle cell proliferation in vitro by leflunomide, a new immunosuppressant, is antagonized by uridine. Immunol Lett 1996 (in press).
108. Zaruba F, Kuta A, Ellis J. Treatment of mycosis fungoides with 6-azauridine. Lancet 1963; 275.
109. Welch AD, Handschumacher RE, Finch SC, Jaffe JJ, Cardoso SS, Calabresi P. A synopsis of recent investigations of 6-azauridine (NSC-32074). Cancer Chemother Rep 1960; 9:39–46.
110. Rada B, Doskocil J. Azo-pyrimidine nucleosides. Pharmacol Ther 1980; 9:171–217.
111. Skoda J. Azapyrimidine nucleosides. In: Sartorelli AC, Johns DG, eds. Handbuch der Experimentellen Pharmacologie. Berlin: 1975:347–372.
112. Handschumacher RE, Pasternak CA. Inhibition of orotidylic acid decarboxylase, a primary site of carcinostasis by 6-azauracil. Biochim Biophys Acta 1958; 30: 451–452.
113. Calabresi P, Turner RW. Beneficial effects of triacetyl-azauridine in psoriasis and mycosis fungoides. Ann Intern Med 1966; 64:352–371.
114. Levine S, Paulus HE. Treatment of psoriatic arthritis with azaribine. Arthritis Rheum 1976; 19:21–28.
115. Dubin HV, Harrell ER. Azaribine in the treatment of psoriasis. Ann Intern Med 1971; 74:797–798.
116. McDonald CJ. Tried and true and new therapy in psoriasis. Consultant 1973; 64–65.
117. Drell W, Welch AD. Azaribine-homocystinemia-thrombosis in historical perspective. Pharmacol Ther 1989; 41:195–206.
118. Drell W, Welch AD. Azaribine-homocystinemia-thrombosis in historical perspective. Pharmacol Ther 1989; 41:195–206.

119. De Groot PG, Willems C, Boers GHJ, Gonsalves MD, Van Aken WG, Van Mourik JA. Endothelial cell dysfunction in homocystinuria. Eur J Clin Invest 1983; 13: 405–410.
120. Clarke R, Daly L, Robinson K, et al. Hyperhomocysteinemia: an independent risk factor for vascular disease. N Engl J Med 1991; 324:1149–1155.
121. Gitel SN, Grieco AJ, Wessler S, Snyderman SE. The thrombogenicity of 6-azauridine. Haemostasis 1979; 8:54–57.
122. Grieco AJ, Homocystinuria: pathogenetic mechanisms. Am J Med Sci 1977; 273:120–132.
123. Ueland PM, Refsum H. Plasma homocysteine, a risk factor for vascular disease: plasma levels in health, disease and drug therapy. J Lab Clin Med 1989; 114: 473–501.
124. Johnson JT, Ferretti GA, Nethery WJ, et al. Oral pilocarpine for post-irradiation xerostomia in patients with head and neck cancer. N Engl J Med 1993; 329:390–395.
125. Keefer RA, Roenigk HH Jr, Hawk WA. Azaribine therapy for psoriasis: evaluation of potential effects on the liver and other organ systems. Arch Dermatol 1975; 111:853–856.
126. Vogler WR, Olansky S. A double-blind study of azaribine in the treatment of psoriasis. Ann Intern Med 1970; 73:951–956.
127. Fox RI, Pintea G, Vasilco R, Paslaru L. Treatment of rheumatoid arthritis (RA) with the pyrimidine synthesis inhibitor triacetyl-6-azauridine plus pyridoxine. Arthritis Rheum 1992; 155:S67.

21

Leflunomide

A New Immunosuppressive Drug

DAVID L. SCOTT
*King's College School of Medicine and Dentistry and
Kings College Hospital, London, England*

VIBEKE STRAND
Stanford University School of Medicine, Stanford, California

I. INTRODUCTION

Leflunomide is a new immunosuppressive drug effective both in experimental models of transplantational and autoimmune diseases and in clinical trials in patients with rheumatoid arthritis (RA). Leflunomide is a low-molecular-weight (270 kDa) isoxasole compound. It is rapidly converted by the opening of the isoxasole ring to produce the active metabolite, A77 1726.

II. MECHANISM OF ACTION

In vitro studies have shown that the active metabolite of leflunomide (A77 1726) inhibits the proliferation of mononuclear cells in a dose-dependent fashion. It reduces the proliferation of human blood mononuclear cells and a variety of transformed human and murine cell lines induced by lipopolysaccharide (LPS), concanavalin A, and phytohemagglutinin (PHA).

The antiproliferative effect of the leflunomide metabolite is due to a change in the de novo synthesis of pyrimidines and is reversed by the addition of uridine to in vitro culture systems. It is also partially reversed by the addition of cytidine. It appears that the metabolite binds to a key enzyme in the pyrimidine pathway, which is termed dihydro-orotate dehydrogenase. As a consequence of enzyme inhibition, there is a reduction in uridine triphosphate (UTP) levels, and to a lesser extent other high-energy phosphates (CTP, ATP, and GTP). In vivo pyrimidine synthesis by lymphocytes and other rapidly dividing cells may therefore be reduced, leading to changes in DNA and RNA synthesis, cell proliferation, and glycosylation.

There may be other actions. There is some evidence that the active metabolite of leflunomide also affects tyrosine kinase. For example, A77 1726 inhibits epidermal growth factor–stimulated proliferation of cultured skin fibroblasts and it also reduces phosphorylation of the epidermal growth factor receptor. In addition, the metabolite of leflunomide may inhibit adhesion lymphocytes to vascular endothelium. Leflunomide is weakly active or inactive in models of inflammation where cyclo-oxygenase or lipoxygenase inhibitors show potent anti-inflammatory effect, but does inhibit the inducible cyclo-oxygenase-2 pathway in stimulated peripheral blood mononuclear cells.

III. EXPERIMENTAL MODELS

Leflunomide has been studied in a variety of experimental models including: adjuvant arthritis, proteoglycan-induced arthritis, collagen arthritis, the MRL/lpr model of SLE, experimental allergic encephalomyelitis, experimental interstitial nephritis, anti-GBM glomerulonephritis, and experimental autoimmune uveitis. In most instances it has been used for prophylaxis, but treatment has also been successful. These features suggest that leflunomide has the propensity to act as an immunomodulator.

This agent has also been tested in models of inflammatory responses, such as the carrageenan paw edema assay and the mouse arachidonic acid ear edema assay. In these models leflunomide is weakly active, suggesting it has slight anti-inflammatory activity, which is less than its immunomodulatory actions.

Detailed information about the mode of action of leflunomide and its effect in experimental models has been provided by Bartlett et al. (1–4).

IV. STUDIES IN HEALTHY VOLUNTEERS

A number of studies have been carried out in healthy volunteers to evaluate the safety, pharmacokinetics, and metabolism of leflunomide in humans. These investigations have included administration of radiolabeled leflunomide, single and multiple dosing studies, and evaluation of the effects of cholestyramine or charcoal on its elimination. When ^{14}C-leflunomide as a single oral dose of 100 mg was administered to three healthy volunteers, approximately 90% was recovered in urine and feces by 28 days. Elimination of the drug occurred slowly with a terminal half-life of 7–8 days. Multiple dosing studies using 100-mg tablets showed that plasma levels accumulated during this time period. The plasma half-life of the active metabolite varied considerably between individuals and the average value was about 11 days. In longer-term studies in patients with rheumatoid arthritis (RA), it was determined that the plasma half-life varied between 14 and 16 days. When leflunomide is taken with food, its absorption is not significant.

Oral administration of either cholestyramine or charcoal led to a rapid clearance of the active metabolite of leflunomide (A77 1726) from the plasma, with reduction of plasma levels by 48–76% within 24 hr. This is a way of dealing with any overdose or adverse effect from the drug.

V. AN OPEN DOSE RANGING STUDY

In the first study 23 patients with active RA were given a loading dose of 50 mg or 100 mg followed by one of three daily dose levels of leflunomide (5 mg, 100 mg, and 25 mg). Treatment lasted 6 weeks. This approach of a loading dose followed by regular daily dosing gave acceptable plasma levels, though a steady state was not reached. The open study paved the way for a randomized placebo-controlled trial.

VI. EFFICACY IN A RANDOMIZED CONTROLLED TRIAL

The most important prospective randomized clinical trial on leflunomide examined 402 patients with active RA (5). There were 398 evaluable cases. There was a placebo group of 102 cases. Patients on active therapy has a loading dose of 50 mg leflunomide followed by daily dosing with 5 mg (95 cases), 100 mg followed by 10 mg daily (100 cases), and 200 mg followed by 25 mg daily (101 cases). Therapy continued for 6 months.

The key efficacy variables included the number of tender joints, the number of swollen joints, pain, patient's and physician's global assessments, erythrocyte sedimentation rate (ESR), C-reactive protein, and the health assessment questionnaire (all in the OMERACT core data set). Other measures included the duration of morning stiffness, grip strength, and rheumatoid factor titer.

The main results are summarized in Tables 1 and 2, including mean changes in individual clinical variables over 6 months as well as the number of responders in both general response indices (the Paulus criteria and the preliminary response criteria of the American College of Rheumatology) and specific measures (conversion to seronegative for rheumatoid factor and return of C-reactive protein to normal).

There were similar improvements in the 10-mg and 25-mg treatment groups in both individual clinical variables and general response indices. By contrast, 5 mg daily of leflunomide was not markedly different from placebo, although the

Table 1 Main Changes in Clinical and Laboratory Variables in a Placebo-Controlled Prospective Trial of Leflunomide

Variable	Placebo ($n = 102$)	5 mg ($n = 95$)	10 mg ($n = 100$)	25 mg ($n = 101$)
Tender joint count	−9.7	−10.5	−13.6	−16.5*
Swollen joint count	−6.5	−7.6	−10.4*	−11.7*
Pain	0.4	0.5	−0.9*	−1.0*
Patient's global assessment	0.5	0.6	1.1*	1.0*
Physician's global assessment	0.6	0.7	1.1*	1.1*
ESR	3.1	4.2	−5.2*	−5.4*
C-reactive protein	5.3	2.4	−14.9*	−9.5*
Health Assessment Questionnaire	−8.1	−5.8	−14.5*	−13.6*

*Significant change compared to placebo ($p < 0.05$).

placebo response was high. Overall 25 mg seemed marginally more effective than 10 mg. The general impression of these results is that leflunomide is effective in treating active RA, that it improves symptoms of joint tenderness, joint swelling, pain, and global disease activity, and that its effect on measures of the acute-phase response (ESR and C-reactive protein) is less marked.

There have been other small prospective randomized studies of RA patients given similar active dose levels without a placebo treatment group (6). The initial 6-month studies have also been extended for longer periods of observation. Findings in these studies are supportive of the main conclusions from the larger investigation of 402 patients. In total 500 patients with active RA have received leflunomide: 414 for at least 6 months and 146 for at least 2 years.

Table 2 Overall Responses to Therapy in a Placebo-Controlled Prospective Trial of Leflunomide

Variable	Placebo ($n = 102$)	5 mg ($n = 95$)	10 mg ($n = 100$)	25 mg ($n = 101$)
Paulus criteria improve by ≥20%	25	32	56*	59*
ACR criteria improve by ≥20%	31	30	52*	60*
Rheumatoid arthritis positivity converts to negative	7	5	14*	19*
C-reactive protein becomes negative	14	9	26	32

*Significant change compared to placebo ($p < 0.05$).

The time course of improvement with leflunomide shows evidence of an effect at 4 weeks. Improvement continues until 16–26 weeks when it stabilizes. Patients respond to leflunomide whether or not they are taking additional therapy with corticosteroids or nonsteroidal anti-inflammatory drugs. They also respond both early and late in the course of their arthritis.

VII. ADVERSE EFFECTS

Analysis of adverse effects in all patients with RA given leflunomide, in both randomized and open-label studies, showed that a variety of reactions occurred that appeared to be dose-related. Reversible alopecia appears to be strongly related to treatment with leflunomide but occurred in less than 10% of all treatment groups. Hypertension and dizziness may be related, especially at higher dosages, but it is difficult to interpret these in a patient population with preexisting hypertensive cardiovascular disease. Allergic reactions, such as rashes and pruritus, occur. There are a number of gastrointestinal adverse effects including abdominal pain, anorexia, dyspepsia, nausea, vomiting, diarrhea, and weight loss. The incidence of these reactions is summarized in Table 3. In addition five cases of vasculitis have been reported in patients taking leflunomide. These include lateral malleolus ulcers, periungual infarcts, digital infarction, digital gangrene, and nodular pulmonary lesions. The relationship between this pattern of vasculitis, a known complication of severe seropositive RA, and leflunomide therapy is uncertain.

There is always a risk that an immunosuppressive drug such as leflunomide can lead to infections. Overall 69 patients have had an infection while treated with leflunomide, but there was no evidence of more infections nor in the type of infections than with placebo therapy. At present, caution is warranted but the risk for infections does not appear excessive.

Table 3 Incidence of Main Adverse Reactions in Clinical Studies with Leflunomide

Adverse reaction	Placebo	5 mg	10 mg	25 mg
All gastrointestinal reactions	3	15	10	12
Weight loss	2	2	4	4
Hypertension	5	3	4	11
Dizziness	0	1	0	3
Rashes and allergic reactions	5	6	4	8
Alopecia	1	1	1	7

Leflunomide is teratogenic in rabbits and rats and should not be used in pregnant women. It must be used with caution in patients with hepatic dysfunction as the active metabolite is extensively protein-bound and is cleared by biliary as well as renal excretion. It should be used with caution and a possible dose reduction in the elderly. It should not be given to adolescents or children. Administration of cholestyramine will rapidly reduce plasma levels of leflunomide within 24–48 hours.

There appears to be a transient dose-related elevation of liver transaminases (ALT, AST) and, to a lesser extent, alkaline phosphatase with leflunomide. Caution is needed when monitoring patients receiving 25 mg daily. In seven cases there has been transient thrombocytopenia, which resolved with continued drug administration. Leukopenia has not been seen.

VIII. CONCLUSIONS

The current results, at the end of the phase II program with leflunomide, show it to be an apparently effective and safe immunomodulatory agent for the treatment of active RA. It can be used together with nonsteroidal anti-inflammatory drugs and corticosteroids. Three questions now need to be addressed: How effective is leflunomide? How safe is leflunomide? Can it be combined with current slow-acting antirheumatic drugs? Studies are currently ongoing throughout Europe and North America that should define its value. These will take several years before they reach fruition. At present, the omens look good, and we must hope the drug fulfills its early promise.

REFERENCES

1. Bartlett RR, Schleyerbach R. Immunopharmacological profile of a novel isoxazol derivative, HWA486, with potential anti-rheumatic activity. I. Disease modifying action on adjuvant arthritis of the rat. Int J Immunopharmacol 1985; 7:7–18.
2. Bartlett RR, Mattar T, Weithmann U, Anagnostopulos H, Popovic S, Schleyerbach R. Leflunomide (HWA 486): a novel immunorestoring drug. In: Lewis AJ, Doherty NS, Ackerman NR, eds. Therapeutic Approaches to Inflammatory Diseases. New York: Elsevier, 1989:215–228.
3. Bartlett RR, Dimitrijevic M, Mattar T, Zielinski T, Germann T, Rüde E, Thoenes GH, Küchle CCA, Schorrlemmer H-U, Bremer E, Finnegan A, Schleyerbach R. Leflunomide (HWA 486), a novel immunomodulating compound for the treatment of autoimmune disorders and reactions leading to transplantation rejection. Agents Actions 1991; 32:10–21.
4. Bartlett RR, Anagnostopulos H, Zielinski T, Mattar T, Schleybach R. Effects of leflunomide on the immune response and models of inflammation. Springer Semin Immunopathol 1993; 14:381–394.

5. Mladenovic V, Domlian Z, Rozman B, Jajic J, Mihajlovic D, Dordevic J, Popvic M, Dimitrijevic M, Zivkovic M, Campion G, Misikic P, Löwe-Freidrich I, Oed Ch, Seifert H, Strand V. Safety and effectiveness of leflunomide in the treatment of patients with active rheumatoid arthritis. Arthritis Rheum 1995; 38:1595–1603.
6. Leflunomide Investigator's Brochure. Weisbaden, Germany: Hoechst AG, 1995.

VII
Future Directions

22
An Assessment of Novel Agents in the Treatment of the Rheumatic Diseases

VIBEKE STRAND
Stanford University School of Medicine, Stanford, California

LEE S. SIMON
Harvard Medical School, Deaconess Hospital, and Dana-Farber Cancer Institute, Boston, Massachusetts

We have attempted to provide the reader with as complete a review as possible of novel therapeutic interventions for autoimmune diseases. Although all of these therapeutics are not biologic agents, most have been designed to specifically interfere within the inflammatory cascade and thereby alter disease progression. However, as quickly as clinical information about these agents becomes available many have already been proven ineffective. Several years are required to bring a promising new treatment to the clinic; hypotheses about its mechanism of action may be superseded by more recent research, even before the first patient is recruited into the first clinical trial. Although many of these interventions have not resulted in significant or sustained clinical benefit, they have served as immunologic probes, revealing previously unrecognized aspects of the underlying disease pathophysiology.

The potential advantages of immunologically active agents in the treatment of autoimmunity include a rapid onset and increased specificity of effect without concomitant immunosuppression. Alternatively, the apparent specificity evident in syngeneic animal models of disease and in vitro assays has rarely translated into meaningful clinical effects in established disease. This failure exemplifies the redundancy of immunologic and cytokine networks. Inhibition of a proinflammatory cytokine, either by prevention of receptor binding or effector cell function may offer transient clinical benefit. Frequently, breakthrough occurs, abrogating clinical effect, either due to an immune response to the administered agent, or more commonly, redundancy in the inflammatory process, thereby bypassing the biologic effect through another cytokine or effector pathway.

As in cancer therapy, recently developed immunologically active agents may serve as induction therapy to gain control of an autoimmune process, allowing subsequent use of less toxic, chronically administered treatments. Acutely, they may ameliorate disease exacerbations, or "flares." Combinations of agents to specifically affect pro-inflammatory cytokine secretion, such as TNFα and IL-1 may offer significant clinical benefit by intervening at several points in the inflammatory process (1).

There are disadvantages to the use of these newer biologically active agents. Products targeted to specific aspects of the immune response may be less toxic, but initial clinical effects may be of short duration, and not maintained with readministration or chronic therapy. Other problems include systemic administration, frequently parenterally and in large doses, which pose additional costs and difficulties to the patient. Substantial costs of treatment, well beyond the hidden costs of developing the new agent, may be difficult to justify.

Long-term toxicities may argue against even dramatic transient benefits. Treatments which resulted in prolonged peripheral CD4 T cell depletion (chimeric anti CD4 monoclonal antibodies (MAbs); CAMPATH 1H MAb) have raised serious concerns about late-occurring infections and secondary malignancies (2–10). Many of the newer and possibly more effective agents, including the IL-1 receptor antagonist (IL-1ra), CAMPATH 1H, and the anti TNFα MAbs, have been associated with a higher incidence of acute treatment-associated infection; their clinical significance is difficult to determine (11–15). It is equally difficult to attribute development of secondary malignancies to these treatments or to the underlying disease process. There are now 15 cases of non-Hodgkin's lymphoma (NHL) and Hodgkin's lymphoma (HL) reported in RA patients receiving methotrexate (16–20), as well as 2 cases of NHL after CAMPATH 1H administration (7), 2 of NHL and one HL following treatment with anti TNFα MAbs (21).

Autoimmune manifestations have also occurred after administration of biologic agents in autoimmune disease, similar to the experience reported following interleukin-2 (IL-2), interferon, granulocyte colony stimulating factor (G-CSF) and granulocyte-monocyte colony stimulating factor (GM-CSF) treatment of cancer or after transplantation. Worsened manifestations of autoimmune disease have been reported following G- and GM-CSF treatment, including vasculitis (22–30), TTP, hemolytic uremic syndrome and vasculitis after CAMPATH-1H (7, and Yocum, personal communication), and an increased incidence of antinuclear antibodies and anti DNA antibodies after anti TNFα MAbs (21,31).

Where should we go from here? Following our experience with MAbs targeted to T cell surface antigens (CD7, anti CD5 immunoconjugate, chimeric anti CD4

MAbs), it is clear that prospective randomized, placebo controlled clinical trials are of critical importance in identifying the potential clinical benefit of these agents. These must be performed early in the clinical development process if sponsors are to accurately assess the potential promise of a new therapeutic. They must be large enough to have sufficient power to determine if the agent can offer clinical benefit. Well-designed and detail-oriented dosing and dose scheduling work are essential, as in vitro and in vivo preclinical data in animals may not translate directly to clinical effects in humans. Initial trials must be conducted in patients who have failed other available treatments and therefore are likely to have refractory and/or longstanding disease. Nonetheless, it is imperative to study a promising agent in a patient population with earlier and potentially responsive disease, once safety has been demonstrated. Most trials have been of short duration; in a disease of twenty to thirty years, 6 months time does little to demonstrate effects of specific therapies on the disease. Outcome measures must be standardized to facilitate comparisons across disease populations and therapeutic agents. Because of known delayed toxicities, long-term follow-up will be required for patients enrolled in clinical trials of these promising immunologically active agents (32,33).

Future treatments should be designed to alter specific disease processes and not be generally immunosuppressive, thereby resulting in as little toxicity as possible. Their use may be required in the setting of "background" therapy with current second-line agents such as methotrexate. There may be few or no comparator treatments with similar onset or duration of effect. Therefore, large multicenter trials, employing placebo as well as active controls, will likely be necessary before the clinical effect of a new therapeutic can be accurately assessed. Finally, these new therapies will be required to demonstrate their pharmacoeconomic justification (34). Although today this appears to be a rather tall order, the proliferation of potential new therapeutic agents for the treatment of autoimmune diseases suggests this will be a goal we will achieve in the foreseeable future.

REFERENCES

1. Strand V. The future use of biologic agents in combination for treatment of rheumatoid arthritis. J Rheumatol 1996; 23:91–96.
2. Moreland LW, Pratt PW, Bucy RP, et al. Treatment of refractory rheumatoid arthritis with a chimeric anti-CD4 monoclonal antibody. Long term follow up of CD4+ cell counts. Arth Rheum 1994; 37:834–838.
3. Moreland LW, Pratt PW, Mayes MD, et al. Double blind placebo controlled multicenter trial using chimeric monoclonal anti-CD4 antibody, cM-T412, in RA patients receiving concomitant methotrexate. Arth Rheum 1995; 38:1581–1588.

4. Van der Lubbe PA, Dijkmans BAC, Markusse HM, et al. A randomized double blind placebo controlled study of CD4 monoclonal antibody therapy in early RA. Arth Rheum 1995; 38:1097–1106.
5. Johnston J, Spreen W. Treatment of rheumatoid arthritis with humanized monoclonal antibody CAMPATH-1H. In: Strand V, Simon L, Pillemer S. eds. Biologic Agents In Autoimmune Diseases III. Arthritis Foundation 1994:55–66.
6. Weinblatt ME, Coblyn J, Maier A, et al. Continued lymphocyte suppression following single dose mAb therapy with CAMPATH 1H: A 20 month follow up. Arth Rheum 1994; 37:S420.
7. Weinblatt, ME. The experience with CAMPATH 1H. In: Strand V, Simon L, Pillemer S, eds. Biologic Agents In Autoimmune Diseases IV. Arthritis Foundation, Atlanta: 1996:161–168.
8. Matteson EL, Yocum DE, St Clair EW, et al. Treatment of active refractory RA with humanized monoclonal antibody CAMPATH-1H administered by daily subcutaneous injection. Arth Rheum 1995; 38:1187–1193.
9. Jendro MC, Ganten T, Mateson EL, et al. Emergence of oligoclonal T cell populations following therapeutic T cell depletion in RA. Arth Rheum 1995; 38:1242–1251.
10. Weinblatt ME, Maddison PJ, Bulpitt KJ, et al. CAMPATH 1H, a humanized monoclonal antibody, in refractory RA. Arth Rheum 1995; 38:1589–1594.
11. Lebsack ME, Paul CC, Martindale JJ, Catalano MA. A dose and regimen ranging study of IL-1 receptor antagonist in patients with rheumatoid arthritis. Arth Rheum 1993; 36:S39.
12. Elliot MJ, Maini RN, Feldmann M, et al. Randomized double-blind comparison of chimeric monoclonal antibody to tumor necrosis factor a (cA2) versus placebo in rheumatoid arthritis. Lancet 1994; 344:1105.
13. Elliot MJ, Maini RN, Feldmann M, et al. Repeated therapy with monoclonal antibody to tumor necrosis factor a (cA2) in patients with rheumatoid arthritis. Lancet 1994; 344:1127.
14. Rankin ECC, Choy EHS, Kassumos D, Kingsley GH, Sopwith AM, et al. The therapeutic effects of an engineered anti-TNF α antibody (CDP 571) in rheumatoid arthritis. Brit J Rheumatol 1995; 34:334–342.
15. Rankin ECC, Choy EHS, Sopwith M, et al. Repeated doses of 10 mg/kg of an engineered human anti-TNFα, CDP571, in RA patients are safe and effective. Arth Rheum 1995; 38:S185.
16. Shiroky JF, Frost A, Skelton JD, et al. Complications of immunosuppression associated with weekly low dose methotrexate. J Rheumatol 1991; 18:1172–1175.
17. Ellman MH, Hurwitz H, Thomas C, et al. Lymphoma developing in a patient with RA taking low dose weekly methotrexate. J Rheumatol 1991; 18:1741–1743.
18. Kamel OW, van de Rijn M, Weiss LM, et al. Brief report: reversible lymphomas associated with EBV occurring during methotrexate therapy for RA and dermatolyositis. NEJM 1993; 328:1317–1321.
19. Zimmer-Galler I, Lie JT. Choroidal infiltrates as the initial manifestation of lymphoma in RA after treatment with low dose methotrexate. Mayo Clin Proc 1994; 69:258–261.

20. Davies JMS, Kremer JM, Furst DE, et al. Lymphomatous changes during methotrexate therapy. Arth Rheum 1995; 38:S2004.
21. Maini RN, Elliott MJ, Long-Fox A, et al. Clinical response of RA to anti-TNFα (cA2) monoclonal antibody is related to administered dose and persistence of circulating antibody. Arth Rheum 1995; 38:S186.
22. Schleisser G, Pralle H, Lohmeyer J. Leukocytoclastic vasculitis complicating G-CSF induced neutrophil recovery in Tg-lymphocytosis with severe neutropenia. Ann Hematol 1992; 65:151–152.
23. Sawamura M, Sakura T. Exacerbation of monoclonal gammopathy in a patient treated with G-CSF. Letter. Annals Int Med 1993; 118:318.
24. Wun T. The Felty syndrome and G-CSF associated thrombocytopenia and severe anemia. Letter. Annals Int Med 1993; 118:318–319.
25. Jain KK. Cutaneous vasculitis associated with G-CSF. J Am Acad Dermatol 1994; 31:213–215.
26. Schwab UM, Corzillius M, Harten P, et al. Treatment of SLE associated neutropenia with granulocyte colony stimulating factor. Arth Rheum 1995; 38:S303.
27. Canvin JMG, Crilly A, Field M. G-CSF treatment associated with cytokine level elevation and exacerbation of arthritis. Arth Rheum 1995; 38:S355.
28. Hazenberg BPC, Van Leeuwen MA, Van Rijswijk MH, et al. Correction of granulocytopenia in Felty's syndrome by GM-CSF. Simultaneous induction of IL-6 release and flare up of the arthritis. Letter. Blood 1989; 74:2769–2770.
29. Joseph G, Neustadt DH, Hamm J, et al. GM-CSF in the treatment of Felty syndrome. Am J Hematol 1991; 37:55–56.
30. Hoekman K, vonBlomberg-van der Flier BME, Wagstaff, et al. Reversible thyroid dysfunction during treatment with GM-CSF. Lancet 1991; 338:541–542.
31. Rankin E, Choy E, Ehrenstein M, et al. Serological effects of repeated doses of an engineered human anti-TNFα antibody, CDP571, in patients with rheumatoid arthritis (RA). Arth Rheum 1995; 38:S279.
32. Strand V, Scott DL, Panayi GS. Evaluating biologic agents in rheumatoid arthritis: a framework for clinical trials. Editorial. J Rheumatol 1994; 21:1390–1392.
33. Strand V. Issues in clinical trials of biologic agents. In: Brooks PM, Furst D. Innovative Treatment Approaches for Rheumatoid Arthritis. Balliere's Clin Rheumatol, vol 9, number 4, Balliere Tyndall, London, 1995:825–835.
34. Strand V. OMERACT II: The biologics perspective. J Rheumatol 1995; 22:1415–1417.

Index

Adjuvant arthritis (AA), oral tolerance in treatment of, 211
Adverse effects of therapeutic agents:
   anti-CD5/RTA immunoconjugate therapy, 15
   anti-ICAM-1 MAb therapy, 160–163
   CAMPATH-1H therapy, 79–80, 296
   in treatment of RA patients, 90
   IFN-Γ in RA, 100
   IVIg, 245–246
   leflunomide, 291-292
Amyotrophic lateral sclerosis, IVIg in treatment of, 235, 241
Anti-CD5/ricin A chain immunoconjugate, 11–24
   anti-CD5/ricin A chain immunoconjugate, 12–15
   CD5 IC-mechanism of action, 13–14
   immunoconjugate toxicity, 14–15
   clinical experience with CD5 IC, 15–21
   controlled trial of CD5 IC, 20–21
[Anti-CD5/ricin A chain immunoconjugate]
   [clinical experience with CD5 IC]
   immunological responses to therapy, 19
   pharmacokinetics, 19–20
   retreatment, 18
   toxicity, 17–18
   uncontrolled clinical trials in RA, 15-16
   future directions for CD5 IC, 21–22
   side effects associated with, 15
Antigen-specific immunotherapy, 205–215, 224–225
   mechanisms of oral tolerance, 206–209
   overview, 205–206
   treatment of organ-specific autoimmune diseases in animals, 209–212
   treatment of autoimmune diseases in humans, 212–215
Antiglobulin responses to cA2, 117

Anti-TNF-α antibody (CDP571), 124–127
  changes in disease activity in placebo,
    1 mg/kg and 10 mg/kg CDP571-
    treated patients, 126
  clinical experience, 124–125
  clinical results, 125–126
Anti-TNF-α MAb, 296
Arthritis Impact Measurement Scales
    (AIMS2), 4
Articular chondrocytes, effects of IFN-Γ
    on metabolism of, 98–99
Assessment of novel agents, 295–299
Autoimmune hematological diseases,
    IVIg in treatment of, 239–240
  autoimmune neutropenia hemolytic
    anemia, 239
  factor VIII inhibitors, 240
  other diseases, 240
Autoimmune neurological diseases, IVIg
    in treatment of, 240–242
  chronic inflammatory demyelinating
    polyneuropathy, 241
  Guillain-Barré syndrome, 235, 241
  multiple sclerosis, 235, 241–242
  myasthenia gravis, 235, 240
  other diseases, 242
Autoreactive T cells, activation of,
    204–205
Azaribine (triacetyl-6-azauridine), 261,
    270, 276–278
Azathioprine, 70, 261, 268–269, 270

Biological agents, evaluation of, 1–10
  data analysis, 7
  disease assessment, 3–6
    articular indices, 3–4
    imaging, 5–6
    patients' views, 4–5
    simple clinical endpoints and com-
      bined indices, 5
  outline of clinical trials methodology
    with biologics, 7, 8
  study design, 2–3
  surrogate laboratory markers, 6–7
Brequinar, 261, 270, 275–276

CAMPATH-1H therapy, 75–82
  clinical studies using, 76–79
    biological activity, 76–77
    clinical studies in RA, 77–78
    multiple sclerosis, 79
    systemic vasculitis, 78–79
  in RA, 83–93
    CAMPATH-1H antigen, 84
    clinical experience in RA, 85–89
    clinical side effects in RA patients,
      89
    humanized CAMPATH-1H anti-
      body, 84
    preclinical data, 84–85
  side effects in, 79–80, 296
CD4 monoclonal antibody therapy,
    55–64
  CD4 MAb as treatment of RA,
    59–60
  immunopharmacology of CD4 MAb,
    57–59
  rationale for CD4 MAb as anti-
    arthritic agents, 56–57
  selection of CD4 MAb, 57
CD4+ T cells, 41–42
  effects of CE9.1 on, 69–73
  effects of chimeric anti-CD4 antibody
    on, 46–47
  in pathogenesis of RA, 42–44, 66–67
  role in inflammation characteristics of
    RA, 155
sCD4/sCD8 ratio, 6
CDP571 (engineered human anti-body
    TNF-α antibody), 124–127
  changes in disease activity in placebo,
    1 mg/kg and 10 mg/kg CDP571-
    treated patients, 126
  clinical experience, 124–125
  clinical results, 125–126
CE9.1 (primatized monoclonal anti-
    body), 65–74
  animal studies, 67–69
  CE9.1 anti-CD4, 67
  effects on CD4 cells and other
    immune functions, 69–73

Index

[CE9.1 (primatized monoclonal antibody)]
  phase I single-dose, dose-finding trial in human RA, 69
  role of CD4 cells in RA, 66–67
Childhood epilepsy, IVIg in treatment of, 242
Chimeric anti-CD4 antibody, 41–53
  experiences for treatment of RA, 44–46
Chlorambucil, 70
Chronic inflammatory demyelinating polyneuropathy (CIDP), IVIg in treatment of, 241
Collagen-induced arthritis, oral tolerance in treatment of, 211
Connective tissue target cells in RA, effects of IFN-Γ on, 98–99
  effect on metabolism of articular chondrocytes, 98–99
  effects on metabolism of synovial fibroblasts, 98
Corticotropin-releasing hormones (CRH), TNF-α as stimulator of, 123
Cyclophosphamide, 5, 70
Cytokines, 1, 7
  present in RA tissues, 107–108

De novo synthesis of purines and pyrimidines, 260–264
Dermatomyositis/polymyositis (DM/PM), IVIg in treatment of, 235, 238
DEXA scans for periarticular osteoporosis, 5
Diabetes. *See* Recent-onset IDDM; Type I diabetes
Digital image analysis of standard radiographs, 5
Disease-modifying antirheumatic drugs (DMARDS), 125, 156–157, 158
  clinical trials of cA2 and, 109

Endothelial adhesion molecules, 115
Epitope mapping of autoantigen, 202–204

Erythrocyte sedimentation rate (ESR), effects of Rhu IL-1R1 on, 135
Experimental autoimmune encephalomyelitis (EAE), 173, 223
  as model for T-cell mediated autoimmune disease, 190–193
  oral tolerance in animals for treatment of, 210–211
  therapeutic effects of TCR peptide on, 175

Factor VIII inhibitors, IVIg in treatment of, 240
Fetal loss, recurrent, IVIg in treatment of, 237
Fibrin D-dimer, 6

Gold, 5
  intramuscular, 70
Graft-versus-host disease, 12
Granulocyte colony stimulating factor (G-CSF), 296
Granulocyte macrophage colony-stimulating factor (GM-CSF), 107
Granulocyte-monocyte colony-stimulating factor (GM-CSF), 296
Guillain-Barré syndrome (GBS), IVIg in treatment of, 235, 241
Gut-associated lymphoid tissue (GALT), 229

Hemolytic anemia, IVIg in treatment of 239
Hodgkin's lymphoma (HL) in RA patients receiving methotrexate, 296
Human antiimmunoconjugate antibody (HAIA) response, 14
  immunological response in CD5 IC therapy, 18
Human anti-mouse antibody (HAMA) response, 14
  CD4+ cells and, 67
  immunological response to CD5 IC therapy, 18
  to MAb, 65

Human autoimmune disease, 201–205
    activation of autoreactive cells,
        204–205
    disease initiation, 201
    epitope mapping of autoantigen,
        202–204
    organ-specific proteins as targets for
        T-cells, 202
Human recombinant IFN-Γ, 95
Hydroxychloroquine, 70

Idiopathic thrombocytopenia purpura
    (ITP), IVIg in treatment of, 235,
    236, 243
Immune system, effect of CD4 MAb on,
    59
Inflammatory bowel disease, 12
Inflammatory cytokines present in RA
    tissues, 107–108
Inhibitors of de novo nucleotide syn-
    thesis, 257–286
    azaribine (triacetyl 6-azauridine), 261,
        270, 276–278
    azathioprine, 261, 268–269, 270
    brequinar, 261, 270, 275–276
    leflunomide, 261, 270, 273–275
    methotrexate, 261, 264–267, 270
    mizoribine, 261, 270, 272–273
    mycophenolic mofetal (MPA), 261,
        268–272
    pathways of de novo synthesis of
        purine and pyrimidines, 260–264
Injectable gold, 5
Insufficient IFN-Γ production in RA, 98
Insulin-dependent diabetes mellitus
    (IDDM). *See* Recent-onset
    IDDM
Interferon-α (IFN-α), 95, 296
Interferon-β (IFN-β), 95
Interferon-Γ (IFN-Γ), 95–105
    effects on connective tissue target cells
        in RA, 98–99
        effect on metabolism of articular
            chondrocytes, 98–99

[Interferon-Γ (IFN-Γ)]
    [effects on connective tissue target
        cells in RA]
        effects on metabolism of synovial
            fibroblasts, 98
    in rheumatoid synovial environment,
        96–98
    therapeutic use in RA, 99–101
        adverse reactions, 100
        doses and schedules, 99–100
        dropouts, 100
        efficacy, 100–101
    TNF-α production stimulated by, 121
Interleukin-1 (IL-1), 1, 107, 131–132
Interleukin-1-α (IL-1-α), 7
Interleukin-2 (IL-2), 7, 296
Interleukin-2 fusion toxin, 25–39
    background, 25–29
    evaluation in autoimmune diseases,
        27–37
        recalcitrant psoriasis, 36–37
        RA, 27–34
        recent-onset IDDM, 34–36
Interleukin-2 receptor, 6
Interleukin-6 (IL-6), 7, 107
Interleukin-8 (IL-8), 107, 115
Intraepithelial lymphocytes (IEL), 229
Intramuscular (IM) gold, 70
Intravenous immunoglobulin (IVIg),
    235–256
    adverse effects, 245–246
    potential mechanisms of action,
        242–245
    treatment of autoimmune hemato-
        logical disease, 239–240
        autoimmune neutropenia/hemolytic
            anemia, 239
        factor VIII inhibitors, 240
        other diseases, 240
    treatment of autoimmune neurological
        disease, 240–242
        chronic inflammatory demyelinating
            polyneuropathy, 241
        Guillain-Barré syndrome, 235, 241
        multiple sclerosis, 235, 241–242

Index

[Intravenous immunoglobulin (IVIg)]
[treatment of autoimmune neurological disease]
  myasthenia gravis, 235, 240
  other diseases, 242
treatment of dermatomyositis/ polymyositis, 235, 238
treatment of idiopathic thrombocytopenia purpura, 235, 236, 243
treatment of Kawasaki disease, 235, 238
treatment of other autoimmune diseases, 242
treatment of recurrent fetal loss, 237
treatment of rheumatoid arthritis/ juvenile rheumatoid arthritis, 235, 238
treatment of systemic lupus erythematosus, 235, 236–237
treatment of vasculitis, 235, 239

Juvenile rheumatoid arthritis, IVIg in treatment of, 235, 238

Kawasaki disease (KD), IVIg in treatment of, 235, 238, 242–243
Keyhole limpet hemocyanin (KLH), 227–228

Lambert-Eaton syndrome, IVIg in treatment of, 242
Leflunomide, 261, 270, 273–275, 287–293
  adverse effects, 291–292
  efficacy in randomized controlled trials, 289–291
  experimental models, 288
  mechanism of action, 287–288
  open dose ranging study, 289
  studies in healthy volunteers, 288–289
Leukemia inhibitory factor (LIF), 107
Lipopolysaccharide (LPS), TNF-α production stimulated by, 121

Macrophage colony-stimulating factor (M-CSF), 107
Magnetic resonance imaging (MRI), 5, 6
Mechanism of oral tolerance, 206–209
  active suppression, 208–209
  anergy, 209
Mesenchymal cells, TNF-α as stimulator of, 123
Methotrexate, 70, 261, 264–267, 270
  Hodgkin's and non-Hodgkin's lymphomas and, 296
Microfocal radiology, 6
Mizoribine, 261, 270, 272–273
Monoclonal antibody to intercellular adhesion molecule-1 (ICAM-1), 155–172
  discussion and conclusions, 167–170
  methods and results, 156–166
    adverse effects, 160–163
    clinical assessment, 158–159
    clinical outcomes, 159–160
    laboratory parameters, 163–166
    MAb preparation and administration, 158
    patients, 156–158
Monoclonal antibodies (MAb), 1, 11–12, 65
Monoclonal anti-TNF (cA2) in RA, 109–110
Mononuclear cells, TNF-α as stimulator of, 122
Multiple sclerosis (MS):
  CAMPATH-1H in treatment of, 79
  IVIg in treatment of, 235, 241–242
  oral tolerance in treatment of, 213, 223
  T-cell-receptor (TCR) peptide therapy for, 173–188
    assessment of effects, 179–183
    clinically important variables, 176–179
    preclinical animal studies, 173–176
Myasthenia gravis (MG):
  IVIg in treatment of, 235, 240
  oral tolerance in treatment of, 223

Mycophenolic mofetil (MPA), 261, 269–272
Myelin basic protein (MBP), 190–192, 223
  epitope mapping and, 202–204

Neopterin, 6
Non-Hodgkin's lymphoma (NHL) in RA patients receiving methotrexate, 296
Nonsteroidal anti-inflammatory drugs (NSAIDS), 109
Novel agents, assessment of, 295–299

Oral gold, 70
Oral tolerance, 201–233
  antigen-specific immunotherapy, 205–215, 224–225
  mechanisms of oral tolerance, 206–209
  overview, 205–206
  treatment of organ-specific autoimmune diseases in animals, 209–212
  treatment of autoimmune diseases in humans, 212–215
  autoimmune disease overview, 201–205, 221–224
  activation of autoreactive cells, 204–205
  disease initiation, 201
  epitope mapping of autoantigen, 202–204
  organ-specific proteins as targets for T-cells, 202
  new directions, 230–231
  possible explanations for tolerance, 228
  potential mechanisms of, 228–230
  tolerance in animals, 209–212, 225
  tolerance in humans, 212–215, 225–228

Penicillamine, 70
Periarticular osteoporosis, DEXA scans for, 5

Primate/human anti-CD4 antibody (IDEC- CE9.1). *See* CE9.1 (primatized monoclonal anti-CD4)
Primatized antibodies, 65–66
Psoriasis, 12
  evaluation of IL-2 fusion toxin in treatment of, 36–37
Purines, 260–264
Pyrimidines, 260–264

Radiology, 5–6
Rapid Assessment of Disease Activity in Rheumatology (RADAR), 4
Recalcitrant psoriasis, evaluation of IL-2 fusion toxin in treatment of, 36–37
  biological effects, 36–37
  safety, 36
Recent-onset IDDM, evaluation of IL-2 fusion toxin in, 34–36
  biological effects, 35–36
  safety, 35
Recombinant human interleukin-1 receptor type 1 (Rhu IL-1R1), 131–139
  effects of treatment with Rhu IL-1R1 and placebo on pain, swelling, and ESR, 135
  effects of treatment with Rhu IL-1R1 or placebo on stiffness, walk time, and global assessment, 136
  results, 133–137
  treatment protocols, 132
Recurrent fetal loss, IVIg in treatment of, 237
Rheumatoid arthritis (RA), 1
  CAMPATH-1H in treatment of, 77–78, 83–93
    CAMPATH-1H antigen, 84
    clinical experience in RA, 85–89
    clinical side effects in RA patients, 89
    humanized CAMPATH-1H antibody, 84
    preclinical data, 84–85

Index

[Rheumatoid arthritis (RA)]
  core data set for assessment, 3
  evaluation of IL-2 fusion toxin in, 27–34
    antiarthritic activity, 31–34
    RA study methods and populations, 27–29
    safety observations, 30–31
  IVIg in treatment of, 235, 238
  oral tolerance in treatment of, 213–214, 223
  See also Anti-CD5/ricin A chain immunoconjugate; CD4 monoclonal antibody therapy; Chimeric anti-CD4 antibody; Inhibitors of de novo nucleotide synthesis; Interferon-Γ (IFN-Γ); Monoclonal antibody to ICAM-1; Recombinant human interleukin-1 receptor type 1 (Rhu IL1-R1); Soluble tumor necrosis factor receptor (sTNFR); T-cell-receptor peptide vaccination; Tumor necrosis factor-α (TNF-α) blockade

Secondary amyloidosis, cA2 in management of, 116
Side effects of therapeutic agents. *See* Adverse effects of therapeutic agents
Slow-acting antirheumatic drugs (SAARDS), 5, 11
Soluble tumor necrosis factor receptor (sTNFR), 141–153
  potential problems with antagonism of TNF, 143–144
  structure and binding of TNF-α to TNF receptors, 142–143
  TNF-α in animal models of RA, 144
  TNF-α antagonists as therapeutic agents in RA, 144–145
  use of Rhu sTNFR:Fc fusion protein in RA, 145–147, 148
Sulfasalazine, 5, 70

Surrrogate laboratory measures of efficacy of biological agents, 6–7
Synovial fibroblasts, effects of IFN-Γ on metabolism of, 98
Systemic lupus erythematosus, 12
  IVIg in treatment of, 235, 236–237
Systemic vasculitis, CAMPATH-1H in treatment of, 78–79

T-cell-receptor (TCR) peptide therapy for multiple sclerosis, 173–188
  assessment of effects, 179–183
    changes in BF frequencies, 180–182
    changes in frequencies of TCR peptide-specific T cells, 179–180
    clinical changes, 183
    toxicity of TCR peptides, 183
  clinically important variables, 176–179
    injection regime, 179
    screening for Vβ bias, 177–178
    selection of TCR epitopes, 178–179
  preclinical animal studies, 173–176
    mechanism, 175–176
    rationale, 173–175
    therapeutic effects on EAE, 175
T-cell-receptor peptide vaccination, 189–199
  EAE as model for T-cell-mediated autoimmune disease, 190–193
  experience with Vβ17 T-cell-receptor peptide vaccine in RA, 193–194
Thrombocytopenia associated with HIV disease, IVIg in treatment of, 240
Toxicity of therapeutic agents:
  CD5/RTA immunoconjugate, 17–18
  TCR peptides, 183
Transforming growth factor-β (TGF-β), 97
Tumor necrosis factor-α (TNF-α), 1, 6, 121–122
  binding to TNF receptors of, 142–143
  function of and role in inflammation and arthritis, 122–124, 141
    basic science, 122–123
    studies in arthritis, 123–124

[Tumor necrosis factor-α (TNF-α)]
  role in pathogenesis of RA, 141–142
  structure of, 142–143
Tumor necrosis factor-α antagonists:
  potential problems with, 143–144
  as therapeutic agents in RA, 144–145
Tumor necrosis factor-α (TNF-α)
    blockade, 107–120
  clinical trials with monoclonal anti-
    TNF(cA2) in RA, 109–114
  open-label trial of cA2, 109–110
  placebo-controlled trial, 110–113
  repeated therapy with cA2, 113
  safety profile of cA2, 113–114
  clinical utility of cA2 and future
    studies, 115–117
  mechanisms of action of cA2, 114–115
  TNF-α in RA, 108–109
Type I diabetes mellitus, 12
  oral tolerance in treatment of, 212, 223

Uveitis:
  oral tolerance in animals for treatment
    of, 211–212
  oral tolerance in humans for treatment
    of, 214, 223

Vasculitis:
  IVIg in treatment of, 235, 239
  See also Systemic vasculitis
Vβ17 T-cell-receptor peptide vaccine in
    RA, 193–194

About the Editors

VIBEKE STRAND has over 15 years of experience, as a clinical investigator and subspecialist in private practice as well as in senior positions in pharmaceutical and biotechnology firms, planning and overseeing clinical trials in autoimmune diseases, AIDS, transplantation and cancer. For the past five years she has consulted to pharmaceutical and biotechnology companies, regarding regulatory and clinical research strategy and clinical development plans to approval. She has worked to promote forums for the discussion of rational product development among industry, FDA and academia; as co-founder and chairman of a biyearly conference, "Biologic Agents in Autoimmune Diseases," and as a member of the organizing committees for three international consensus conferences on outcome measures for rheumatoid arthritis clinical trials (OMERACT). She is a Clinical Associate Professor in the Division of Immunology at Stanford University, California.

DAVID L. SCOTT is a Reader in Rheumatology and Honorary Consultant Rheumatologist at King's College School of Medicine and Dentistry, and King's College Hospital, London, England. The author or coauthor of over 150 journal articles, he is a Fellow of the Royal College of Physicians. Dr. Scott received the M.D. degree with Distinction (1981) from the University of Leeds, England.

LEE S. SIMON is an Associate Professor of Medicine at the Harvard Medical School, Boston, Massachusetts, and Assistant to the Chairman, Department of Medicine, for Undergraduate Medical Education, Assistant to the President, Deaconess Hospital, for Medical Education, and Director of Rehabilitative Services, Deaconess Hospital, and the Dana-Farber Cancer Institute, Boston, Massachusetts. Dr. Simon has over 15 years' experience as a clinical investigator, in designing clinical trials and developing new assay methodologies for studying bone turnover, and has continued to broaden an understanding of the actions of anti-inflammatory drugs. In his role as an educator he has fostered the maturation of forums for the free exchange of information between academia, clinical investigators, industry, regulatory agencies, and government about the design of new therapies for inflammatory and autoimmune diseases. He has served on the organizing committees of three biyearly conferences entitled "Biologic Agents in Autoimmune Diseases" and served as an Associate Editor of the proceedings from two of these meetings. He is a member of several scientific advisory boards for both traditional pharmaceutical manufacturers and for a biotechnology firm. He is the Chairman of Education for the American College of Rheumatology.